Property of
DeVry Tech. Library
Woodbridge, NJ

D0119657

INTERFACING TECHNIQUES IN DIGITAL DESIGN WITH EMPHASIS ON MICROPROCESSORS

Ronald L. Krutz, Ph.D.

Director, Computer Engineering Center
Mellon Institute
Carnegie Mellon University

WILEY

John Wiley & Sons

New York · Chichester · Brisbane · Toronto · Singapore

Copyright © 1988, by John Wiley & Sons, Inc.

All rights reserved. Published simultaneously in Canada.

Reproduction or translation of any part of
this work beyond that permitted by Sections
107 and 108 of the 1976 United States Copyright
Act without the permission of the copyright
owner is unlawful. Requests for permission
or further information should be addressed to
the Permissions Department, John Wiley & Sons.

Library of Congress Cataloging in Publication Data:

Krutz, Ronald L., 1938–
    Interfacing techniques in digital design with
emphasis on microprocessors.

    Includes index.
    1. Interface circuits.   2. Microprocessors.
3. Digital electronics.   I. Title.

TK7868.I58K78   1988   621.398   87-25227
ISBN 0-471-08289-9

Printed in the United States of America

10 9 8 7 6 5 4 3 2 1

To Hilda, Sheri, and Lisa

The challenges are simpler—
the failures are fewer—
the rewards are greater

. . . . because of you

ABOUT THE AUTHOR

Ronald L. Krutz is Associate Director of Mellon Institute, Carnegie Mellon University. He is also Director of the Mellon Institute Computer Engineering Center. Prior to this position, Dr. Krutz was on the faculty of the Carnegie Mellon Electrical Engineering Department and held senior management and research positions with the Singer Corporate Research & Development Laboratory and Gulf Research Laboratories. He is a Registered Professional Engineer in Pennsylvania and is the author of a videotape course on microprocessors (1984) as well as two other textbooks: *Microprocessors and Logic Design* (1980) and *Microprocessors for Managers* (1983).

PREFACE

Microprocessor interfacing technology is evolving as dynamically as the VLSI components it supports. The techniques of interfacing are based on fundamental electrical engineering principles, advanced microfabrication rules, and computer hardware and software design concepts. It is the aim of this book to tie together these different areas and develop interfacing skills based on practical design experience and meaningful, relevant examples.

Since an increasing number of microcomputer peripheral circuits are being integrated onto higher density chips, a complete coverage of interfacing should include an introduction to basic on-chip device and logic circuit design principles. This approach develops an understanding of the on- and off-chip drive requirements that have to be considered by the designer. Following a discussion of microprocessor interfacing methodologies, terms, and definitions in Chapter 1, the fundamentals of MOS and CMOS logic devices and their interfacing are presented in the first part of Chapter 2. Then transmission line principles and formulas are developed in detail and related to on- and off-chip line-driving applications. Useful transmission line design rules are derived and illustrated with examples using commonly available components. Chapter 2 concludes with a review of noise sources in digital circuits and presents the corresponding shielding and grounding techniques to minimize the associated problems.

Chapter 3 is devoted to the analysis of bus architectures and the interfacing to these buses. Specific sections include detailed coverage of dynamic RAM interface design, EEPROM system design, and HSCMOS/NMOS interfacing. Building on the detail of Chapter 3, the material in Chapter 4 covers serial and parallel interface standards and gives specifications for the most popular of these standards. Detailed examples using available components illustrate the application of serial and parallel interface standards.

As industry standards, the architecture of the IBM PC and the later architectures using the 80386 microprocessor are the main topics of Chapter 5. An introductory summary of the IBM Personal System/2 family is also provided. Through these architectural media, the interfacing material developed in the preceding chapter is illustrated, reviewed, and extended. Details of 80386 memory and I/O interfacing are specifically covered in depth. Included in this coverage are 80386 bus timing, pipelined and non-pipelined cycles, and cache memory interfacing. This approach provides valuable exposure to the student who will more than likely encounter these architectures in many applications.

Realizing that the "real world" is analog in nature, the conversion and interfacing of the analog signals to digital form is explained in detail in Chapter 6. The reverse technology, digital-to-analog conversion, is also developed.

In light of the emphasis on Computer Integrated Manufacturing (CIM) and Computer-Aided Engineering (CAE) in most industries, the topic of local area networks (LAN's) in Chapter 7 is especially relevant. In particular, the ISO-OSI standards and the related MAP standard for manufacturing system communication as presented in Chapter 7 are essential knowledge for the student now and in the future. Also, other token passing and CSMA/CD LAN standards are defined and discussed in depth.

Chapter 8 includes material not usually associated with interfacing—software. In most actual design and interfacing problems, however, software development is the most critical and costly area to be addressed. In particular, software engineering techniques and languages such as C and PL/M-86 that are used to develop interfaces are summarized and relevant examples are presented.

RONALD L. KRUTZ

CONTENTS

1 MICROPROCESSOR INTERFACING: AN OVERVIEW 1

1.1 Interfacing Layers 1
    1.1.1 Other Considerations 2

1.2 Definitions 2
    1.2.1 Buses 3
    1.2.2 Memory Characteristics 4

1.3 Costs 5

1.4 Interfacing 6

2 CIRCUIT AND ELECTRICAL INTERFACING CONSIDERATIONS 7

2.1 Internal Characteristics of MOSFET Digital Systems 7
    2.1.1 MOSFET Circuits 9
        NOR/NAND Circuits 11
        Pass Transistor 11
        CMOS Structure 13
        Interface Between the Real World and Digital Systems 13

2.2 Transmission Line Considerations 14
    2.2.1 Effects of Characteristic Impedance on Signal Propagation 18
    2.2.2 Application of Transmission Line Characteristics to MOS
        Circuits 25

2.3 MOS Memory as Transmission Line Load 28
    2.3.1 Other Methods of Avoiding Oscillation Problems 29

2.4 Noise 31
    2.4.1 Noise Sources 31
    2.4.2 Shielding Techniques 32
    2.4.3 Summary of Noise Sources, Effects, and Noise Reduction
        Techniques 34
    2.4.4 Interconnection Wiring Noise Reduction 34
    2.4.5 Grounding Connections 38

3 BUS INTERFACING 41

3.1   Delays   41

3.2   Asynchronous Buses   43
 3.2.1   Asynchronous Bus Read   43
 3.2.2   Asynchronous Bus Write   45

3.3   Synchronous Buses   45
 3.3.1   Synchronous Bus Read and Write   45
 3.3.2   Synchronous Bus Utilizing WAIT State (Semisynchronous)   46
 3.3.3   Thirty-Two-Bit Bus Standards   48

3.4   Memory and Peripheral Interfacing   49
 3.4.1   Dynamic RAM   49
  Power Supply   50
  Memory Cycles   54
  Timing Estimates   56
  Detailed Dynamic RAM Design Example   57
 3.4.2   High Speed CMOS Logic Interfacing   72
  NMOS to HSCMOS   72
  HSCMOS to NMOS   73
  Noise Margin   74
  Microprocessor Bus Interfacing Example   74
  Timing Characteristics   76
 3.4.3   Electrically Erasable Programmable Read Only Memory
  (EEPROM)   79
  Processor Interface   79
  Read Mode   79
  Byte Write/Erase   80
  Chip Erase   82
  8298 Integrated Controller   83

4 COMMUNICATIONS AND DATA TRANSFER 86

4.1   Serial Communications   86
 4.1.1   Asynchronous Transmission   86
  20 Milliampere Current Loop   88
  RS-232-C   88
  RS-232-C Specification Limits   92
  Typical RS-232-C Connection and Timing   93
  RS-422-A, RS-423-A, RS-449   95
  RS-422-A   95
  RS-423-A   96
  RS-449   97
  RS-485   99

4.1.2   Synchronous Transmission   102
HDLC/SDLC Data Link Control   102

4.2   IEEE Standard Digital Interface for Programmable Instrumentation,
IEEE STD 488-1978, Parallel/Asynchronous   115
4.2.1   Terminology and Definitions   116
4.2.2   Logic Levels   118
4.2.3   Drivers and Receivers   118
4.2.4   Example Circuits   119
4.2.5   Handshake Timing   122
4.2.6   General Interface Management Bus   130

4.3   Microprocessor Parallel Communications   132
4.3.1   Addressing   133
4.3.2   Basic I/O Circuit   134
4.3.3   Strobed I/O Circuits   135
4.3.4   Pseudobidirectional I/O Port   137

5   IBM PERSONAL COMPUTER AND 80386 PROCESSOR INTERFACING, ASPECTS OF IBM PERSONAL SYSTEM/2 FAMILY   139

5.1   IBM PC   139
5.1.1   System Board/RAM Interface   140
5.1.2   I/O Channel   143
General Interface   143
5.1.3   SDLC Controller (8273) as Used on SDLC
Communications Adapter   153
5.1.4   Printer Adapter   156
5.1.5   General RAM Interface   158
5.1.6   General ROM Interface   160
5.1.7   IBM PC Summary   160

5.2   IBM Personal System/2 Summary   160

5.3   80386 Microprocessor   162
5.3.1   80386 General Description   163
5.3.2   80386 Structure   163

5.4   80386 Bus Interface   164
5.4.1   Bus Timing   165
5.4.2   Non-Pipelined Read Cycle   165
5.4.3   Non-Pipelined Write Cycle   167
5.4.4   Pipelined Cycles   168
5.4.5   Interrupt Acknowledge Cycle   169

5.4.6   Timing Considerations   172
         READY# Timing   172
         READ Timing   172
         WRITE Timing   172
5.4.7   LOCK Cycle   173
5.4.8   Reset Timing   173

5.5   Interfacing to Memory   175
5.5.1   Dynamic RAM Subsystem Design   176
5.5.2   Dynamic Memory Refresh   177

5.6   80386 Cache Memory Interfacing   179
5.6.1   Organization of Cache Memory   180
5.6.2   Cache Data Integrity   183
5.6.3   Typical Cache Performance   183
5.6.4   Example Design   183

5.7   I/O Interfacing   186
5.7.1   General I/O Interface   187

6   ANALOG INTERFACING TO MICROCOMPUTERS   192

6.1   Analog-to-Digital Converters   192
6.1.1   Approach   192
6.1.2   Errors and Accuracy   193
6.1.3   Temperature Effects   198
6.1.4   Methods of A/D Conversion   200
         Counter Converter   200
           Timing   200
         Successive Approximation A/D Converter   201
         Flash Converter   202
         Integrating A/D Converters   205
           Single-Slope (Single-Ramp)   205
           Dual Ramp (Dual Slope) Converter   206
6.1.5   Example Error Calculations   209

6.2   Digital-To-Analog (D/A) Converters   209
6.2.1   Weighted Current Source   210
6.2.2   R-2R Network D/A Converter   211

6.3   Interfacing A/D and D/A Converters   213
6.3.1   Data Acquisition Systems   213
6.3.2   D/A Converter Interface   215
6.3.3   Additional A/D Interface Example   216

7   Local Area Networks   217

7.1   Topology   217
　　　7.1.1   Star   218
　　　7.1.2   Ring   218
　　　7.1.3   Bus   219

7.2   Related Communications Terminology   219
　　　7.2.1   Frequency Division Multiplexing   220
　　　7.2.2   Time Division Multiplexing   220
　　　7.2.3   Space Division Multiplexing   220
　　　7.2.4   Broadband and Baseband Approaches   221

7.3   Architecture   222
　　　7.3.1   International Standards Organization OSI Model   223
　　　7.3.2   Medium Access Options   227
　　　　　　Token Passing   227
　　　　　　　Token Ring   228
　　　　　　　Token Bus   229
　　　　　　　MAP   233
　　　　　　　TOP   240
　　　　　　CSMA/CD   241
　　　　　　　Carrier Sensing   242
　　　　　　　Detection of Collision   243
　　　　　　　Retransmission   243
　　　　　　　Ethernet Message Format   244
　　　　　　　Preamble   245
　　　　　　　Destination Address   245
　　　　　　　Source Address   245
　　　　　　　Type Field   245
　　　　　　　Data Field   245
　　　　　　　Frame (Packet) Check Sequence   245
　　　　　　　Bit Encoding   246
　　　　　　　Typical Implementation   246
　　　　　　　Some Products   251

8   SOFTWARE INTERFACING   252

8.1   Modular Programming   252

8.2   Structured Programming   254

8.3   Programming Style   255

8.4   BASIC   257

8.5   FORTRAN   259

8.6   PASCAL   261

8.7   Ada   262

8.8   LISP   264
      8.8.1   Definitions   265
      8.8.2   Examples   265
      8.8.3   Control   267

8.9   PL/M-86   268
      8.9.1   Background   268
      8.9.2   Description   268
      8.9.3   Identifiers   268
      8.9.4   Data Types   269
      8.9.5   Operators   271
      8.9.6   Declarations   273
      8.9.7   Initialization   274
      8.9.8   Utilization of Data and Initial Attributes   274
      8.9.9   Procedures   275
      8.9.10   Attributes   277
      8.9.11   PL/M-86 Flow Control   281
      8.9.12   PL/M-86 Built-In Procedures   288
      8.9.13   Examples   291
      8.9.14   Conclusion   297

8.10   The C Programming Language   297
      8.10.1   General Description   298
      8.10.2   Variables   298
      8.10.3   Preprocessor   300
      8.10.4   Operators   301
      8.10.5   Program Structure and Control   301
      8.10.6   Library Functions   302
      8.10.7   Reference   303
      8.10.8   Interfacing Example   303
      8.10.9   Conclusion   322

Appendix A   ASCII CHARACTER SET   323

Appendix B   INTERFACE-RELATED READINGS   324
             Real-Time Operating System Needs   324
             DIM: A Network Compatible Intermediate Interface
             Standard   332
             VME Bus Interfacing: A Case Study   359

INDEX   379

1

MICROPROCESSOR INTERFACING: AN OVERVIEW

Interfacing is a term that applies across a broad range of electronic implementations. It relates to systems as well as to individual transistors. Because of this breadth, interfacing involves many technologies that are encompassed by the disciplines of electrical engineering and computer engineering. Since these technologies are usually separated into subdisciplines with their own areas of emphasis, their integration into a reasonably concise and unified treatment is desirable. The treatment of interfacing in this book is directed toward the integration of these various subdisciplines.

1.1

INTERFACING LAYERS

Interfacing usually involves effectively traversing a boundary from one entity to another. In the field of electronics, the entities can be viewed in a hierarchical fashion from a system, subsystem, component, and transistor level. Boundaries have to be traversed in all these levels. In doing so, opposite ends of the spectrum, such as the effects of electrons in motion and instruction execution times, may have to be considered.

In a microprocessor-based system, a number of levels can be defined arbitrarily to attempt to structure the interfacing hierarchy. One possible layering is

electrical(physical)

signal

logic

protocol

code

algorithmic/heuristic

Note that this model includes software implementations as part of the hierarchy. In microprocessor-based systems, especially real-time data acquisition and control applications, the software "interface" to the equipment to be controlled is a costly and critical portion of the total implementation. This aspect, denoted by the code and algorithmic/heuristic levels, is treated in Chapter 8.

Protocol interfacing is studied through communications and local area network standards and implementations. Interfacing at the logic level is addressed through circuit applications, bus interfacing, and data transfer. Signal interfacing techniques are developed in electrical interfacing, bus interfacing, and data transfer techniques. The physical interfacing techniques are developed through the electrical and analog interfacing coverages. All the levels are summarized and illustrated by means of the IBM Personal Computer family and 80386 microprocessor-based material.

1.1.1 OTHER CONSIDERATIONS

In the majority of interfacing applications, timing considerations are critical and must be incorporated into the overall design. In many real-time control systems, the exact response times of all the elements to be controlled may not be available prior to system installation. This unavailability may be due to the sheer number of controlled elements or to their inability to be measured until the total system is assembled and tuned. In such cases, hardware and software accommodations must be made during the specification, design, and implementation phases of the project to handle these timing variables. This "handling" must include the ability to make some changes in the software without requiring radical rewrites or time-consuming modifications of the interface.

Theoretically, since software is more easily changed than hardware, required adaptations to the control system should be made through code changes. If the software was designed using modern software engineering principles, this approach is feasible. Even assembly language coding that is required for time-critical portions of the software can be designed and implemented in a structured manner. Examples of good real-time software implementations for the C and PL/M 86 languages are given in Chapter 8.

Before proceeding to the specifics of interfacing, some fundamental microprocessor-related definitions, characteristics, and economics are provided.

1.2

DEFINITIONS

Strictly speaking, a *microprocessor* is composed of an arithmetic logic unit (ALU), control circuitry, a small amount of scratch pad memory, and, in some cases, timing (clock) circuits. A block diagram of a microprocessor is given in

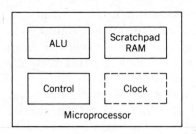

FIGURE 1.1 Microprocessor block diagram.

Figure 1.1. On the other hand, a *microcomputer* is made up of a microprocessor, program memory that is usually a type of read only memory (ROM), read/write random access memory (RAM), and a peripheral interface that provides communication to external devices. If the timing (clock) circuits are not contained in the microprocessor, then they must also be provided for in the microcomputer. A general microcomputer configuration is shown in Figure 1.2.

1.2.1 BUSES

The component elements of the microcomputer are interconnected by wiring paths known as *buses*. There are three types of buses: *address, data,* and *control*. The address bus carries binary-coded patterns (addresses) from the microprocessor to the ROM, RAM, and peripheral interface. The addresses identify unique locations that will be the source or destination of information transferred to or from the microprocessor. This information travels over the bidirectional data bus. The information falls into the four general categories of instructions, control words, status indicators, and data. Instructions contained in ROM are executed by the microprocessor, but intelligent peripheral

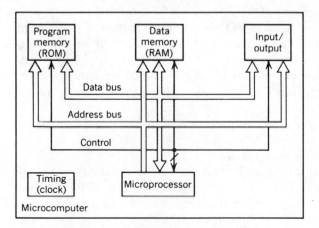

FIGURE 1.2 General microcomputer configuration.

devices that contain their own processors can execute instructions initiated by the microprocessor. Control words usually originate in the microprocessor and are transmitted to intelligent peripheral devices. Data are available from all elements of the microcomputer and travel on the data bus. RAM is usually the primary repository of data present in the microcomputer.

Even though the term random access memory (RAM) applies to read/write memory, it is also a characteristic of ROM. Random access means that all units of data contained in the memory can be accessed in essentially the same amount of time. An X-Y coincident matrix is a good model for this type of memory. Alternately, a memory can be *serial access,* which means that the time to access data contained in the memory is a function of its location. A magnetic tape wound on a cassette is a good example of a serial access memory. The amount of time to acquire a piece of data that is stored near the center of the tape spool is greater than that needed to retrieve data stored on the edge of the spool.

The control bus contains a variety of signals such as reset, interrupt, halt, acknowledges, and so on. The reset places the microcomputer in a known state to begin or resume operation. Interrupt signals are used to interrupt the current program flow and cause the microprocessor to execute instructions in a different part of the program. The interrupt is normally used to service emergency or high priority requirements. Acknowledge signals provide information as to the status of the microprocessor. For example, an interrupt acknowledge line may be provided to provide confirmation that an interrupt has been recognized by the microprocessor.

1.2.2   MEMORY CHARACTERISTICS

Read/write random access memory (RAM) is usually *volatile,* that is, it loses its contents when power is lost. RAM provides for very high speed storage and retrieval of data (100 ns or less). ROM, which is essentially an array of fusible links, provides storage that is *nonvolatile* but cannot be modified by the microprocessor. Thus a program or data in ROM cannot be changed but are maintained if power is lost. A ROM is programmed by a mask in the final stage of chip production by the semiconductor manufacturer. Thus, once defined and produced, the contents cannot be modified. Use of a ROM is usually reserved for proven, debugged programs that have to be duplicated in high volume (thousands or more).

A PROM or programmable read only memory is similar to a ROM but can be programmed once at the user's site. Again, it is not modifiable after the initial programming and is nonvolatile.

In an attempt to achieve nonvolatility and read/write capability in a single device, EPROMS and EEPROMS have been developed. The EPROM (Erasable, Programmable, Read Only Memory) provides nonvolatility with a type of read/write capability. The EPROM can be erased by shining ultraviolet light through a quartz window on the EPROM package for approximately 15–20 min. This procedure erases the entire contents of the memory. To write information into the EPROM, the chip is placed into a special programming fixture and electrically programmed. Then it can be placed into the microcomputer system. An EEPROM (Electrically Erasable Programmable Read Only Memory) can be modified electrically in situ. The write time (20 ms) is much slower than with a conventional RAM, but the EEPROM is nonvolatile and, thus, permits resident programs to be changed locally, or even remotely, through a telephone line connection.

1.3

COSTS

A general hardware cost breakdown of a microcomputer system is as follows:

Processor (CPU)	15%
Peripheral interface (I/O)	20%
Memory	65%

If we look at the specific, average costs of a microcomputer system produced for moderate volume (<200), the following numbers are typical:

Processor, I/O, memory, other components	500
Circuit boards, connectors, assembly	160
Miscellaneous	100
Total	760

An interesting perspective is obtained by calculating related software development costs. Assumptions made are

1. Software is developed "custom" for the application on the microcomputer.
2. Software cost is approximately $8–$10 per line of debugged code.
3. 10–20 lines of debugged code are produced per day.
4. Software costs can be amortized over all microcomputer systems produced.

With these assumptions, a 64K byte program (assuming 4 memory bytes per line of code) development cost is calculated as:

$$64\text{K bytes} \times \frac{1 \text{ line of code}}{4 \text{ bytes}} \times \$8/\text{line} = \$128 \text{ K}$$

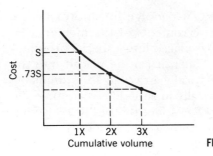

FIGURE 1.3 Semiconductor learning curve.

For one microcomputer system, the disparity between hardware and software costs is obvious. This is reflected by the increased labor intensity of software over hardware. A simple break-even comparison to determine the number of units (n) to be sold to bring the software cost per unit in line with the hardware cost per unit is as follows.

$$\frac{128,000}{n} \text{ units} = 760$$

$$n = \frac{128,000}{760} = 169 \text{ units}$$

Thus, as software development can be amortized over a large number of units and sold in ROMs, PROMs, and so on, as *firmware,* some software can be made to follow the learning curve of computer hardware. This curve describes the situation that for each doubling of the cumulative volume of devices sold, the cost of development is reduced by 27%. The learning curve is shown in Figure 1.3.

1.4

INTERFACING

With this brief background, we can proceed to Chapter 2 to investigate the circuit-related and electrical considerations that impact the interfacing task.

FOR FURTHER READING

Editors, *Electronics,* Special Commemorative Issue, V-1, 53, No. 9, April 17, 1980.
Editors, *Intel OEM Systems Handbook,* No. 210941-003, 1985.
Evans, Christopher, *The Micro Millennium,* Washington Square Press, New York, 1979.
Krutz, R. L., *Microprocessors for Managers: A Decision-Maker's Guide,* Holt, Rinehart, and Winston, 1983.

2

CIRCUIT AND ELECTRICAL INTERFACING CONSIDERATIONS

In order to perform useful work, digital systems receive information, operate on it, and provide information as an output. The operations may be trivial or complex, but if the input and output functions are not implemented properly, the system is useless. The boundary between the digital system and the outside world is called the *interface*. For example, the integrated circuit (IC) chips that connect a microprocessor's data output to a printer or display are called an output interface. Looking at interfaces in a hierarchical fashion, there can also be interfaces between subsystems in a digital system. In fact, in *Very Large Scale Integration (VLSI)* chips, the communications and associated interfaces among digital subsystems are the most important considerations in the design process and may take up to 70% or more of the chip area. These types of interfaces are illustrated in Figure 2.1.

2.1

INTERNAL CHARACTERISTICS OF MOSFET DIGITAL SYSTEMS

The *Metal Oxide Semiconductor Field Effect Transistor (MOSFET)* is the fundamental building block in VLSI circuits and, therefore, will be emphasized in this book. It is the basic unit of the high density CMOS and NMOS technologies that is used to fabricate today's microprocessors and interface devices. The symbol for the MOSFET and the names of its three terminals are given in Figure 2.2. In practice, the MOSFET acts as a switch connecting the source and drain. The switch is either on (closed) or off (open) depending on the voltage applied between the gate and source. If the *gate-to-source voltage, V_{gs},* exceeds a particular voltage known as the *threshold voltage, V_{th},* the switch is closed. Otherwise, the switch is open. In reality, the MOSFET switch is not simply open (infinite impedance) or closed (zero impedance). Typical "on" impedances are in the range of hundreds or thousands of ohms while "off" impedances have values in millions of ohms (Megaohms). Figure 2.3 illus-

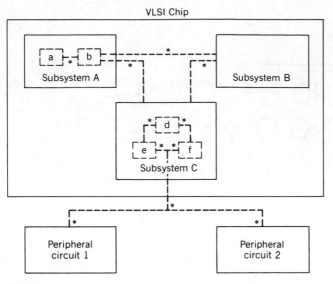

*Interfaces

FIGURE 2.1 VLSI interfaces.

Source   FIGURE 2.2 MOSFET symbol.

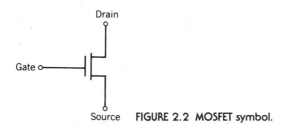

(a)   (b)

FIGURE 2.3 MOSFET operation: (a) o switch, $V_{gs} < V_{th}$; (b) closed switch, $V_{gs} \geq V_{th}$.

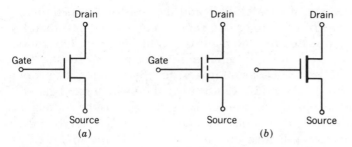

FIGURE 2.4 MOSFET transistor symbols: (a) enhancement mode;
(b) depletion mode.

trates the switch operation of a MOSFET. If the threshold voltage, V_{th}, is greater than zero, the MOSFET is called an *enhancement mode* MOSFET. Conversely, a MOSFET with a threshold voltage less than zero is termed a *depletion mode* MOSFET. Symbols for both types of MOSFETs are shown in Figure 2.4. Two different symbols can be used for the depletion mode device; they are given in Figure 2.4b.

2.1.1 MOSFET CIRCUITS

A digital inverter is one of the most basic circuits in a VLSI device. In its simplest form, it consists of an enhancement mode MOSFET and a "pull-up" resistor as shown in Figure 2.5. In Figure 2.5, V_{DD} is the power supply voltage, typically +5 volts, V_i is the input voltage, and V_o is the output voltage. If we use the switch analogy, when $V_i > V_{th}$ for a fixed V_{DD}, the switch is on (closed) and V_o is approximately at 0 volt. Thus, for a logic 1 input ($V_i > V_{th}$), $V_o = 0$. When $V_i < V_{th}$, the switch is off (open) and V_o is "pulled up" through resistor R to V_{DD}. If V_o is tied to a high impedance input in the following stage, V_o will be approximately the same as V_{DD}. If $V_i < V_{th}$ is considered a logic 0 input, V_o will be a logic 1. The circuit, therefore, does implement a digital inverter.

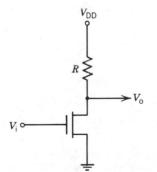

FIGURE 2.5 Digital inverter.

Since the enhancement mode MOSFET and the resistor R function as a voltage divider, the value of R has to be large enough to make V_o less than V_{th} when the MOSFET is on. Because the materials available in IC fabrication require relatively large areas to implement the inverter pull-up resister R, a depletion mode MOSFET is used as a pull-up resistor. By selecting the proper geometry, a depletion mode MOSFET can be turned on with a fixed gate-to-source voltage and act as pull-up resistor. By tying the gate to the source, V_{gs} will be 0 V. Recalling that V_{th} of a depletion mode MOSFET is less than 0 V, we find that a V_{gs} of zero will turn on the transistor. If the MOSFET is designed properly, the ratio of the current through the MOSFET and the voltage drop across it (V_{ds}/I_{ds}) will provide the desired pull-up resistance. The circuit diagram of a digital inverter utilizing a depletion mode MOSFET as a pull-up resistor is shown in Figure 2.6. In order to discuss the relative areas of Q_1 and Q_2 as implemented in silicon, we can refer to the simplified physical layout of an *n-channel* (*NMOS*) MOS transistor in Figure 2.7. In this type of transistor, a positive voltage on the gate with respect to the source (V_{gs}) induces negative charges in the channel region. These negative charges then provide a connecting path in the channel between the *n*-type source and drain regions. A *p-channel* (*PMOS*) MOS transistor is constructed in a similar manner, but with the *p*-type and *n*-type materials reversed. A negative charge on the gate with respect to the source induces positive charges in the channel region of a PMOS transistor.

The length-to-width ratio, L/W, of a MOSFET is usually referred to by the symbol Z. If a digital inverter is to drive another inverter with its output, the length-to-width ratio, Z_{pu}, of the depletion mode pull-up transistor is usually designed to be 4 times that of Z_{pd}, the length-to-width ratio of the enhancement mode pull-down transistor. In other words,

$$\frac{Z_{pu}}{Z_{pd}} = \frac{(L_{pu}/W_{pu})}{(L_{pd}/W_{pd})} = 4:1 \tag{2.1}$$

This ratio ensures that the inverter input level that initiates the inverter out-

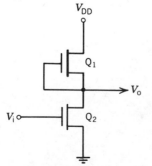

FIGURE 2.6 Digital inverter with depletion mode pull-up: Q_1, depletion-mode MOSFET; Q_2, enhancement mode MOSFET.

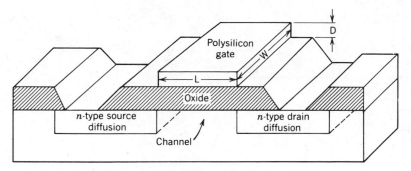

FIGURE 2.7 Simplified physical layout of an *n*-channel MOSFET.

put transition from one logic level to another lies approximately halfway between ground and V_{DD}.

NOR/NAND CIRCUITS

Other logic circuits can be fabricated by extending the basic inverter circuit. An MOS NOR logic circuit is given in Figure 2.8 and an MOS NAND circuit in Figure 2.9. From Figure 2.8, if either input A or B or both A and B are at a logic 1 level, output T is pulled down to ground. If A and B are at a logic 0 level, T is pulled up to V_{DD}, a logic 1. Thus the NOR function is implemented. Similarly, the output of the NAND circuit in Figure 2.9 is pulled down to a logic 0 or ground level only if inputs A and B are simultaneously at a logic 1 level.

PASS TRANSISTOR

Another useful MOSFET circuit utilizes the *pass* transistor. In this application, MOSFETs are connected in series as switches to propagate signals.

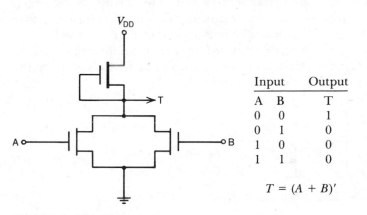

Input		Output
A	B	T
0	0	1
0	1	0
1	0	0
1	1	0

$$T = (A + B)'$$

FIGURE 2.8 MOS NOR circuit.

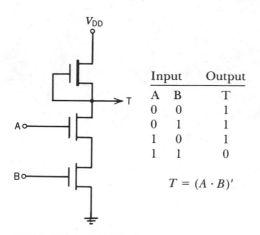

Input		Output
A	B	T
0	0	1
0	1	1
1	0	1
1	1	0

$$T = (A \cdot B)'$$

FIGURE 2.9 MOS NAND circuit.

Figure 2.10 illustrates a number of pass transistor stages that are used to shift binary data.

Since pass transistors act as series resistances and capacitances to ground, the propagating signal is delayed through each stage. In order to restore the waveshape of signals transmitted through a series of pass transistors, inverter stages are inserted in the pass transistor chain. The general rule of thumb is to insert an inverter stage when the cumulative pass transistor stage delays are equal to an inverter stage delay. Typically, this results in an inverter being inserted after three or four pass transistors, as illustrated in Figure 2.11.

Unlike inverters, pass transistors have no static power dissipation and are, therefore, advantageous to use. Logic implemented with pass transistors is referred to as *steering* logic, whereas logic implemented with the inverter gate is known as *ratio* logic.

When an inverter is used as a level restorer between pass transistors, the ratio Z_{pu}/Z_{pd} must be increased from $4:1$ to $8:1$. This change is necessary, since the voltage drop across the pass transistor(s) reduces the inverter gate input voltage. Thus the driven enhancement mode MOSFET is turned on less than it would be if driven directly from the output of an inverting gate. A value of 8 for Z_{pu}/Z_{pd} of the driven inverter ensures that its V_o is equal to the V_o of the inverter preceding the pass transistor.

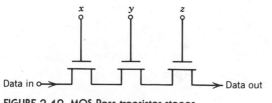

FIGURE 2.10 MOS Pass transistor stages.

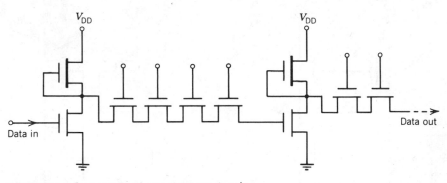

FIGURE 2.11 Restoration of pass transistor signals.

CMOS STRUCTURE

As NMOS integrated circuits have become denser and operate at higher speeds, the power dissipation in these circuits has emerged as a serious limitation to achieving higher densities. An enhanced version of *CMOS (Complementary MOS)*, a relatively old technology, has proven to be the vehicle for achieving high speed, high complexity microprocessors and interface chips.

In a basic CMOS inverter, the NMOS and PMOS transistors are connected in series between power and ground so that, except for switching times when the transistors are changing state, only one transistor of the pair is on at any given time. Thus no current (except small leakage) current flows from the power supply to ground through the CMOS circuit. On the average, therefore, relatively little current flows in a CMOS gate and the gate appears as a primarily capacitive load. This characteristic reduces the power to be dissipated in the circuit and, consequently, allows for closer packing of the circuit elements.

INTERFACE BETWEEN THE REAL WORLD AND DIGITAL SYSTEMS

The MOS devices appear as primarily capacitive loads. Therefore, driving these devices will result in delays, particularly when sending signals off-chip. For example, consider a VLSI chip providing data to a peripheral device. The load capacitance to be driven off chip may be very large relative to the gate capacitance of the on-chip inverter. One technique for driving a large off-chip load is through a series of increasingly larger (in current driving capability) inverter gates. Another approach is to use a buffer gate that can supply approximately 4 times the current required in one stage. This is accomplished by tying the gate of the depletion mode pull-up transistor to the complement of the logic level applied to the gate of the enhancement mode pull-down transistor. This connection scheme turns on the transistor more than tying the

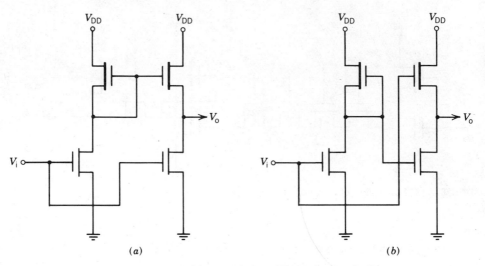

FIGURE 2.12 High current buffers: (a) inverting buffer; (b) noninverting buffer.

gate of the depletion mode pull-up to its source as in a standard MOS inverter. Figure 2.12 shows inverting and noninverting versions of this high-current buffer connection. High-current buffers can also be used in sequential stages to drive large current loads.

2.2

TRANSMISSION LINE CONSIDERATIONS

Another effect that influences interfacing is that of the *RF transmission line*. For example, consider the MOS driver model in Figure 2.13.

Since the line from the driver to the load exhibits a series component of

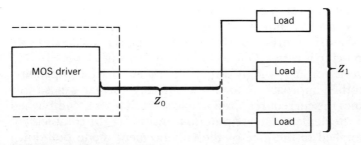

FIGURE 2.13 Typical MOS driver model.

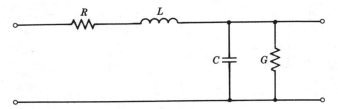

FIGURE 2.14 Transmission line segment, one representation.

resistance and inductance and a parallel component of capacitance and conductance, it can be modeled by a distributed transmission line equivalent circuit segment of the form given in Figure 2.14. The form of the segment shown in Figure 2.14 is not the only possible representation of the transmission line. There are other alternatives, one of which is given in Figure 2.15.

The values of the distributed elements in the transmission line representations are given in per unit length. If we use the representation of Figure 2.14, the transmission line segments can be joined to represent a distributed transmission line. This model ignores second order loss terms, such as skin effect, and is given in Figure 2.16.

When a digital circuit switches at high frequencies (> 100 kHz), the driven lines can be modeled to a first-order approximation by Figure 2.16. An important parameter of the transmission line relative to the driving circuit is the input impedance, Z_i, of the transmission line. Assuming a transmission line of infinite length, we can redraw Figure 2.16 as shown in Figure 2.17, letting $Z_1 = R + j\omega L$ and $Z_2 = \dfrac{1}{(G + j\omega L)}$.

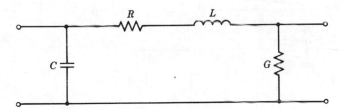

FIGURE 2.15 Transmission line segment, another representation.

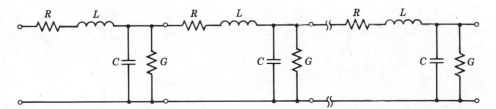

FIGURE 2.16 Distributed transmission line model. (R, L, C, and G are per unit length.)

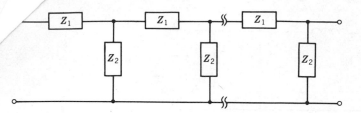

FIGURE 2.17 Reduced distributed transmission line model.

Since the transmission line is assumed to be infinite in length, the impedance looking into any section of the line is the same. This impedance is called the *characteristic impedance*, Z_o. Thus any of the line models given in Figure 2.18 represent the infinite transmission line.

The equivalent circuit of Figure 2.18b can be used to calculate Z_o in terms of Z_1 and Z_2, the per unit length parameters of the transmission line.

$$Z_i = Z_1 + \frac{Z_2 Z_o}{Z_2 + Z_o} \tag{2.2}$$

$$Z_i(Z_2 + Z_o) = Z_1(Z_2 + Z_o) + Z_2 Z_o \tag{2.3}$$

$$Z_i Z_2 + Z_i Z_o = Z_1 Z_2 + Z_1 Z_o + Z_2 Z_o \tag{2.4}$$

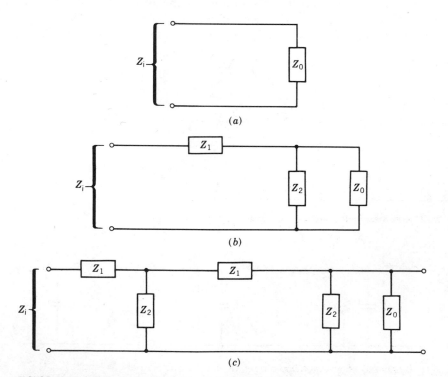

(a)

(b)

(c)

FIGURE 2.18 Infinite transmission line equivalent model.

Noting from Figure 2.18*a* that $Z_i = Z_o$, we can rewrite equation (2.4) as

$$Z_o Z_2 + Z_o^2 = Z_1 Z_2 + Z_1 Z_o + Z_2 Z_o \tag{2.5}$$

or

$$Z_o^2 - Z_1 Z_o - Z_1 Z_2 = 0 \tag{2.6}$$

Solving for

$$Z_o = \frac{Z_1 \pm \sqrt{Z_1^2 + 4Z_1 Z_2}}{2} \tag{2.7}$$

and substituting for Z_1 and Z_2 yields

$$Z_o = \{(R + j\omega L) \pm [(R + j\omega L)^2 + 4(R + j\omega L)/(G + j\omega C)]^{1/2}\}/2 \tag{2.8}$$
$$Z_o = (R + j\omega L)/2 \pm [(R + j\omega L)^2 + 4(R + j\omega L)/(G + j\omega C)]^{1/2}/2 \tag{2.9}$$

Recalling that the values of R, L, and C are in per unit length, as the length of the section is decreased toward zero in the limit, we find that the ratio term $(R + j\omega l)/(G + j\omega C)$ becomes the only term of any significance in equation (2.9). Then the expression for Z_o reduces to

$$Z_o = \frac{\sqrt{4(R + j\omega L)/(G + j\omega C)}}{2} \tag{2.10}$$

or

$$Z_o = \sqrt{\frac{R + j\omega L}{G + j\omega C}} \tag{2.11}$$

A general plot of equation (2.11) is given in Figure 2.19.

Since digital signals consist of pulses in the megahertz range, the flat part of the curve in Figure 2.19 applies. Thus, for our interests,

$$Z_o = \sqrt{\frac{L}{C}} \tag{2.12}$$

where L is the inductance per unit length and C is the capacitance per unit length of the transmission line. The significance of this result is that, in general, a transmission line presents a characteristic impedance of $Z_o = \sqrt{L/C}$ to a digital circuit and this value is independent of the line length. To a good approximation, a transmission line being driven by a digital circuit can be represented by a resistor that equals $\sqrt{L/C}$ Ω across the output of the circuit. As an example, a twisted pair of 24 AWG wire has a characteristic impedance of 96 Ω and can be approximated to a digital driver as shown in Figure 2.20.

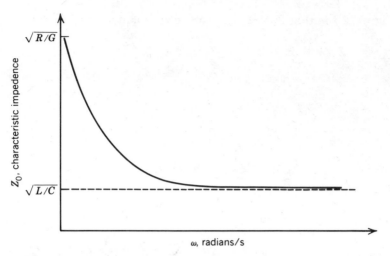

FIGURE 2.19 General plot of characteristic impedance.

Similarly, RG 58.7/U coaxial cable has a characteristic impedance of 75 Ω and appears to a digital circuit as represented in Figure 2.21.

An important point to note is that not only do external wires and coaxial cables exhibit transmission line properties to digital circuits, but also printed circuit traces and on board wiring have these characteristics.

2.2.1   EFFECTS OF CHARACTERISTIC IMPEDANCE ON SIGNAL PROPAGATION

Without going into the derivation, it can be shown that on a lossless line (zero attenuation), there is a phase shift of the signals in the megahertz range equal to $\omega\sqrt{LC}$ radians per unit length. With this information, the propagation velocity, v_p can be calculated. This velocity is equal to $\omega/\omega\sqrt{LC}$ or $1/\sqrt{LC}$ unit lengths/second. Thus the expression for v_p is independent of the transmitted signal frequency in this ideal case of a lossless line. For most digital circuits, v_p is approximately 0.169 m/ns. The propagation time, then, is $\frac{1}{0.169}$ or 5.9 ns/m.

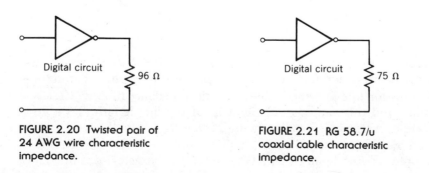

FIGURE 2.20 Twisted pair of 24 AWG wire characteristic impedance.

FIGURE 2.21 RG 58.7/u coaxial cable characteristic impedance.

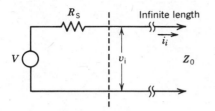

FIGURE 2.22 Ideal transmission line with no load.

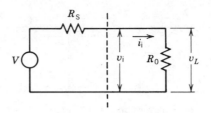

FIGURE 2.23 Finite length transmission line terminated in its characteristic resistance.

With the information developed to this point, a general transmission line model for a digital circuit driving a line can be analyzed to determine the effects on the transmitted signal. Consider the circuit in Figure 2.22, where V is the source voltage and R_s is the source resistance.

Since the transmission line is assumed to be infinite in length, the voltage source "sees" the characteristic impedance, Z_o, when it looks into the line. Equivalently, a resistance of Z_o can be inserted across a finite lossless line at any distance from the source and the line will still appear as an infinite transmission line, since it sees its characteristic impedance. This situation is shown in Figure 2.23.

As a consequence the input voltage, v_i, of the circuit of Figure 2.22 is equal to the input voltage of the circuit of Figure 2.23, as are the corresponding input currents, labeled i_i.

There are no impedance mismatches when the transmission line is terminated in its characteristic impedance. Therefore, all the power transmitted down the line is absorbed by the load and no reflections are generated back toward the source. In the steady state, the voltage, v_L, at the load is

$$v_L = V \left[\frac{R_o}{R_o + R_s} \right] \tag{2.13}$$

If the source impedance, R_s, is very much less than R_o,

$$v_L \simeq V \tag{2.14}$$

in the matched load impedance case.

Now assume that the load impedance, R_L, is larger than R_o. This situation will result in the reflection of a portion of the initial voltage wave back toward the source when the initial voltage wave arrives at the load. In order to calculate the fraction of the initial wave that is reflected at the source, Figure 2.24 is used.

Summing currents at node n yields

$$i_n = i_r + i_l \tag{2.15}$$

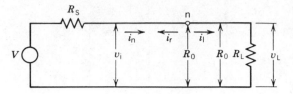

FIGURE 2.24 Transmission line with load.

where

i_n is the current into node n
i_r is the reflected current
i_l is the load current

Now,

$$i_n = \frac{v_n}{R_o} \tag{2.16}$$

$$i_r = \frac{v_r}{R_o} \tag{2.17}$$

$$i_l = \frac{v_L}{R_L} \tag{2.18}$$

where

v_n is the voltage at node n
v_r is the reflected voltage
v_L is the load voltage
R_o is the line characteristic impedance
R_L is the load resistance

Substituting into equation (2.15), we obtain

$$\frac{v_n}{R_o} = \frac{v_r}{R_o} + \frac{v_L}{R_L} \tag{2.19}$$

The voltage at the load when the first voltage wave arrives is equal to the voltage, v_n, at node n plus the reflected voltage, v_r. Rewriting equation (2.19) with these substitutions produces

$$\frac{v_n}{R_o} = \frac{v_r}{R_o} + \frac{v_n + v_r}{R_L} \tag{2.20}$$

Solving for the percentage, v_r/v_n of the originally transmitted wave that is reflected yields

$$\frac{v_n}{R_o} = \frac{v_r}{R_o} + \frac{v_n}{R_L} + \frac{v_r}{R_L} \tag{2.21}$$

$$\frac{v_n}{R_o} - \frac{v_n}{R_L} = \frac{v_r}{R_o} + \frac{v_r}{R_L} \tag{2.22}$$

$$v_n \left(\frac{1}{R_o} - \frac{1}{R_L}\right) = v_r \left(\frac{1}{R_o} + \frac{1}{R_L}\right) \tag{2.23}$$

$$\frac{v_r}{v_n} = \frac{(1/R_o - 1/R_L)}{(1/R_o + 1/R_L)} \tag{2.24}$$

$$\frac{v_r}{v_n} = \frac{[(R_L - R_o)/R_L R_o]}{[(R_L + R_o)/R_L R_o]} \tag{2.25}$$

$$\frac{v_r}{v_n} = \frac{R_L - R_o}{R_L + R_o} \tag{2.26}$$

Equation (2.26) gives the fraction of the initial wave that is reflected back to the source from the load as a function of the characteristic impedance of the line and the load resistance. This fraction is known as the *reflection coefficient*. From equation (2.26), it is seen that for an open-circuit termination of a transmission line ($R_L = \text{inf}$), the reflection coefficient is $+1$. Conversely, for a short-circuited transmission line, the reflection coefficient is -1. A reflection coefficient of $+1$ indicates that the reflected signal adds to the amplitude of the original transmitted signal, while a coefficient of -1 describes the situation where the reflected signal subtracts from the original transmitted signal. When the transmission line is terminated with a resistance equal to the characteristic impedance, the reflection coefficient is zero and no reflections occur. Let us examine some specific cases using different values of R_o and R_L.

To illustrate the situation where $R_L > R_o$, let $R_L = 5R_o$. Substituting into equation (2.26) gives

$$\frac{v_r}{v_n} = \frac{5R_o - R_o}{5R_o + R_o} \tag{2.27}$$

$$= \frac{4R_o}{6R_o} = \frac{2}{3} \tag{2.28}$$

Thus $\frac{2}{3}$ of the initial transmitted wave is reflected from the load because of the impedance mismatch. Since the sign of v_r/v_n is positive, the reflected wave will add to the amplitude of the initial transmitted wave. As this wave returns along the line, it will be absorbed or reflected at the source according to equation (2.26), except that R_L is replaced by R_s, the source resistance. In this particular case, if $R_s = R_o$, the wave will be absorbed by the source and no further reflections will occur. If $R_s \neq R_o$, reflections will occur back to the load. The steady state voltage at the load, after reflections have dampened out, can be easily calculated from

$$v_L = V\left[\frac{R_L}{R_L + R_s}\right] \tag{2.29}$$

With $R_L = 5R_o$ and $R_s = R_o$,

$$v_L = V\left[\frac{5R_o}{5R_o + R_o}\right] = \frac{5}{6}V \tag{2.30}$$

Insight can be obtained into the characteristics of this circuit by looking at voltage v_i as a function of time. In the subsequent developments, v_i is equivalent to v_n and the two will be used interchangeably. When the voltage V is first applied, the voltage source sees a voltage divider as shown in Figure 2.25. Thus

$$v_i = V\left[\frac{R_o}{R_o + R_o}\right] = \frac{V}{2} \tag{2.31}$$

When the wave reaches the load, a reflection, v_r, with an amplitude equal to $\frac{2}{3}$ of v_i or $V/3$ is generated, since $v_r/v_n = \frac{2}{3}$. This reflected wave travels back toward the source. Then the instantaneous voltage, v_L, at the load at this time is

$$v_L = v_i + \left(\frac{2}{3}\right)(v_i) \tag{2.32}$$

$$= \frac{V}{2} + \left(\frac{2}{3}\right)\left(\frac{V}{2}\right) \tag{2.33}$$

$$= \frac{V}{2} + \frac{V}{3} \tag{2.34}$$

$$= \frac{5}{6}V \tag{2.35}$$

The reflected wave traveling back toward the source sees a termination of $R_s = R_o$ and no further reflections occur. The amplitude of the voltage at the source is then

$$v_i + v_r \tag{2.36}$$

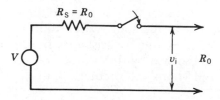

FIGURE 2.25 Initial environment for v_i calculation.

or

$$\frac{V}{2} + \frac{V}{3} = \frac{5}{6} V \tag{2.37}$$

This voltage is equal to the load voltage and steady state values are achieved.

Reflections also occur when R_L is less than R_o. As an example, let $R_L = R_o/5$ and $R_s = R_o$. The steady state value of voltage at the load is

$$V_L = V\left[\frac{R_L}{R_L + R_o}\right] \tag{2.38}$$

$$= V\left[\frac{(R_o/5)}{(R_o/5 + R_o)}\right] \tag{2.39}$$

$$= V\left[\frac{(R_o/5)}{(6R_o/5)}\right] \tag{2.40}$$

$$= \frac{V}{6} \tag{2.41}$$

$$\frac{v_r}{v_n} = \frac{R_L - R_o}{R_L + R_o} \tag{2.42}$$

$$= \frac{(R_o/5 - R_o)}{(R_o/5 + R_o)} \tag{2.43}$$

$$= \frac{(-4/5R_o)}{(6/5R_o)} \tag{2.44}$$

$$= -\frac{2}{3} \tag{2.45}$$

The minus sign for v_r/v_n indicates that the phasing of the reflected wave is such that it subtracts from the amplitude of the initially transmitted wave. As before, at the initial moment of the application of V to the circuit,

$$v_i = \frac{V}{2} \tag{2.46}$$

Upon encountering the unbalanced load of $R_L = R_o/5$, a reflection, v_r, is generated such that

$$v_r = v_n\left(-\frac{2}{3}\right) \tag{2.47}$$

$$= \frac{V}{2}\left(-\frac{2}{3}\right) \tag{2.48}$$

$$= -\frac{V}{3} \tag{2.49}$$

The voltage, v_L, at the load is

$$v_L = v_i + v_n \tag{2.50}$$

$$= \frac{V}{2} + \left(-\frac{V}{3}\right) \tag{2.51}$$

$$= \frac{V}{6} \tag{2.52}$$

At the source, the reflected wave and the initial source voltage sum as follows:

$$v_i + v_r \tag{2.53}$$

or

$$\frac{V}{2} + \left(-\frac{V}{3}\right) = \frac{V}{6} \tag{2.54}$$

As before, since $R_s = R_o$, no reflections occur and a steady state situation is established.

In the previous examples, R_s was set to equal R_o. If $R_s \neq R_o$, reflections would occur and be transmitted to the load. If $R_L \neq R_o$, the reflected wave would then be transmitted back to the source, and so on, until the reflections subsided and steady state was reached. The instantaneous voltage at any point in the circuit would be the sum of all of the voltage waves at that point at the time of measurement.

Since there are two variables, R_s and R_L, that can have values less than, equal to, or greater than R_o, a summary of each condition with associated advantages/disadvantages will be presented.

1. $R_s = R_o$, $R_L = R_o$—matched load and source resistance. Steady state voltage at load $\cong V/2$. The waveform transmitted to the receiver has minimum distortion in relation to other methods and essentially no oscillations. Allows for use of low tolerance parallel termination resistors around R_o value. Consumes power at load through the parallel termination resistor.

2. $R_s = R_o$, $R_L > R_o$—essentially a series termination, since $R_L > R_o$. The voltage wave is reflected from the load and back to the source. Since, at the source, $R_s = R_o$, no further reflection occurs. Steady state voltage at load is higher than with the fully matched case ($R_s = R_o = R_L$). Also, series termination has reduced power consumption over parallel termination. Power is consumed only when the signal is on the line.

3. $R_s = R_o$, $R_L < R_o$—negative reflections occur subtracting from the initial transmitted voltage. For $R_L \ll R_o$, reflected wave all but cancels transmitted wave, resulting in a low dc level at the load. This circuit is not useful for data transmission.

4. $R_s < R_o, R_L = R_o$—a parallel termination providing excellent transmission characteristics with little or no oscillation. The voltage at load is higher than with fully matched case. Again, power is consumed through parallel terminating resistance.

5. $R_s < R_o, R_L > R_o$—essentially unterminated. Oscillations occur around steady state value and eventually settle out. They may cause problems with thresholding-type receivers.

6. $R_s < R_o, R_L < R_o$—varying waveforms at different portions of transmission line. Not used.

7. $R_s > R_o, R_L = R_o$—delivers low amplitude voltages to the load relative to other useful cases. Low distortion and essentially no oscillations are produced.

8. $R_s > R_o, R_L > R_o$—waveform at load rises exponentially to steady state value, causing a delay in sensing the transmitted signal.

9. $R_s > R_o, R_L < R_o$—oscillation produced and very low amplitude voltage available at load. Not used.

Of the nine cases discussed, the four most useful are

1. $R_s = R_o, R_L = R_o$
2. $R_s = R_o, R_L > R_o$
3. $R_s < R_o, R_L = R_o$
4. $R_s < R_o, R_L > R_o$

Graphical representations of these four cases are given in Figures 2.27 through 2.30 for the transmission line circuit of Figure 2.26. In these figures, τ is the one-way transmission time from the source to the load.

2.2.2 APPLICATION OF TRANSMISSION LINE CHARACTERISTICS TO MOS CIRCUITS

Recalling a typical MOS driver as shown in Figure 2.31, we can apply the transmission line model. In this model, there is a transmission line section with characteristic impedance Z_o and a number of loads attached to the transmission line forming other line segments called *stubs*. Since the MOS loads are

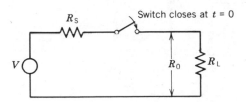

FIGURE 2.26 Transmission line circuit.

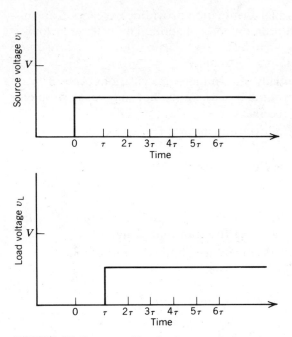

FIGURE 2.27 Source and load voltages with $R_s = R_o = R_L$.

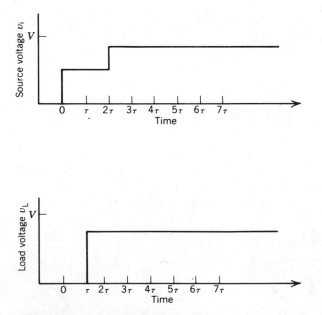

FIGURE 2.28 Source and load voltages with $R_s = R_o$, $R_L > R_o$.

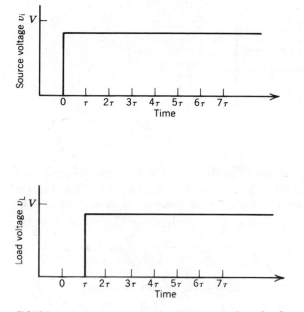

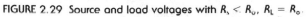

FIGURE 2.29 Source and load voltages with $R_s < R_o$, $R_L = R_o$.

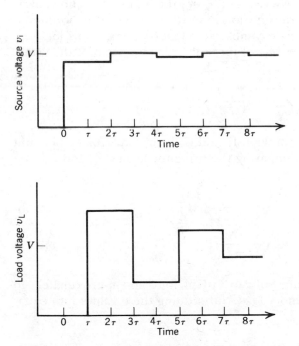

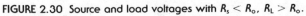

FIGURE 2.30 Source and load voltages with $R_s < R_o$, $R_L > R_o$.

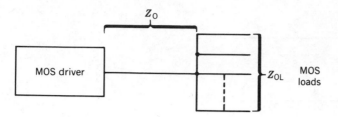

FIGURE 2.31 MOS driver and loads.

primarily capacitive, the characteristic impedance of the transmission line will be lowered and propagation delay will be increased. To minimize this effect, the transmission line length should be kept as short as possible. This rule applies to lengths of interconnect on a chip, printed circuit board, or off-board wiring runs.

2.3

MOS MEMORY AS TRANSMISSION LINE LOAD

Recall from equation (2.12) that the characteristic impedance, Z_o, of a transmission line is equal to $\sqrt{L/C}$, where L and C are the inductance and capacitance per unit length of the transmission line. In most cases the additional inductance of the load is negligible compared to that of the transmission line, so the characteristic impedance, Z_{oL}, of the stub or load portion of the transmission line is

$$Z_{oL} = \sqrt{\frac{L}{C + C_1}} \tag{2.55}$$

where L and C are as defined previously and C_1 is the capacitance per unit length of the attached load. Comparing the two impedances, Z_o and Z_{oL}, we find that

$$\frac{Z_{oL}}{Z_o} = \sqrt{\frac{L}{C + C_1}} \bigg/ \sqrt{\frac{L}{C}} \tag{2.56}$$

$$= \sqrt{\frac{C}{C + C_1}} \tag{2.57}$$

Typical values are in the range of 2 to 3 pF/in. for C and approximately 40 pF/in. for C_1 for an MOS memory load. Substituting these values into equation (2.57) yields

$$\frac{Z_{oL}}{Z_o} = \sqrt{\tfrac{2}{42}} = \sqrt{\tfrac{1}{21}} = 0.218 \tag{2.58}$$

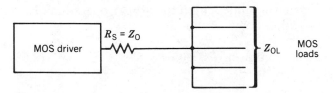

FIGURE 2.32 Series terminated MOS driving circuit.

Equation (2.58) indicates that the characteristic impedance Z_{oL}, of the loaded section or stub is approximately one-fifth of the characteristic impedance, Z_o, of the connection portion of the transmission line. This approximate calculation shows that unless the load is extremely close to the driver, an impedance mismatch exists and some type of termination is required. If we review the four practical termination options of

1. $R_s = R_o, R_L = R_o$ $\hspace{3cm}$ (2.59)
2. $R_s = R_o, R_L > R_o$ $\hspace{3cm}$ (2.60)
3. $R_s < R_o, R_L = R_o$ $\hspace{3cm}$ (2.61)
4. $R_s < R_o, R_L > R_o$ $\hspace{3cm}$ (2.62)

the series termination case, option 2, is a good choice for this application. Less power is consumed than in the parallel termination case, and reflections are absorbed after one round trip from the source to the load and return. Since the series termination resistor is approximating the source impedance, it should be placed physically as close to the MOS driver as possible. Figure 2.32 illustrates the series termination case.

2.3.1   OTHER METHODS OF AVOIDING OSCILLATION PROBLEMS

Another technique that is useful in eliminating problems of signal reflections is controlling the length of the transmission line, when feasible. Recall that the period of oscillation of a transmission line not terminated with the characteristic impedance Z_o is 2τ, where τ is the one-way transmission time. The harmful effect of this condition is that a single transmitted zero-to-one level change at the source will appear as one or more pulses at the load end of the line.

If the load was a clock or gating input to a register, the register would receive multiple clock pulses and function incorrectly. Figure 2.30 illustrates both the source and load voltages for this situation. A popular approach is to limit the maximum length of the transmission line segment, where feasible, to a value such that the 2τ pulse width caused by the oscillation is less than the minimum acceptable pulse width of the driven logic circuit input. For example, if the minimum clock input pulse width required by a register is 50 ns, 2τ must be less than 50 ns.

If l is the maximum transmission line length for proper operation of the register and a typical propagation time is 5.9 ns/m,

$$2\tau = 2(5.9 \text{ ns/m})l < 50 \text{ ns} \tag{2.63}$$

solving for l, yields

$$l < \frac{50 \text{ ns}}{2(5.9 \text{ ns/m})} = 4.24 \text{ m} \tag{2.64}$$

Thus, if the length of the transmission line is kept below approximately 4 m, the pulses caused by the oscillation will not be sensed by the clock input of the register.

A second method, aimed at eliminating the effect of oscillations on non-clocking digital inputs such as data inputs, is to delay the activation of the clock input to the device. If the clock input is not pulsed until the oscillations on the data inputs have dampened and the voltage at the load has reached a steady state value, the data lines will be read only when they are in a stable condition. In a periodic or synchronous system, the minimum clock period satisfying this condition is given by the general relationship

$$t_c > k\tau \tag{2.65}$$

where

t_c is the clock period
k is an odd integer
τ is the one-way propagation time

For the circuit in Figure 2.33 with a data line length of 1 m and a signal propagation time of 0.169 m/ns, t_c is calculated as

$$t_c > k\tau = k(5.9 \text{ ns/m}) (1 \text{ m}) \tag{2.66}$$
$$= k(5.9 \text{ ns}) \tag{2.67}$$

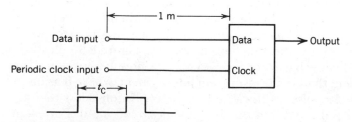

FIGURE 2.33 Data input entered by periodic clock signal.

If three round trips are required for the originally transmitted signal to reach steady state at the data input,

$$t_c > 3(5.9 \text{ ns}) = 17.7 \text{ ns} \tag{2.68}$$

The clock period, t_c, must be greater than 17.7 ns or the clock frequency, $1/t_c$, must be less than 56 MHz.

2.4
NOISE

In a sometimes mysterious manner, electrical disturbances called *noise* can cause faulty operation or even destruction of components in a digital system. Noise can be generated external to the digital system or induced by portions of the system itself.

2.4.1   NOISE SOURCES

Some typical noise sources are motors, generators, relays, and, in general, any device that switches large currents. A circuit of this type is shown in Figure 2.34. The inductance, L, in the wires associated with such devices produces a voltage equal to $L di/dt$, where di/dt is the rate of change of current during the switching interval. For example, an inductance of 100 nH in a line switching a current at a rate of 10 amperes/ns produces a voltage of 1000 volts. The current rise time of 10 amperes/ns in a switching circuit is not extraordinary. Thus the 1000-V noise spike generated during the switching of the current can cause problems in digital circuits through the power supply lines. This type of noise is reduced by the use of filters or transient suppressors in the power supply lines.

Noise from spark discharges and arcs that is radiated into a digital circuit is known as radio frequency interference or (*RFI*) or electromagnetic pulse

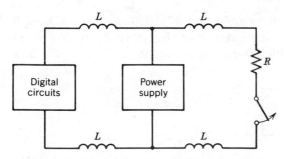

FIGURE 2.34 Current switching affecting digital circuits.

TABLE 2.1 Approximate Near Field Distance of
Influence for Sample Radiated Frequencies

Radiated Frequency (MHz)	Approximate Near Field Distance of Influence (ft)
1	170
5	34
10	17
20	8
50	3
100	2

(*EMP*). Radiation of this type consists of time-varying electric (E) and mag-netic (H) fields. The ratio E/H is called the *wave impedance*. The value of E/H at a great distance from the source of radiation is a constant 377 Ω. This number is known as the *far field* value. Conversely, the *near field* that extends approximately $\frac{1}{6}$ of a wavelength from the source, has a E/H value much greater or less than 377 Ω. Table 2.1 gives the near field distance of influence from the source for some sample values of radiated frequencies.

Looking at the distances in Table 2.1, we see that radiated noise interfer-ence within the general area of the digital system in all likelihood can be attributed to a near field; therefore, corresponding shielding action should be taken.

2.4.2   SHIELDING TECHNIQUES

The proper shielding method chosen is determined by the value of the E/H ratio. If E/H is much greater than 377 ohms, then the major source of inter-ference is an E field. Conversely, if E/H is much less than 377 Ω, the H field is the culprit. The E field sources are linear in nature and coupling to the digital circuit is primarily capacitive. The H fields emanate from loops, and coupling to the circuit is inductive. Figures 2.35 and 2.36 illustrate the E and H field coupling, respectively.

Binding posts and long pins on printed circuit boards can be sources of E field interference. Loops conducting current in circuits can generate an H field. The basic mechanism for shielding is to dissipate the field energy before

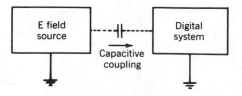

FIGURE 2.35 E field capacitive coupling.

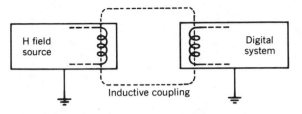

FIGURE 2.36 H field inductive coupling.

it reaches the digital circuit. This reduction in EMI field strength can be accomplished through absorption losses (I^2R) in the shielding material and losses caused by reradiation of the field from the shielding material. Absorption losses are effective in reducing the inductively coupled H field while reradiation reduces the strength of the capacitively coupled E field. For example, a grounded metal Faraday shield can be used to reduce E field interference in a circuit board while a ground plane can be used to reduce H field inductive coupling. These methods are illustrated in Figures 2.37 and 2.38, respectively. With respect to the relative material parameters and frequency of the EMI field, the absorption and reradiation losses can be increased, thus improving shielding, as shown in Table 2.2.

The table shows that a material that has a low permeability and high conductivity is useful in reducing E field radiation. Also, lower E field frequencies are easier to shield from than higher frequencies. For H fields, higher frequencies are easier to shield and thick materials with high permeability and high conductivity are most effective. Generally, high permeability materials are not among the highest in conductivity, so a compromise must be made.

Another important factor in the effectiveness of shields is the absence or presence of discontinuities in the shielding material. Openings or holes in the material cause induced currents to flow around them and, in many cases, reduce the ability of the material to shield the EMI field. Single large openings are less desirable than a series of evenly spaced, smaller openings that permit currents to flow between them. Seams in cabinets and connections between cabinets fall into the same category and must be considered in the overall shielding strategy.

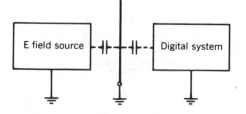

FIGURE 2.37 Grounded metal (Faraday) shield to reduce capacitive coupling.

TABLE 2.2 Maximizing EMI Field Losses

Field Losses	μ	σ	ω	t
E field				
loss increase (reradiation)	Low	High	Low	—
H field				
loss increase (absorption)	High	High	High	High

Note: μ is permeability, σ is conductivity, ω is the frequency of EMI, and t is thickness.

2.4.3 SUMMARY OF NOISE SOURCES, EFFECTS, AND NOISE REDUCTION TECHNIQUES

Table 2.3 summarizes the noise sources, corresponding effects, and noise reduction techniques relative to digital circuits. Because of the wide variety of combinations involved, the information in Table 2.3 has to be general in nature.

2.4.4 INTERCONNECTION WIRING NOISE REDUCTION

The two most commonly used physical wiring means of connecting digital circuits are coaxial cable and twisted pair. A cross section of coaxial cable is shown in Figure 2.39. If the current that flows in the center conductor is equal to the current flowing in the shield, the resulting magnetic fields from both conductors will effectively cancel each other at points outside the cable. The cable shield conductor should be connected to the circuit to ensure that the two currents are equal. A schematic of the proper connection is given in Figure 2.40. In many cases, the ground at digital circuit A is not at the same potential as the ground at digital circuit B. Thus a ground loop is introduced as shown in Figure 2.41. Usually, this type of loop can be tolerated. In ex-

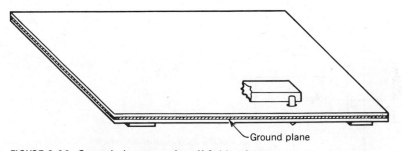

Ground plane

FIGURE 2.38 Ground plane to reduce H field radiation.

TABLE 2.3 Summary of Noise Sources, Effects, and Noise Reduction Techniques Applicable to Digital Circuits

Type of noise	Supply voltage line transients	Electromagnetic interference (EMI) EMP/RFI	Electrostatic discharge ESD	Ground noise
Source	Voltage generated by line inductance ($v = Ldi/dt$) as a result of current switching	Spark discharges, arcs	Static voltage discharge	Unwanted currents in ground line caused by potential difference among grounds in a circuit
Method of Introduction into Circuit	Conduction through power lines	Radiation	Physical contact	Conduction through ground lines
Means of Noise Reduction	Transient suppressors, power line filters, circuit layout, good grounding techniques, on-board voltage regulation	Circuit layout, shielding, reduce area of current loops (i.e., reduce loop inductance), coaxial cable	Transient suppressors, buffering circuits between cpu and keyboards, and so on	Coaxial cable, circuit layout, ground plane, opto-isolators
Typical Symptoms	Intermittent operation, software malfunctions	Intermittent operation, unrepeatable incorrect operation	Software malfunctions, damage or destruction of components, input part damage	Intermittent operation, circuit damage, software malfunction

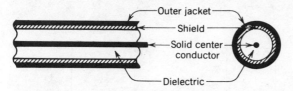

FIGURE 2.39 Coaxial cable.

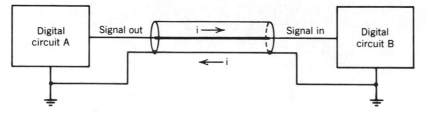

FIGURE 2.40 Proper coaxial cable connection for EMI shielding.

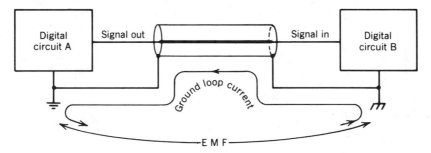

FIGURE 2.41 Ground loop introduction in digital circuits.

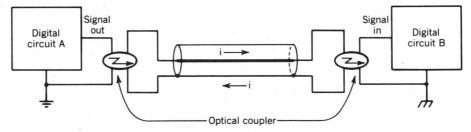

FIGURE 2.42 Optical coupling to break the ground loop.

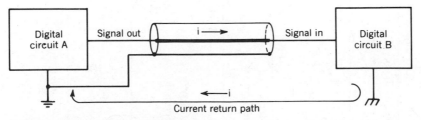

FIGURE 2.43 Noisy ground return path.

tremely noisy situations, optical coupling can be used to break the ground loop as illustrated in Figure 2.42. If the coaxial cable shield is connected to ground at only one end, electrostatic shielding is accomplished, but no magnetic shielding is provided. In this instance, the current return path is through the ground system as shown in Figure 2.43 and can result in relatively large ground current noise due to the potential difference between the two ground points.

The twisted pair provides almost as good magnetic interference protection as a coaxial cable when frequencies are 1 MHz or less. Since the twisted

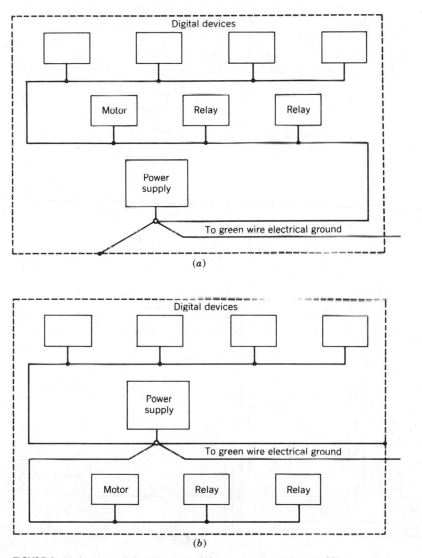

FIGURE 2.44 Incorrect (*a*) and correct (*b*) way to connect ground lines.

pair consists of two wires twisted around each other, magnetic field generation and reception are essentially canceled as with the coaxial cable. The
twisted pair does not provide any electrostatic shielding, however.

2.4.5   GROUNDING CONNECTIONS

In digital circuits, there are usually two main classes of grounds, "clean" and
"noisy." A "clean" ground is that associated with the conventional integrated

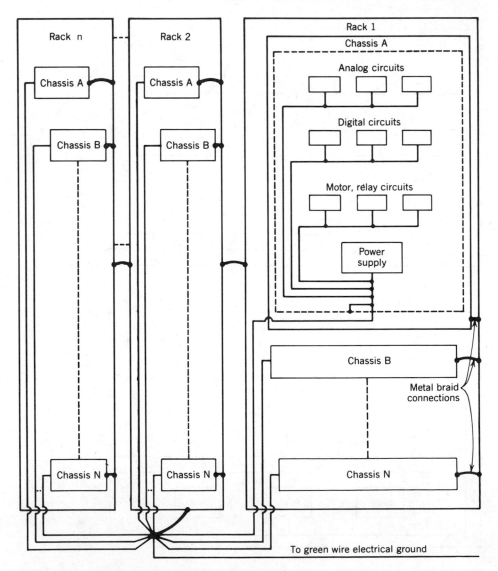

FIGURE 2.45 Proper grounding technique for multiple racks.

circuit devices while "noisy" grounds are those connected to motors, relays, and so on. It is good practice not to have currents from both types of devices running through the same ground. Figure 2.44a shows the improper grounding method, and Figure 2.44b shows the recommended technique.

If analog circuits are in the system, they should also have a separate ground line connected to the power supply ground. Note that in the correct grounding method, the metal shield or case of the system is connected to signal ground.

From a systems point of view, where there may be multiple racks of equipment with individual chassis within the rack, the proper grounding hierarchy is given in Figure 2.45.

PROBLEMS

2.1 The clocked data input to a digital circuit is not properly terminated and oscillations occur when step changes appear at the input. The clock input to the circuit has a frequency of 43.5 MHz. If the line length from the clock source is 0.5 m and three round trips are required for the originally transmitted signal to reach steady state at the data input, will the data line be clocked after it has reached a stable condition? If not, what clock frequency will allow the stable data input?

2.2 The characteristic impedance, Z_0, of different conductor geometries can be calculated as follows:

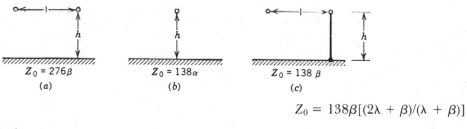

$$Z_0 = 138\beta[(2\lambda + \beta)/(\lambda + \beta)]$$

where

$\alpha = \log(4h/d)$

$\beta = \log(2l/d)$

$\lambda = \log(2h/l)$

d is the conductor diameter

For what values of l and h would the characteristic impedances of the conductor configurations in (a) and (b) be equal, assuming that both configurations are utilizing conductors of the same diameter?

FOR FURTHER READING

Morrison, Ralph, *Grounding and Shielding Techniques in Instrumentation*, 2nd. ed., John
    Wiley & Sons, Inc., New York, 1977.
Gray, J. P. *VLSI 81*, Academic Press, New York, 1981.
Williams, J., and S. Dendinger, "Use Motor-Drive IC to Solve Tricky Design Prob-
    lems," *EDN*, May 12, 1983.

3

BUS INTERFACING

The paths or buses over which various types of information are passed among microcomputer-based system elements deserve separate attention because of their critical impact upon overall system operation and performance. Buses are generally categorized in terms of their timing protocol, that is, *asynchronous, synchronous,* and *synchronous utilizing a WAIT state, (semisynchronous).* The general procedures and operations that occur in microprocessor interfacing to peripheral devices and memories are really a subset of the general bus discussion. In addition to the address, data, and control buses associated with microcomputers, handshaking and arbitration (bus conflict resolution) buses are necessary to implement the various bus alternatives. Fundamental considerations associated with all these aspects of bus interfacing are the delays involved with the bus transactions. An important aspect of bus interfacing involves memories in microprocessor systems. Specific examples in these areas are covered in the latter part of this chapter.

3.1

DELAYS

In general, there are three types of delays associated with signals in a digital system: *logic, capacitive,* and *transit time.* A *logic* delay (sometimes known as *propagation* delay) is the time required for the output of a digital element to switch with respect to the input. Typical TTL gate delays are around 3–6 ns. *Skew,* or the time difference between two signals originating from the same or coincident sources but experiencing different delays, primarily is the result of logic delay. An example of skew caused by different logic delays is shown in Figure 3.1.

Capacitive delay is caused by capacitive loading on the output of a logic element. Figure 3.2 is a typical plot of propagation delay versus load capacitance for the low-power Schottky TTL family. From Figure 3.2 it can be seen that the average delay time increases at a rate of 0.08 ns/pF and has a value of approximately 5 ns for a 15-pF load.

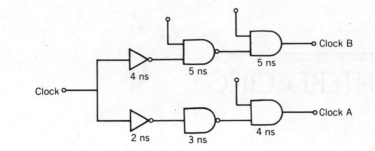

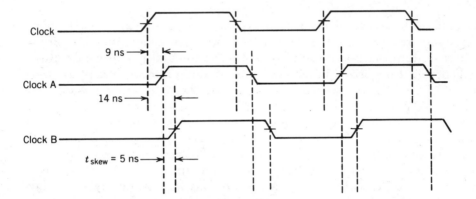

FIGURE 3.1 Skew caused by logic delays.

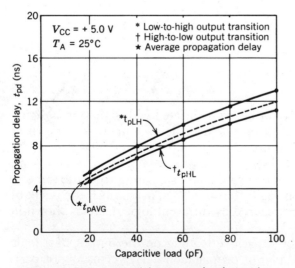

FIGURE 3.2 Propagation delays versus load capacitance for LSTTL: * Low-to-high output transition; † high-to-low output transition; ★ average propagation delay.

Transit time delays are the time necessary for the digital signal to travel along a wire or printed circuit board trace. From transmission line theory, the worst case transit time delay per unit length, tp(ns), is equal to $\sqrt{LC}$, where L is inductance per unit length and C is capacitance per unit length and device capacitance. A typical value of tp is 2 ns/ft.

Thus, in considering driving a bus through gates, decoders, and so on, the total of all the delays must be considered along with the resulting skewing of signals. A good example is the presentation of an address to the bus. Because of varying delays in the address line paths, skewing of the address bits may occur at the bus. Setup time allows for skew and other delays as discussed.

3.2

ASYNCHRONOUS BUSES

With an asynchronous bus, the transaction time on the bus is a function of the individual device rather than of a fixed clock rate as in a synchronous bus. In an asynchronous bus system, the handshaking signals are usually referred to as Master and Slave signals. These signals are generated by appropriate elements taking part in a bus transaction.

An advantage of the asynchronous bus is that both fast and slow devices can be mixed on the bus. Thus the bus speed is a function of the devices that are communicating at the time. Newer, faster devices can be added to the bus, increasing the speed as technology improves. When signals are applied to an asynchronous bus by a Master device, a *sync* pulse must also be presented on the bus. The Slave then responds with *acknowledge* signals. In general, then, asynchronous buses operate with handshaking occurring between two devices in acquiring data from the bus. Because of this handshaking requirement, a time-out safety feature must be built in so as to abort a bus cycle in the event of a failure that results in no handshaking response.

3.2.1 ASYNCHRONOUS BUS READ

A typical asynchronous bus Read transaction is illustrated in Figure 3.3.

In the Read operation, the Master places the address on the bus and then allows for the address lines to stabilize. The Master then goes active, requesting data from the Slave. The Slave places the data on the bus and, after allowing for the data setup time, goes active to indicate to the Master that data are available for reading. The Master accepts the data and indicates this by going inactive. The data must be held stable for a hold time after being read by the Master. The Slave acknowledges the Master's receipt of the data by also going to the inactive state. Data are then removed from the data bus.

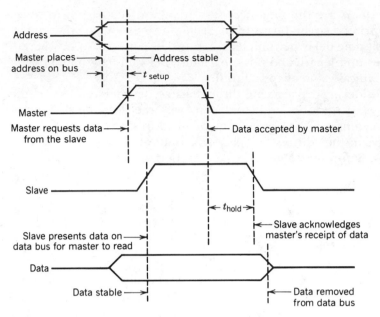

FIGURE 3.3 Typical asynchronous bus read.

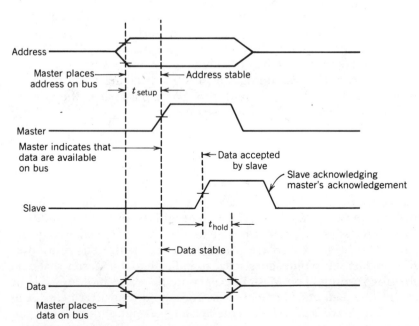

FIGURE 3.4 Typical asynchronous bus write.

3.2.2 ASYNCHRONOUS BUS WRITE

For an asynchronous bus write, as shown in Figure 3.4, the bus Master places the addresses and data on the bus. After a setup time delay to allow the address and data lines to become stable, the Master signal becomes active, indicating to the Slave that data are available on the bus. The Slave then accepts the data and goes active to indicate that it has the data. The data have to be held stable for a period equal to the hold time after the Slave accepts the data. The Master then goes to the inactive state, acknowledging the Slave's receipt of the data. The Slave, in turn, returns to the inactive state, acknowledging the Master's acknowledgment.

The type of bus protocol described for the asynchronous bus read and write just described is referred to as *fully interlocked* because of the assertions and acknowledgments between the Master and Slave.

3.3

SYNCHRONOUS BUSES

A synchronous bus operates as its name implies, in fixed clock intervals. Thus requests for data from a Slave by the Master must be honored within a fixed period of time or "garbage" will be read from the bus. This clocked operation of a bus results in faster operation, since all elements on the system are in synchronism and operate within known times in interactions with each other.

In many instances synchronous buses have fewer lines and are easier to understand and test. The rate of propagation of the signals is particularly critical in synchronous bus designs. Since the clock is the key timing signal in a synchronous bus, its arrival at various points along the bus is critical. The clock can either be designed to travel along the bus with inherent delays occurring before it reaches devices at the bus extremes or it can be designed to arrive simultaneously at each connector point. The latter, of course, is accompanied by increased cost but results in a higher-speed bus.

Synchronous buses use partial handshaking, since a device can take more than one clock cycle to respond to a master. Thus, if the time of one clock cycle is not enough for a slave device to accommodate the master, it must increase its time to respond by multiples of the clock cycle. This can be a disadvantage with respect to the asynchronous bus, since the asynchronous bus takes the exact time required and no more.

3.3.1 SYNCHRONOUS BUS READ AND WRITE

A synchronous bus Read is shown in Figure 3.5. The setup and hold conditions apply as discussed in the asynchronous protocol. The example utilizes a

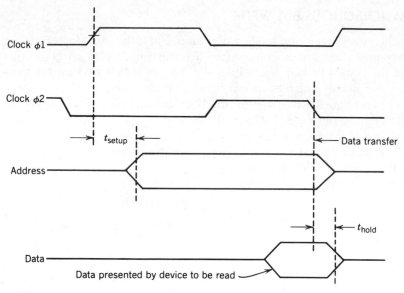

FIGURE 3.5 Synchronous bus read.

nonoverlapping two-phase clock with addresses presented on phase 1 active and data transfers occurring on phase 2 active. Similarly, a synchronous bus write is given in Figure 3.6.

3.3.2   SYNCHRONOUS BUS UTILIZING WAIT STATE (SEMISYNCHRONOUS)

This type of bus is a synchronous bus that supports "slow" devices by providing for a WAIT state. The WAIT state is essentially a stretching of the clock time to allow a slow device to respond. A good example is a slow memory coupled to a microprocessor. Figure 3.7 shows a microprocessor whose synchronous read operation requires that any memory used with it have an access time of 500 ns or less. Further assume that the microprocessor utilizes a two-phase clock and that memory is addressed during phase 1 and data are transferred during phase 2. Under these circumstances, an Erasable Programmable Read Only Memory (EPROM) with an access time of 1 microsecond could not normally be used. But, if the bus allows for a WAIT state, the slow memory could be used. Typically, the microprocessor has a READY input that, when activated, extends phase 2 to allow the slow memory to respond to an access. The extension of the phase 2 clock by the active low READY line is depicted in Figure 3.8.

Because of the use of dynamic registers in the microprocessor, and their refreshing requirements, phase 2 width cannot be extended arbitrarily. A

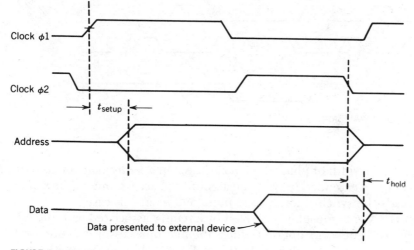

FIGURE 3.6 Synchronous bus write.

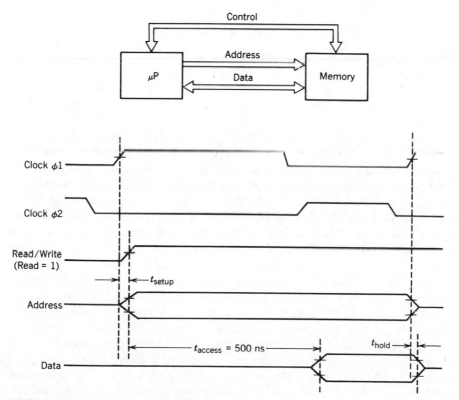

FIGURE 3.7 Synchronous bus with 500-ns access time.

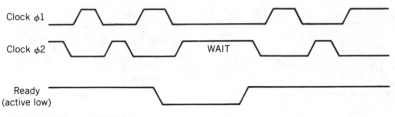

Clock φ1

Clock φ2   WAIT

Ready
(active low)

FIGURE 3.8 Phase 2 WAIT state.

typical maximum width of phase 2 is 5 μs or less. On a semisynchronous basis, phase 2 should be extended only by the amount the access time of the slowest device exceeds the synchronous access time (500 ns in this case).

The READY signal must be generated by circuitry associated with the slow memory system. The WAIT option reduces the microprocessor execution rate only when it accesses a slow device. In this example, when devices with access times of 500 ns or less are used, operation is the same as with a synchronous bus.

3.3.3   THIRTY-TWO-BIT BUS STANDARDS

At this time there are six major 32-bit buses, not all standards. They are Multibus II, FutureBus, FastBus, NuBus, VersaBus, and the VME-32 bus. FastBus is an IEEE standard used in the specialty nuclear experimentation area and VersaBus has been superceded by VME-32. Thus there are four major 32-bit buses aimed at general use.

Trying to preempt a proliferation of incompatible 32-bit buses, the IEEE Computer Society Microprocessor Standard Committee established the P896 FutureBus Standardization Committee in 1978. Some early agreements resulting from P896 deliberations were that the FutureBus would utilize handshaking, be based on Eurocards, use 100-mA drivers, and use a single 96-pin connector. While the P896 work progressed at a relatively slow pace, a number of 32-bit buses began to emerge from the chip manufacturers, utilizing some of the P896 work. These buses were the VME bus (Motorola), NuBus (Western Digital/Texas Instruments), and Multibus II (Intel).

The VME-32 bit bus was derived from the earlier 16-bit VME bus and is closely associated with the 68000 family of microprocessors. Because of its 16-bit lineage, the VME-32 does not provide for individual strobes for the four groups of bytes that are present on the bus. Thus 24-bit transfers cannot be accomplished cleanly. To transfer a 24-bit word, 32-bit words aligned at odd memory locations must be transmitted by sending a single byte followed by the remaining 3 bytes. The VME bus was developed in 1981 and is, therefore, the oldest of the 32-bit microprocessor buses; VME-32 is based on double-size Eurocards, is handshaken, and has a maximum speed of 152 Mb/s for a single read transfer. Because of its relationship with the Motorola 68000 family of processors, VME-32 is, and will continue to be, a popular bus.

The NuBus is a synchronous 32-bit bus that was derived from a Texas Instrument/Western Digital synchronous bus that was, in turn, adapted from an MIT handshaken bus. NuBus is similar to the Multibus II in many respects. It can transmit 8-, 16-, 24-, and 32-bit words, utilizes a triple size card, and performs a single read transfer at a rate of 106 Mb/s. The NuBus was used in the Texas Instruments' NuMachine and will probably not be widely used or supported.

The P896 Future Bus specification was finalized in 1984. It has the highest speed of all four buses under discussion with a single read transfer rate of 200 Mb/s. It is costly to implement and uses ECL drivers. There is scanty support for the bus at the present time. It is based on a triple Eurocard, has individual byte strobes, and uses three-way handshaking.

Multibus II, being pushed by Intel Corporation, is a popular bus and will be widely supported. A double-size Eurocard is preferred for this bus. The single read transfer rate of the synchronous (10 MHz) Multibus II is 106 Mb/s. A Command and Service Module (CSM) is needed on the Multibus II for initialization and clocking functions.

Of the buses discussed, VME-32 and Multibus II will vie for popularity and will coexist. The Intel and National Semiconductor family of microprocessors are more easily utilized on a Multibus II system while the VME-32 bus is oriented toward the 68000 family of processors.

3.4

MEMORY AND PERIPHERAL INTERFACING

In this section, interfacing to dynamic RAM and Erasable Programmable ROM is discussed. Examples include Low Power Schottky TTL and High Speed CMOS devices. These techniques are then extended to peripheral devices in general.

3.4.1 DYNAMIC RAM

The considerations discussed in the previous sections of this chapter apply to interfacing dynamic RAMs to microprocessors. As an example, interfacing the Intel 2118 16K dynamic RAM to a microprocessor will be discussed. A block diagram of the 2118 is given in Figure 3.9.

The 2118 is essentially a 128×128 matrix of memory cells that are accessed by row addresses presented during the RAS (Row Address Strobe) and column addresses presented during the CAS (Column Address Strobe) cycles. The 16K memory space is addressed by having RAS active while row

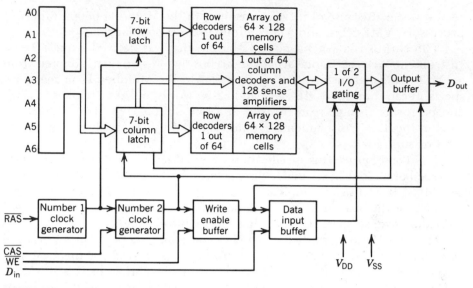

FIGURE 3.9 Intel 2118 Dynamic RAM.

addresses are presented on input lines A_0–A_6 and, similarly, column addresses are presented on the same input lines while CAS is active. Note that RAS can remain active while CAS is active, since the RAS and CAS circuits are independent. Thus successive CAS cycles can be originated without reapplying RAS to implement block transfers of data known as Page Mode. If CAS is not applied following a RAS active, a refresh-only operation is implemented. The correspondence between the input address lines and the absolute 14 address bits required to access the 16K of memory is shown in Table 3.1. Accessed data appear at the D_{out} pin. Pinouts of the 2118 are given in Figure 3.10.

POWER SUPPLY

A positive 5 V on V_{ss} is required to operate the 2118, with ground on V_{DD}. The negative voltage usually required for such memories is developed internally by means of a back-bias generator. The use of the back-bias technique

```
        ┌───∪───┐
N/C ⊏ 1        16 ⊐ Vss
Din ⊏ 2        15 ⊐ CAS
 WE ⊏ 3        14 ⊐ Dout
RAS ⊏ 4   2118 13 ⊐ A6
 A0 ⊏ 5        12 ⊐ A3
 A2 ⊏ 6        11 ⊐ A4
 A1 ⊏ 7        10 ⊐ A5
VDD ⊏ 8         9 ⊐ N/C
        └───────┘
```

FIGURE 3.10 2118 Pinouts.

TABLE 3.1 Multiplexed Address Inputs

	Input Address	Absolute Row Address	Absolute Column Address
$\overline{\text{RAS}}$	$A_0 \oplus A_2$	A_6	—
	$A_1 \oplus A_2$	A_5	—
	A_2	A_4	—
	A_3	A_3	—
	A_4	A_2	—
	A_5	A_1	—
	A_6	A_0	—
$\overline{\text{CAS}}$	A_0	—	A_{13}
	A_1	—	A_{12}
	A_2	—	A_9
	A_3	—	A_{11}
	A_4	—	A_{10}
	A_5	—	A_8
	A_6	—	A_7

produces additional current requirements from the power supply during turn-on. This occurs because time is required after turn-on for the internally produced negative bias voltage to be generated, during which period all MOS transistors in the device will be on and conducting. At power-on with $\overline{\text{RAS}}$ and $\overline{\text{CAS}}$ equal to 0 V, the current (I_{DD}) characteirstic of the 2118 is as depicted in Figure 3.11. Figure 3.12 illustrates an alternative power-on characteristic that is obtained by having $\overline{\text{RAS}}$ and $\overline{\text{CAS}}$ track V_{DD} through pull-up resistors from the $\overline{\text{RAS}}$ and $\overline{\text{CAS}}$ inputs to V_{DD}. This configuration significantly reduces the power-on current load to the power supply.

Power consumed by the 2118 is a function of the three types of currents drawn by the memory chip. These currents are the operating current (I_{DDO}),

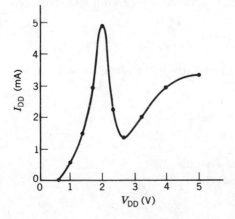

FIGURE 3.11 Typical 2118 power-on current with $\overline{\text{RAS}} = \overline{\text{CAS}} = 0$.

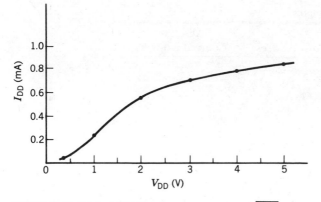

<u>FIGURE</u> 3.12 Typical 2118 power-on current with $\overline{RAS}$ and $\overline{CAS}$ tracking V_{DD}.

standby current (I_{DDS}), and refresh current (I_{DDR}). Devices not selected by RAS draw standby current. These three currents are calculated as follows:

$$I_{DDO} = K(I_{DD2} + I_{DDLO})$$

where

I_{DD2} = operating V_{DD} supply current

I_{DDLO} = output load current (I_{OL}) plus leakage (I_{Lk}) to n other devices = $(I_{OL} + nI_{Lk})$

K = number of active devices selected by both $\overline{CAS}$ and $\overline{RAS}$ at one instance

$$I_{DDS} = M(I_{DD1})$$

where

M = number of inactive devices, that is, devices receiving $\overline{CAS}$-only cycles and not selected by $\overline{RAS}$.

I_{DD1} = V_{DD} supply current.

The standby mode retains data but does not support reading or writing activities

$$I_{DDR} = N(I_{DD3})(t_{RC}/t_{REF})(128)$$

where

N = total number of devices in the system

I_{DD3} = V_{DD} supply current, $\overline{RAS}$-only cycle

t_{RC} = refresh cycle time

t_{REF} = time between refresh cycles

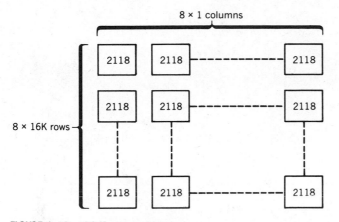

FIGURE 3.13 128 K memory array.

Total power, P_T, is the product of the maximum value of the supply voltage, V_{DD}, and the sum of I_{DDO}, I_{DDS}, and I_{DDR}.

$$P_T = (V_{DD,max})(I_{DDO} + I_{DDS} + I_{DDR})$$

As an example, let us calculate the power consumed by a 128K-byte memory array comprised of 2118's. The memory array will be organized as in Figure 3.13. Thus there will be 64 memory devices, making $N = 64$. For any row accessed, 8 devices will be active at any one time ($K = 8$). The number of standby chips, M, is equal to N-K or 56. From data sheets, values of the other required currents and times can be obtained. Typical values are given as follows for this example.

$$t_{REF} = 2 \text{ ms}$$
$$t_{RC} = 350 \text{ ns}$$
$$I_{DD1} = 2.5 \text{ mA}$$
$$I_{DD2} = 22 \text{ mA}$$
$$I_{DD3} = 15 \text{ mA}$$
$$I_{Lk} = 10 \ \mu A$$
$$I_{OL} = 250 \ \mu A$$

Inserting these values in the appropriate equations and assuming n (number of inactive devices sharing output data lines and drawing leakage current) = 3 yields:

$$I_{DDO} = 8[22 \text{ mA} + 250 \ \mu A + 3(10 \ \mu A)]$$
$$= 178.24 \text{ mA}$$
$$I_{DDS} = 56(2.5 \text{ mA})$$
$$= 140 \text{ mA}$$

$$I_{DDR} = (64 \times 15 \text{ mA}) \left(\frac{350 \text{ ns}}{2 \text{ ms}}\right) (128)$$
$$= (960 \text{ mA})(0.0224)$$
$$= 21.50 \text{ mA}$$

Total power, P_T is then calculated as

$$P_T = V_{DD,max}(178.24 \text{ mA} + 140 \text{ mA} + 21.50 \text{ mA})$$
$$= V_{DD,max}(339.74 \text{ mA})$$

If $V_{DD,max} = 5.5 \text{ V}$

$$P_T = (5.5 \text{ V})(0.339 \text{ A})$$
$$= 1.8 \text{ W}$$

Summarizing, we find that 1.8 W of power are required for the 128K × 8 memory array comprised of the 64 2118 chips.

MEMORY CYCLES

Memory cycles begin with a negative transition on the $\overline{\text{RAS}}$ line. The different types of cycles are illustrated in Figures 3.14 through 3.16. In the figures, WE is the Write Enable signal that is active low.

All memory cycles are initiated by a high-to-low transition of the RAS signal. For a Read cycle, a CAS high-to-low transition occurs as shown in Figure 3.14. RAS and CAS are TTL compatible signals and have minimum pulse widths as specified in data sheets. Also, for a Read cycle, WE remains high. The D_{out} line is at a high impedance state until valid data appear at the output.

In Write cycles, the data in (D_{in}) must be valid at or prior to WE or CAS going low, whichever is the latest. One type of Write cycle, a Read-Modify-Write cycle, reads data and then the addressed memory location contents are modified. This cycle is accomplished by WE going low while RAS and CAS are low as shown in Figure 3.14. The value on D_{out} is read as usual and is not affected during the Modify-Write portion of the cycle. A Delayed Write cycle is similar to a Read-Modify-Write except that timing constraints for reading are not met and D_{out} is undefined. The Early Write Cycle, also shown in Figure 3.14, is implemented by WE going low before CAS goes low. D_{out} stays in the high impedance state and D_{in} is written into the selected bit location.

The basic operation needed to refresh the 2118 is a cycle at each of the 128 row addresses. This can be accomplished by any of the memory cycles such as Read, Write, and so on. Write cycles will, however, cause the addressed memory cell to change state while refreshing of the other cells is occurring. The best refresh mode is the RAS Only cycle as outlined in Figure 3.15. In this mode, multiple rows of memory devices can be accessed without

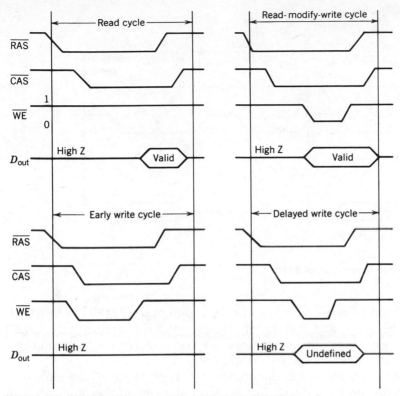

FIGURE 3.14 Read and write cycles of the 2118 RAM.

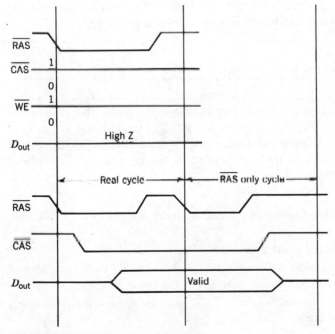

FIGURE 3.15 Refresh cycles of the 2118.

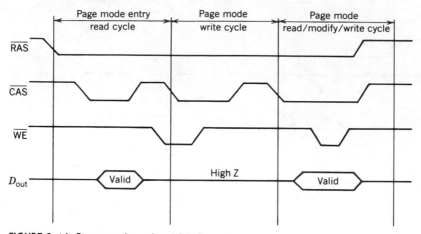

FIGURE 3.16 Page mode cycle of 2118.

bus contention problems. The data out (D_{out}) pins of the 2118's can be wired-ORed together. The READ cycle refresh, also shown in Figure 3.15, should only be used whenever the memory system is composed of single row devices. Otherwise, bus contention problems will occur.

For block read and write operations, the Page Mode Cycles shown in Figure 3.16 should be utilized. These cycles allow multiple columns of an addressed row to be accessed with one activation of the RAS signal and multiple activations of the CAS signal. Thus Page Mode timing is a function of the repetitive CAS signal pulse width and internal precharge timing.

TIMING ESTIMATES

In the development of memory systems, there are many instances where exact timing information is not explicitly available for particular devices or associated wiring. Since this information is useful in developing worst case timing estimates for the memory system, some "rules of thumb" with regard to TTL devices have evolved that are useful in timing calculations. These rules are listed as follows:

1. 0.08 ns/pF are added to the propagation delay of devices for capacitive loads.
2. Traces on PC boards add 2 to 3 pF/in.
3. Minimum propagation delay of a TTL device can be estimated as $\frac{1}{2}$ of the typical propagation delay time given on the data sheet.
4. Maximum skew between Schottky TTL devices in the same package is approximately 0.5 ns.

DETAILED DYNAMIC RAM DESIGN EXAMPLE

With the background material presented on the 2118 dynamic RAM, it is valuable to perform a detailed memory system design using the 2118 part and the 8202A dynamic RAM controller. The controller serves as the interface between the 8086 family of microprocessors and the RAM chips and can directly address and drive up to 64K of RAM. It also provides address multiplexing and refresh control. In this example, the 8202A will be used with the 8088 microprocessor.

A block diagram of the 8202A is given in Figure 3.17 and its pin configuration is shown in Figure 3.18. Descriptions of the pin functions are detailed in Table 3.2. Characteristic of the 8086 family of microprocessors, the 8088 supports minimum parts count implementation (minimum) or large system (maximum) designs. Because the 40-pin package of the 8088 limits the signals

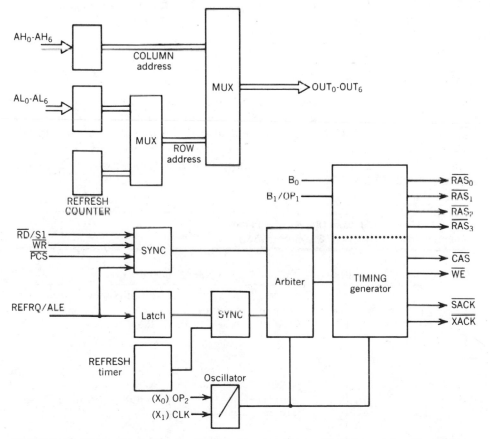

FIGURE 3.17 8202A block diagram. Reprinted by permission of Intel Corporation, Copyright 1986.

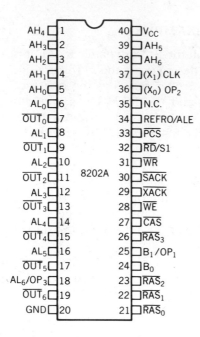

FIGURE 3.18 8202A Pin outs. Reprinted by permission of Intel Corporation, Copyright 1986.

that can be provided, a MN/$\overline{\text{MX}}$ strap pin (pin 33) is provided to permit choice of the required design. In the MN (minimum) mode, pin 33 is tied to V_{cc} and bus control signals are generated directly on pins 24–31 and pin 34. These signals include ALE (address latch enable) $\overline{\text{WR}}$ (Write), and $\overline{\text{INTA}}$ (interrupt acknowledge). Thus, in the minimum mode, the microprocessor generates the $\overline{\text{RD}}$ and $\overline{\text{WR}}$ signals directly.

TABLE 3.2 8202A Pin Functions (Reprinted with the permission of Intel Corporation, Copyright, 1986.)

Symbol	Pin No.	Type	Name and Function
AL_0	6	I	Address Low: CPU address inputs used to generate
AL_1	8	I	memory row address.
AL_2	10	I	AL_6/OP_3 used to select 4K RAM mode.
AL_3	12	I	
AL_4	14	I	
AL_5	16	I	
AL_6/OP_3	18	I	
AH_0	5	I	Address High: CPU address inputs used to gener-
AH_1	4	I	ate memory column address.
AH_2	3	I	
AH_3	2	I	
AH_4	1	I	
AH_5	39	I	

TABLE 3.2 (Continued)

Symbol	Pin No.	Type	Name and Function
AH_6	38	I	
B_0	24	I	Bank Select Inputs: Used to gate the appropriate
B_1/OP_1	25	I	RAS_0–RAS_3 output for a memory cycle. B_1/OP_1 option used to select the Advanced Read Mode.
$\overline{PCS}$	33	I	Protected Chip Select: Used to enable the memory read and write inputs. Once a cycle is started, it will not abort even if $\overline{PCS}$ goes inactive before cycle completion.
$\overline{WR}$	31	I	Memory Write Request.
$\overline{RD}/S1$	32	I	Memory Read Request: S1 function used in Advanced Read mode selected by OP_1 (pin 25).
$\overline{REFRQ}/$ ALE	34	I	External Refresh Request: ALE function used in Advanced Read mode, selected by OP_1 (pin 25).
$\overline{OUT_0}$	7	O	Output of the Multiplexer: These outputs are de-
$\overline{OUT_1}$	9	O	signed to drive the addresses of the Dynamic
$\overline{OUT_2}$	11	O	RAM array. For 4K RAM operation, $\overline{OUT_6}$ is
$\overline{OUT_3}$	13	O	designed to drive the 2104A $\overline{CS}$ input. (Note that
$\overline{OUT_4}$	15	O	the $\overline{OUT}_{0-6}$ pins do not require inverters or driv-
$\overline{OUT_5}$	17	O	ers for proper operation.
$\overline{OUT_6}$	19	O	
$\overline{WE}$	28	O	Write Enable: Drives the Write Enable inputs of the Dynamic RAM array.
$\overline{CAS}$	27	O	Column Address Strobe: This output is used to latch the Column Address into the Dynamic RAM array.
$\overline{RAS_0}$	21	O	Row Address Strobe: Used to latch the Row Ad-
$\overline{RAS_1}$	22	O	dress into the bank of dynamic RAMs, selected by
$\overline{RAS_2}$	23	O	the 8202A Bank Select pins (B_0, B_1/OP_1).
$\overline{RAS_3}$	26	O	
$\overline{XACK}$	29	O	Transfer Acknowledge: This output is a strobe indicating valid data during a read cycle or data written during a write cycle. $\overline{XACK}$ can be used to latch valid data from the RAM array.
$\overline{SACK}$	30	O	System Acknowledge: This output indicates the beginning of a memory access cycle. It can be used as an advanced transfer acknowledge to eliminate wait states. (Note: If a memory access request is made during a refresh cycle, $\overline{SACK}$ is delayed until $\overline{XACK}$ in the memory access cycle).
$(X_0)\ OP_2$	36	I/O	Oscillator Inputs: These inputs are designed for a
$(X_1)\ CLK$	37	I/O	quartz crystal to control the frequency of the oscillator. If X_0/OP_2 is connected to a $1K\Omega$ resistor pulled to +12 V then X_1/CLK becomes a TTL input for an external clock.
N.C.	35		Reserved for future use.
V_{CC}	40		Power Supply: +5 V.
GND	20		Ground.

Tying pin 33 to ground puts the 8088 into the maximum mode. This mode normally uses an 8288 bus controller that decodes status lines $\overline{S0}$ $\overline{S1}$ and $\overline{S2}$ appearing on pins 26, 27, and 28, respectively. The 8288, in turn, generates $\overline{RD}$ and $\overline{WR}$ signals for the 8202A. By having the 8288 generate the bus control signals, other pins of the 8088 can be allocated for large system implementations. Also, the 8288 has larger sink and source current capability.

A third, alternate, configuration, utilizes TTL gates and flip-flops to decode 8088 status outputs $\overline{S1}$, ($DT/\overline{R}$), and ALE to generate $\overline{RD}$ and $\overline{WR}$ signals for the 8202A. The alternate configuration can be used in either the minimum or maximum mode. This configuration can be used to reduce memory speed requirements by generating $\overline{RD}$ and $\overline{WR}$ signals for the 8202A earlier in the timing cycle. In the following example, the minimum configuration will be used.

Refreshing of the 2118 and dynamic RAMs, in general, can be accomplished either synchronously or asynchronously. In the asynchronous or distributed mode, as it is sometimes called, a refresh cycle is requested by the memory controller. Since the refresh request may conflict with the microprocessor's access to RAM, the controller must incorporate logic to arbitrate among these requests. If either an access or refresh cycle is in progress, it must go to completion. Therefore, if a refresh cycle is in progress, a WAIT state in the microprocessor must be initiated and held until the refresh cycle has been completed. A refresh cycle is usually requested by the controller every 12–15 μs.

In the synchronous (hidden) refresh mode, the refreshing of dynamic RAM occurs when the microprocessor is known not to be accessing memory. In some cases, this design can permit refreshing without any need for CPU WAIT states.

Basic Design. A basic memory system functional diagram using the 8088 in the minimum mode is shown in Figure 3.19.

Read Cycle (Minimum Mode). When a READ signal is placed on the RD line by the microprocessor, it is detected by the 8202A. As a result, $\overline{RAS}$ and $\overline{CAS}$ signals are generated by the controller and are used by the RAM chips to latch the address from the 8088. Data are then made available at the selected RAM data output pins and latched into the 8282 data latch by the $\overline{XACK}$ strobe generated by the 8202A. The latched data are read into the CPU during the same Read cycle. The timing diagram for this Read cycle is given in Figure 3.20.

In order to calculate the minimum speed requirement for the 2118 dynamic RAM in the proposed system design, timing information is needed for the 8202A, 2118, and 8282 latch. The parameters of interest in the Read cycle for the 8202A are t_p, the clock period and t_{cc}, the $\overline{RD}$ to $\overline{CAS}$ delay. The time, t_p, is a function of the rate at which the 8202A is set to run.

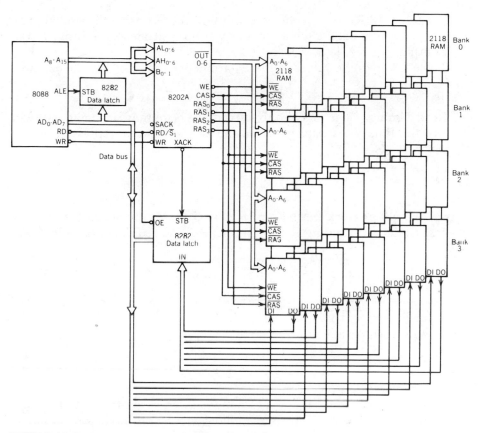

FIGURE 3.19 Basic memory system (minimum mode).

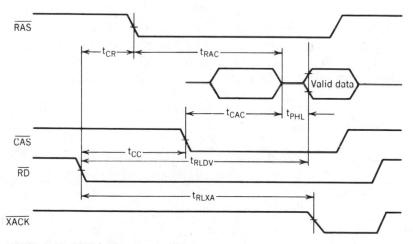

FIGURE 3.20 READ cycle timing diagram.

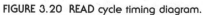

For the 2118 RAM, t_{cac}, the access time from $\overline{CAS}$ active is needed. The time, t_{PHL}, for the 8282 latch is the delay from a strobe (STB) input to the latch output.

Table 3.3 gives the values of these parameters to be used in the example calculations.

Read Timing.   The time required from the beginning of a memory Read cycle to the time valid data are available is t_{RLDV} and is expressed by the equation

$$t_{RLDV} = t_{cc,max} + t_{cac} + t_{PHL}$$

Thus, for a 40-ns clock period and a 2118-3 RAM, the shortest time that external circuitry (including RAM) can make the stored data available to the CPU is

$$t_{RLDV} = 245 + 55 + 45$$
$$= 345 \text{ ns}$$

For a 54-ns clock period with the same RAM,

$$t_{RLDV} = 301 + 55 + 45 = 401 \text{ ns}$$

With no WAIT states, the basic Read timing is given in Figure 3.21. To function without requiring WAIT states, the following relationship in the minimum mode must hold:

$$t_{RLDV(CPU)} \leq 2t_{CLCL(8088)} - t_{CLRL(8088)} - t_{DVCL(8088)}$$

In the relationship, t_{CLCL} is the 8088 clock cycle period, t_{CLRL} is the $\overline{RD}$ signal active delay, and t_{DVCL} is the setup time for data into the 8088. Thus $t_{RLDV(CPU)}$ is the maximum time that the CPU can allow for valid data to be accessed from addressed memory. The CPU timing parameters are also shown in Figure 3.21.

For the 8088 operating with a 5-MHz clock, the maximum value of $t_{RLDV(CPU)}$ is calculated as:

$$t_{RLDV(CPU)} \leq 2(200 \text{ ns}) - 165 \text{ ns} - 30 \text{ ns}$$
$$\leq 400 \text{ ns} - 165 \text{ ns} - 30 \text{ ns}$$
$$\leq 205 \text{ ns}$$

Since previous calculations for the 2118-3 show that $t_{RLDV(CPU)} = 345$ ns for a 40-ns 8202A clock frequency, WAIT states must be inserted to allow the 8088

TABLE 3.3 Timing Parameters for Design Example

8202A		8202A t_{cc}	
Min Clock Period t_p Min	Max Clock Period t_p max	Min $3\,t_p + 25$	Max $4\,t_p + 85$
40	54	At 40 ns clock period = 145 ns	At 40 ns clock period = 245 ns
		At 54 ns clock period = 187 ns	At 54 ns clock period = 301 ns

2118			8282 t_{PHL} STB to Output Delay
t_{cac}			
2118-3	2118-4	2118-7	
55	65	80	45

$t_{DS,MIN(2118)} = 0$ for all RAMs of 2118 family, $t_{DH}(2118\text{-}3) = 25$ ns, $t_{DH}(2118\text{-}7) = 45$ ns

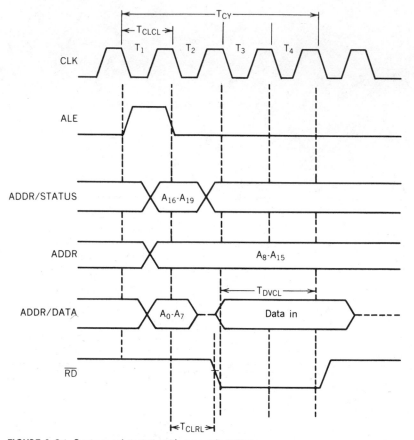

FIGURE 3.21 Basic read timing with no wait states.

to operate with the 2118-3. Each WAIT state is equal to the CPU clock period (200 ns for a 5-MHz clock), so the number of WAIT states required is

$$\frac{345 - 205}{200} = 0.7$$

Rounding off, one WAIT state is needed. With one WAIT state, the basic Read timing is as shown in Figure 3.22.

Alternate Configuration Read. It is interesting to consider an implementation of the alternate configuration utilizing TTL devices between the 8088 and the 8202A. As noted earlier, this implementation allows the use of slower memories, since the $\overline{RD}$ and $\overline{WR}$ signals to the 8202A are generated earlier in the cycle. This approach can be used with the 8088 in either the minimum or maximum mode. A block diagram of the alternate configuration is given in Figure 3.23.

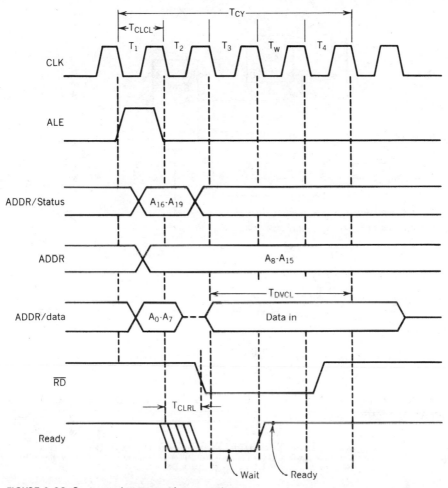

FIGURE 3.22 Basic read timing with one wait state.

The t_{RLDV} calculation for the alternate configuration is

$$t_{RLDV(ALE)} = 2t_{CLCL}(8088) + t_{CHCL}(8088) - t_{CHLL}(8088)$$
$$- t_{DVCL}(8088) - t_{PHL}(74S74)$$

where

t_{CLCL} is the 8088 clock cycle period

t_{CHCL} is the 8088 clock high time

t_{CHLL} is the 8088 ALE active delay

t_{DVCL} is the setup time for data into the 8088

t_{PHL} is the clock to output delay of the 74S74 D-type flip flop

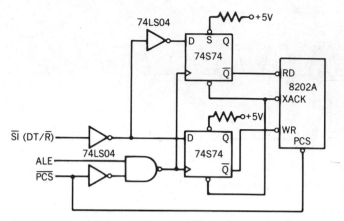

FIGURE 3.23 Alternate configuration block diagram.

For the 8088 with a 5-MHz clock rate, the parameter values in nanoseconds are

$$t_{CLCL} = 200$$
$$t_{CHCL} = 68.67$$
$$t_{CHLL} = 85$$
$$t_{DVCL} = 30$$

The value of t_{PHL} for the 74S74 is 9 ns. Substituting in the equation for $t_{RLDV}(ALT)$ yields:

$$t_{RLDV}(ALT) = 2(200) + 68.67 - 85 - 30 - 9$$
$$= 344.67 \text{ ns} \cong 345 \text{ ns}$$

The alternate mode, therefore, gives the memory system more time to respond to the CPU and no WAIT states would be required for a 2118-3 RAM and a 5-MHz 8088 with a 40-ns clock period for the 8202A.

Write Cycle (Minimum Mode).   Dynamic RAM cycles are of three varieties:

1. Early write—defined by $\overline{WE}$ signal being active before $\overline{CAS}$ by at least the write setup time (t_{wcs}) of the RAM. The RAM data output remains in the high impedance state until the end of the cycle. Thus data setup (t_{ds}) and data hold (t_{DH}) times are measured relative to $\overline{CAS}$ going active (Figure 3.24a).

2. Delayed write—occurs when $\overline{WE}$ signal becomes active less than the write setup time (t_{wcs}) before or after $\overline{CAS}$ is active (Figure 3.24b). In this case, the data output from the RAM can leave the high impedance

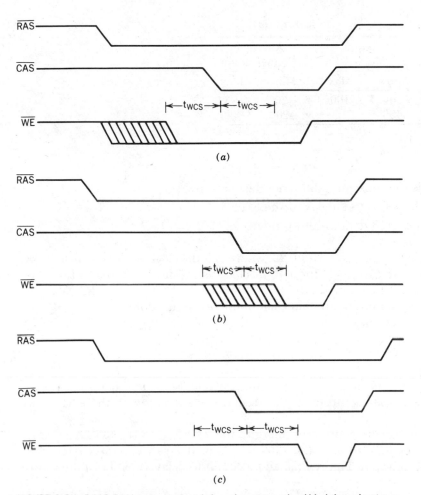

FIGURE 3.24 2118 RAM write cycles: (a) early write cycle; (b) delayed write cycle; (c) read-modify-write cycle.

state and should be prevented from affecting the data bus. This isolation can be easily accomplished through the RAM data latch.

3. Read-modify-write—timing wherein $\overline{WE}$ becomes active a time greater than the write set-up time (t_{wcs}) after $\overline{CAS}$ is active, Figure 3.24c. Memory data are modified during this type of cycle. The D_{out} is the information to be read and is not changed during the modify-write portion of the cycle.

The early-write cycle can be used if there is adequate RAM setup and hold time. For this to be true:

$$t_{\text{WLDV,min(8088)}} + t_{\text{CC,min(8202)}} = t_{\text{DS,min(2118)}}$$

TABLE 3.4 8086 t_{WLDV} and t_{DHADV} Values (5 MHz)

8086 Values	Minimum Mode		Alternate Configuration	
	Min	Max	Min	Max
t_{WLDV}	-100	100	-70	85
t_{DHADV}	420		420	

where

t_{WLDV} is the time for valid write data set-up

t_{CC} is the $\overline{WR}$ to $\overline{CAS}$ delay of the 8202A

t_{DS} is the 2118 data in set-up time

Recalling that $t_{CC,min(8202)} = 3t_p + 25$, where t_p is the clock period of the 8202A, we can calculate $t_{CC,min(8202)}$ for a 40-ns clock as 145 ns. From the 2118-3 data sheet, $t_{DS,min} = 0$ ns. For the 5-MHz 8088, Table 3.4 provides the values of t_{WLDV} for the minimum and alternate mode configurations.

Substituting, we obtain

$$-100 + 145 = t_{DS,min(2118)} \qquad \text{or} \qquad t_{DS,min(2118)} = 45 \text{ ns}$$

Thus there are 45 ns of setup margin available. Since the 2118-3 RAM requires 0 ns of data setup time, this design can succeed with the early-write option.

If there is inadequate setup margin, the $\overline{WR}$ input to the 8202A must be delayed or the $\overline{WE}$ output of the 8202A must be delayed. Circuits to delay the $\overline{WR}$ input are given in Figure 3.25 and a circuit to delay $\overline{WE}$ is shown in Figure 3.26.

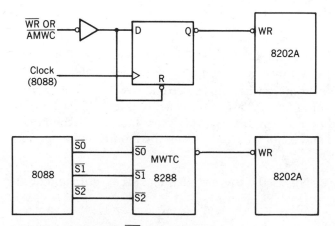

FIGURE 3.25 Delaying $\overline{WR}$ input of 8202A.

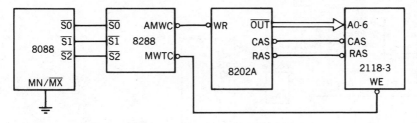

FIGURE 3.26 Delaying $\overline{WE}$ input to RAM.

In addition to the data setup time specification, the data hold time after data are valid (t_{DHADV}) requirement must be met. These values are also listed in Table 3.4. The RAM hold time is met by satisfying:

$$t_{DHADV,min} \geq t_{WLDV,max} + t_{CC,max} + t_{DH}$$

where the new variable, t_{DH}, is the RAM data in hold time and, from the 2118-3 data sheet, is equal to 25 ns. Substituting, we obtain

$$t_{DHADV,min} \geq 100 + (4t_p + 85) + 25$$

For an 8202A value of 40 ns for t_p,

$$t_{DHADV,min} \geq 100 + 245 + 25$$
$$\geq 370 \text{ ns}$$

From Table 3.4, $t_{DHADV,min} = 420$ ns. There is a 50-ns margin in the hold-time requirement. If more delay occurred in the data lines as a result of adding additional drivers, the hold-time margin would increase. Adding delay in the $\overline{WR}$ input to the 8202A, however, would decrease the hold margin.

Chip Select.   The 8202A Dynamic Ram controller has a protected chip select input (PCS) that enables the memory read and write inputs. $\overline{PCS}$ must be stable before $\overline{RD}$ and $\overline{WR}$ inputs to the 8202A become active. $\overline{PCS}$ is normally generated by address decoding. Generation of $\overline{PCS}$ for the minimum mode is shown in Figure 3.27.

Refreshing.   To maintain data integrity, each row of the 128 rows in the 2118 RAM must be refreshed every 2 milliseconds. A row selected by the low-order $\overline{RAS}$ addresses (A_0-A_6) is refreshed by any Read or Write memory cycle. During a Write cycle, however, the addressed memory cell in a selected row will undergo a Write operation, possibly changing the data contents.

The recommended refresh cycle is the $\overline{RAS}$-Only cycle, particularly since it can be utilized with Wired OR outputs without incurring bus contention.

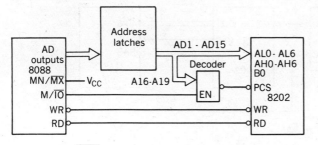

FIGURE 3.27 $\overline{PCS}$ Generation for 8202A minimum mode.

For multiple rows of dynamic RAMs, Read, Write, or Read-Modify-Write cycles will result in bus contention. In the $\overline{RAS}$-Only refresh mode, the D_{out} pin of the RAM remains in the high impedance state.

The 8202A has an on-board refresh timer that generates an internal refresh cycle every 12–16 μsec to ensure that all rows of the RAM are refreshed every 2 ms. An external request for refresh placed on the REFRQ input pin of the 8202A will reset the 8202A refresh timer. As long as external refresh requests occur prior to the interval of 8202A internal refresh request generation, the external timing will prevail. Otherwise, refresh timing will be controlled by the internal refresh timer. If a memory request by the CPU coincides with an external refresh request, the memory request will be completed first.

A refresh cycle begins with the placing of the refresh row address on the $\overline{OUT}$ 0–6 pins of the 8202A. The address is provided by the internal refresh counter that is incremented after every refresh cycle. Then the Row Address strobe signals ($\overline{RAS}$ 0–3) are activated. Refreshing occurs, and the cycle is completed by deactivating all signals. Precharging then occurs to prepare for a CPU memory cycle and the 8202A becomes idle. Timing for the refresh cycle is illustrated in Figure 3.28.

It is important that the minimum pulse width specifications (t_{RAS} and t_{CAS}) are met for $\overline{RAS}$ and $\overline{CAS}$, respectively. Similarly, a new cycle cannot begin until a minimum precharge time, t_{RP}, has been satisfied. For the 2118-3 RAM,

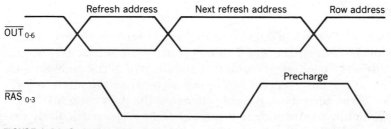

FIGURE 3.28 Refresh cycle timing.

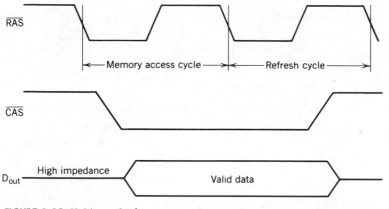

FIGURE 3.29 Hidden refresh.

these minimum values are

$$t_{RAS} = 115 \text{ ns}$$
$$t_{CAS} = 55 \text{ ns}$$
$$t_{RP} = 110 \text{ ns}$$

As an option, a "hidden refresh" of the 2118 can occur. Under hidden refresh, a refresh cycle can occur while maintaining valid data at the D_{out} pin. Hidden refresh is initiated by holding $\overline{CAS}$ active (low) and bringing $\overline{RAS}$ high and then, after a precharge time, t_{RP}, executing a $\overline{RAS}$-Only refresh cycle with $\overline{CAS}$ low. This timing sequence is shown in Figure 3.29.

Since the 8088 utilizes instruction prefetch (advanced fetching of instructions and their placement in an instruction queue), the opportunity for conflict between memory and refresh cycles is reduced. Therefore, there is not a significant advantage in using hidden refresh with the 8086 family of microprocessors.

For the 8085-type microprocessor, the hidden refresh is useful in reducing refresh interference and the circuit of Figure 3.30 will implement hidden

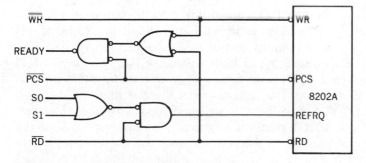

FIGURE 3.30 Hidden refresh for 8085A.

refresh if the following equation is satisfied:

$$4T \geq 2t_{RC,max} + t_{CR,max}$$

where

T is the clock period of the 8085A

t_{RC} is the memory cycle time of the 8202A and is equal to $(10t_p\text{-}30)$ minimum and $12\ t_p$ maximum, where t_p is the 8202A clock period.

t_{CR} is the $\overline{RD}$ or $\overline{WR}$ to $\overline{RAS}$ delay and is equal to $(t_p + 30)$ minimum and $(2t_p + 70)$ maximum.

3.4.2 HIGH SPEED CMOS LOGIC INTERFACING

Traditionally, low-power Schottky TTL (LSTTL) has been used external to NMOS microprocessors to interface to peripheral chips and memories. CMOS offered a low-power alternative but was lacking in speed. High-speed CMOS (HSCMOS) is a third approach that offers the speed of LSTTL along with the low-power consumption of CMOS. Various semiconductor manufacturers' versions of high speed CMOS have become the technology of choice in the implementation of 32-bit microprocessors such as the 68020 and the 80386. Interfacing to the 80386 is covered in detail in Chapter 5.

HSCMOS utilizes silicon gate technology as opposed to metal gate technology of conventional CMOS. A standard HSCMOS gate can sink up to 4 mA while maintaining $V_{OL(max)} \geq 0.4$ V. Typically, a 74HC00 HSCMOS two-input NAND gate can drive up to 10 LSTTL inputs. Also, since HSCMOS inputs only draw 1 μA of current, NMOS devices can drive an increased number of such inputs over LSTTL inputs. HSCMOS devices also compare favorably with LSTTL circuits in terms of propagation delay. The following sections describe interfacing HSCMOS to the inputs and outputs of single-power supply (+5V) NMOS microprocessors and other single-voltage NMOS peripherals.

NMOS TO HSCMOS

Figure 3.31 is the general TTL-compatible NMOS output circuit of a typical microprocessor. The minimum output high voltage, V_{OH} (min) of this circuit is 2.4 V while the maximum output low voltage, $V_{OL(MAX)}$ is 0.4 V. Correspondingly, the minimum input high voltage, $V_{IH(min)}$ of a HSCMOS circuit is 3.5 V. The maximum low voltage, $V_{IL(max)}$ of the HSCMOS input is 1.0 V. Since the NMOS output low voltage of 0.4 V is less than the required HSCMOS level of 1.0 V, compatibility is achieved for low inputs. However, the high output of the NMOS circuit is less than the required 3.5 V of the HCMOS gate. The output of the NMOS circuit must, therefore, be pulled up with a resistor to +5 V to ensure that a high output of the NMOS device exceeds 3.5 V. The resistor is shown dotted in Figure 3.31.

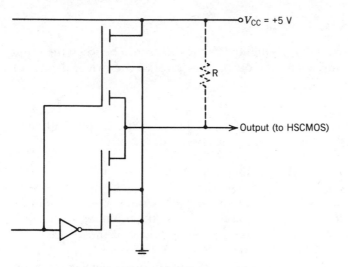

FIGURE 3.31 NMOS output circuit (TTL compatible).

HSCMOS TO NMOS

From the HSCMOS circuit output voltage specification and the NMOS TTL compatible input specification, a HSCMOS device can directly interface to an NMOS device. A typical HSCMOS output with $V_{CC} = +5$ V has $V_{OH(min)} = 4.95$ V and $V_{OL(max)} = 0.05$ V, both able to meet the TTL compatible NMOS input voltage requirements of $V_{IH(min)} = 2.0$ V and $V_{IL(max)} = 0.8$ V, respectively. Figure 3.32 illustrates a typical NMOS input.

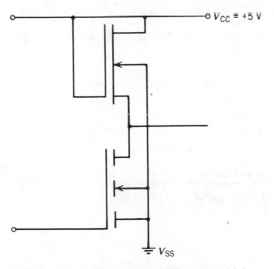

FIGURE 3.32 NMOS input circuit (TTL compatible).

NOISE MARGIN

The low noise margin, V_{NL}, is defined as the difference between the input low level of a driven device and the output low level of the driver device, or

$$V_{NL} = V_{IL} - V_{OL}$$

The high noise margin, V_{NH}, is the difference between the output high level of a device and the input high level of the driven device or:

$$V_{HL} = V_{OH} - V_{IH}$$

The noise margin is a measure of the noise voltage that may occur on an input line that is in either a low or high state and can cause the input logic voltage to enter a region that may initiate a transition on the output of the device to the opposite logic level. For 5-Volt HSCMOS, $V_{NL} = 0.95$ V and $V_{NH} = 1.45$ V. LSTTL has $V_{NL} = 0.40$ V, and $V_{NH} = 0.70$ V. Conventional, metal gate CMOS has $V_{NL} = V_{NH} = 1.45$ V.

Table 3.5 gives the interface noise margins for some mixed logic families.

MICROPROCESSOR BUS INTERFACING EXAMPLE

Interfacing to a microprocessor bus requires different types of circuits, depending upon the device to be coupled to the bus. For example, an input port, when selected, must be connected to the microprocessor data bus. When the port is deselected, it must be isolated from the data bus. This connection/isolation can be accomplished by a three-state buffer that is connected to the bus when it is selected (enabled), or isolated by presenting a high impedance to the bus when it is deselected. Some typical tristate buffers, as they are called, in the high speed CMOS logic family are the 74HC240 and the 74HC244 octal buffers as shown in Figure 3.33. These buffers are unidirectional, that is, data can travel through the buffers in only one direction when the buffers are selected.

Since many devices exchange information with the microprocessor data bus in both directions, a bidirectional bus transceiver is required as an interface. One such device is the 74HC243 HSCMOS noninverting bus transceiver that is shown in Figure 3.34. An example application of the 74HC243 trans-

TABLE 3.5 Noise Margins for Some Mixed Logic Families

Logic Families	$V_{NL} = V_{IL} - V_{OL}$	$V_{NH} = V_{OH} - V_{IH}$
NMOS to LSTTL	0.40 V	0.40 V
HSCMOS to NMOS	0.75 V	2.95 V
NMOS to HSCMOS[a]	0.60 V	1.10 V
LSTTL to NMOS	0.40 V	0.40 V

[a] *Note:* Requires pull-up resistor.

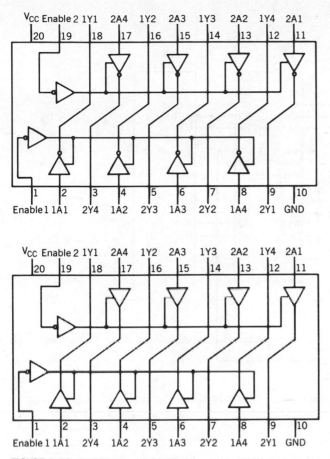

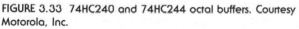

FIGURE 3.33 74HC240 and 74HC244 octal buffers. Courtesy Motorola, Inc.

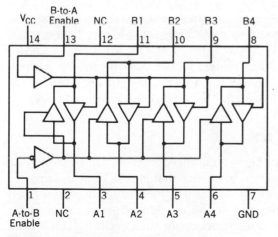

FIGURE 3.34 74HC243 octal bus transceiver.
Courtesy Motorola, Inc.

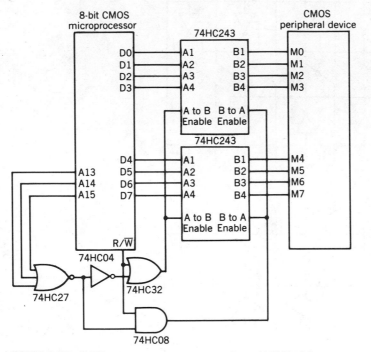

FIGURE 3.35 CMOS microprocessor system utilizing 74HC243 bus transceivers.

ceiver as a bidirectional interface in an all CMOS microprocessor system is illustrated in Figure 3.35.

Continuing with the all CMOS microprocessor system example, we will develop address decoding using HSCMOS components. Decoders use the CPU address outputs to generate unique, mutually exclusive active outputs on decoder output lines that correspond to a specific input address. Diagrams of two HSCMOS decoders, the 74HC138 and the 74HC154, are given in Figure 3.36 along with corresponding truth tables. Figure 3.37 illustrates address decoding using the 74HC138 to decode the address lines of a CMOS microprocessor to generate a chip enable signal for a CMOS peripheral device.

TIMING CHARACTERISTICS

When utilizing interface chips between a microprocessor and memory or peripherals, timing must be calculated to ensure that excessive delays are not introduced. Since most microprocessors operate synchronously with a fixed time allowed to read or write data, the delays introduced by interface chips must be accounted for. A HSCMOS microprocessor such as the MC146805E2 operating at 1 MHz has a 25-ns worst case allowed for address decoding. Propagation delay for the 74HC138 decoder is 20 ns, thus making it compatible with the CMOS CPU.

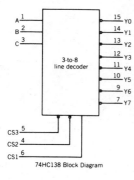

74HC138 Block Diagram

Truth Table

Inputs						Outputs							
Chip select			Output select										
CS1	CS2	CS3	C	B	A	T0	Y1	Y2	Y3	Y4	Y5	Y6	Y7
X	X	H	X	X	X	H	H	H	H	H	H	H	H
X	H	X	X	X	X	H	H	H	H	H	H	H	H
L	X	X	X	X	X	H	H	H	H	H	H	H	H
H	L	L	L	L	L	L	H	H	H	H	H	H	H
H	L	L	L	L	H	H	L	H	H	H	H	H	H
H	L	L	L	H	L	H	H	L	H	H	H	H	H
H	L	L	L	H	H	H	H	H	L	H	H	H	H
H	L	L	H	L	L	H	H	H	H	L	H	H	H
H	L	L	H	L	H	H	H	H	H	H	L	H	H
H	L	L	H	H	L	H	H	H	H	H	H	L	H
H	L	L	H	H	H	H	H	H	H	H	H	H	L

H = high level (steady state)
L = low level (steady state)
X = don't care

74HC154 Block Diagram

Truth Table

CS1	CS2	Data inputs				Selected output at logic "0"
Inhibit		D	C	B	A	
0	0	0	0	0	0	S0
0	0	0	0	0	1	S1
0	0	0	0	1	0	S2
0	0	0	0	1	1	S3
0	0	0	1	0	0	S4
0	0	0	1	0	1	S5
0	0	0	1	1	0	S6
0	0	0	1	1	1	S7
0	0	1	0	0	0	S8
0	0	1	0	0	1	S9
0	0	1	0	1	0	S10
0	0	1	0	1	1	S11
0	0	1	1	0	0	S12
0	0	1	1	0	1	S13
0	0	1	1	1	0	S14
0	0	1	1	1	1	S15
1	X	X	X	X	X	All outputs = 1
X	1	X	X	X	X	

X = don't care

FIGURE 3.36 Two HSCMOS decoders. Courtesy Motorola, Inc.

In data bus buffering of a HSCMOS system, the following delays are typically incurred:

Device(s)	Delay (ns)
Transceiver associated control logic	45
Transceiver (74HC243)	30
Three-state disable time (74HC243)	25

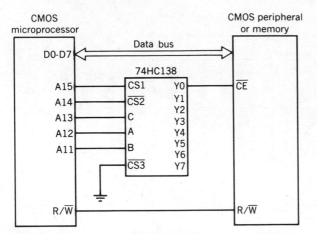

FIGURE 3.37 Address decoding using a 74HC138 HSCMOS decoder.

These delays total 100 ns. Typically, for the MC146805E2 HSCMOS microprocessor operating at 5 MHz with $V_{DD} = 5$ V, the access time allotted for peripherals or memory is 365 ns. Thus a 365 ns − 100 ns or 265-ns delay can be allotted to the memory or peripheral device to provide its data.

In many applications, external data must be captured and held for a period of time before the microprocessor can read it. In a typical application, a MC74HC374 noninverting or MC74HC574 inverting octal D-type flip-flop can perform this latching function. Figure 3.38 illustrates this function with

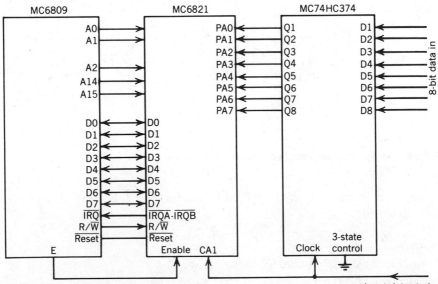

FIGURE 3.38 Latched data input to microprocessor.

the NMOS MC6809 microprocessor and the NMOS MC6821 peripheral interface adapter.

3.4.3   ELECTRICALLY ERASABLE PROGRAMMABLE READ ONLY MEMORY (EEPROM)

The EEPROM is a static (does not require refreshing), nonvolatile, read/write memory. It can be reprogrammed in situ by applying appropriate voltage pulses on the order of 21 V for a period of 10 ms. EEPROMS support the erasure of individual bytes of memory instead of requiring a bulk erase of all memory as in earlier devices. Access times of EEPROMS are in the 200-ns range, thus eliminating the need for a processor WAIT state in the read cycle.

PROCESSOR INTERFACE

A controller is required between a processor and an EEPROM. As a specific example, the Intel 2816 EEPROM will be used to illustrate the controller requirements. The 2816 is organized as a 2-k × 8-bit memory with pinouts as shown in Figure 3.39.

READ MODE

In the read mode, the address bits select data in a particular location in the 2816 that are to be presented on the data output pins. For the selected data to be placed on the data output pins, both CE (chip enable) and OE

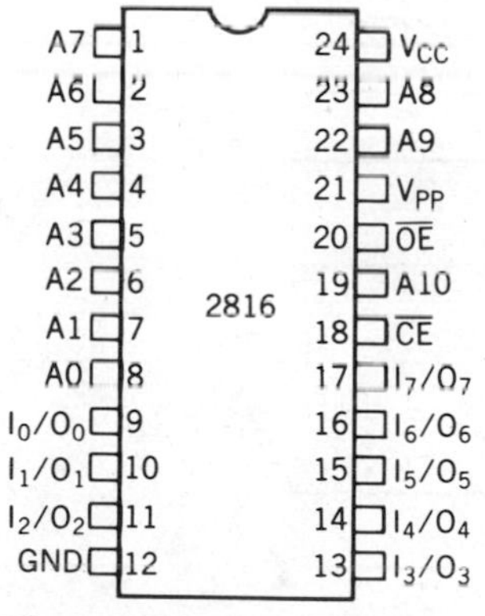

A7	1		24	V_{CC}
A6	2		23	A8
A5	3		22	A9
A4	4		21	V_{PP}
A3	5		20	$\overline{OE}$
A2	6	2816	19	A10
A1	7		18	$\overline{CE}$
A0	8		17	I_7/O_7
I_0/O_0	9		16	I_6/O_6
I_1/O_1	10		15	I_5/O_5
I_2/O_2	11		14	I_4/O_4
GND	12		13	I_3/O_3

Pin Designations

A_0–A_{10}	Addresses
I_0–I_7	Data inputs
O_0–O_7	Data outputs
$\overline{OE}$	Output enable
$\overline{CE}$	Chip enable
V_{pp}	Program voltage

FIGURE 3.39 2816 Pinouts.

(output enable) must be active low. Otherwise, the data output lines are in the high impedance state of the three-state output drivers. The OE line is normally connected to the microprocessor RD (read) line. This eliminates contention that may occur when only the CE line is used to select or deselect the memory chip. The situation may arise where one chip is going from a logic state to the high impedance mode while another chip is undergoing the reverse transition. When this occurs, the data pins may be driven improperly from two sources. Since the RD line becomes active much after the CE line, this condition cannot occur.

The 2816 does not need to utilize the wait state if used with microprocessors that can operate with the memory access time of 250 ns that the 2816 possesses. During a Read of the 2816, only 5 volts DC are needed. This is because V_{pp}, the programming pin must be between 4 and 6 V dc during a Read cycle. Read cycle timing is shown in Figure 3.40.

BYTE WRITE/ERASE

To change a byte of data in the 2816, generally the data must first be erased and then the new data written. A special case occurs when bits within a byte to be changed will only have transitions from 1's to 0's. In this case, no erase of the byte has to occur before the new data are entered. There are six modes of operation of the 2816, namely read, standby, byte erase, byte write, chip erase, and erase/write (E/W) inhibit. The pins involved and appropriate select voltages are presented in Figure 3.41.

The byte write and erase modes are essentially identical, except for the data input pins. A byte erase occurs when the data input pins are all at a logic

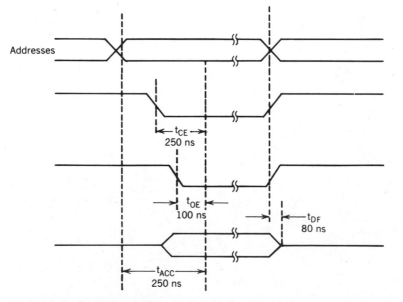

FIGURE 3.40 Read cycle timing.

	Pin			
Mode	$\overline{CE}$	$\overline{OE}$	V_{pp}	Inputs/Outputs
Read	Low	Low	+4 to +6	D_{out}
Standby	High	Don't care	+4 to +6	High Z
Byte erase	Low	High	+21	D_{in} = High
Byte write	Low	High	+21	D_{in}
Chip erase	Low	+9 to +15 V	+21	D_{in} = High
E/W inhibit	High	Don't care	Don't care	High Z

FIGURE 3.41 2816 mode selection.

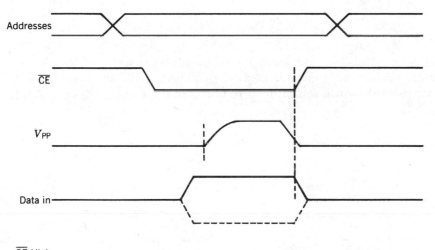

FIGURE 3.42 Byte write/erase timing.

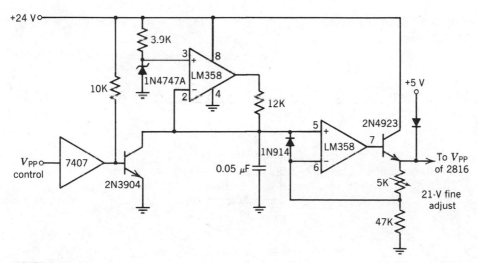

FIGURE 3.43 V_{pp} circuit.

1 or high level. Otherwise, the data pattern presented on the data input pins are written into the memory byte. To erase a byte, V_{pp} must also be pulsed to 21 volts through an exponential. Figure 3.42 shows the appropriate signals and relative timing for a byte erase. Byte write is identical except that data-in are the bits to be entered, not all 1's as in byte erase. In Figure 3.42, V_{pp} must rise through the exponential and be held for a minimum of 9 ms; V_{pp} can fall to zero as quickly as possible. A circuit for implementing V_{pp} is given in Figure 3.43.

CHIP ERASE

To implement the erasure of the complete chip, +9 to +15 volts must be applied to the output enable (OE) pin while V_{pp} is pulsed to +21 V. The current at the OE is only the leakage current of approximately 10 μA. The chip erase timing diagram is illustrated in Figure 3.44, and the circuit to generate the different levels of OE as required in different modes is given in Figure 3.45.

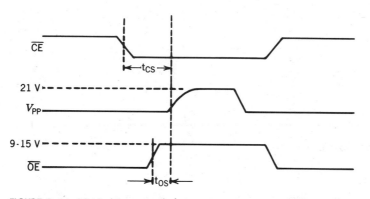

FIGURE 3.44 2816 chip erase timing.

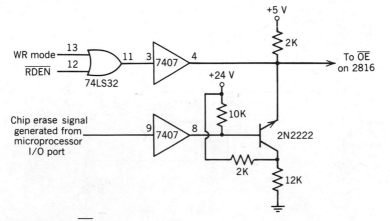

FIGURE 3.45 $\overline{\text{OE}}$ generation for 2816.

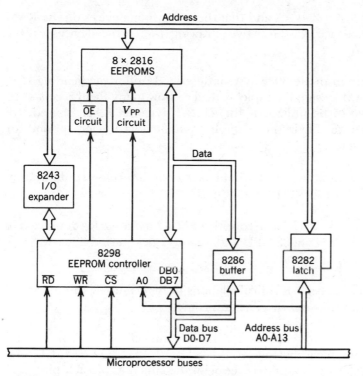

Address

FIGURE 3.46 General 2816 interface using 8298 controller.

8298 INTEGRATED CONTROLLER

The Intel 8298 EEPROM controller integrates many of the discrete circuits necessary to control reading, writing, and erasing of Intel EEPROMS. The 8298 can support up to 16K bytes of EEPROM. The V_{pp} and OE signals must be generated by circuits external to the 8298. A general diagram of the 8298 used in a 16K × 8 EEPROM memory interface with direct access capability is shown in Figure 3.46. For detailed information on the 8298, refer to appropriate Intel manuals.

PROBLEMS

3.1 Calculate the power consumed by a 64K × 16-bit dynamic RAM memory array comprised of 2118-3 devices. From the 2118 data sheet, the following values were obtained:

$$I_{DD1,max} = 2 \text{ mA}, \qquad I_{DD2,max} = 27 \text{ mA}$$
$$I_{DD3,max} = 18 \text{ mA}, \qquad I_{Lk} = 10 \text{ } \mu A$$
$$t_{REF} = 2 \text{ ms}, \qquad t_{RC} = 235 \text{ ns}, \qquad V_{DD} = 5V \pm 10\%$$

Also, assume that $I_{OL} = 250$ μA and that there are four devices on the RAM output line. One device will be active, drawing I_{OL}; three devices will be inactive, drawing I_{Lk}.

3.2   In the minimum mode 8088 system described under the heading "Detailed Dynamic RAM Design Example," calculate how many Read access WAIT states, if any, have to be utilized if the 2118-7 dynamic RAM is used. All needed values are available in the example. Assume a 54-ns clock period for the 8202A controller.

3.3   In Problem 3.2, what effect would using the alternate configuration have on the number of WAIT states required?

3.4   Can the early-write cycle be used with a 2118-7 dynamic RAM utilized in conjunction with a 5-MHz 8088 CPU and a 54-ns clock period 8202A? Show work to support your conclusions.

3.5   From Problem 3.4, perform calculations to determine if the data hold time after data valid (t_{DHADV}) specification is met.

3.6   The 2187, 8-K $\times$ 8 dynamic RAM has on-chip refresh circuitry to simplify its use. Its pins have the following labels: A0-A12, IO0-IO7, CE, OE, WE, RDY, REFEN, VCC, and GND. Explain the function of each pin.

3.7   Compare a single-bus microcomputer system architecture to a distributed architecture with local and global memories. Cite the advantages and disadvantages of each.

3.8   The 2817A (2-K $\times$ 8) EEPROM is similar to the 2816A except that the 2817A has on-chip address latches for direct microprocessor interfacing. Draw the general interface of a 2817A to an 8088 microprocessor.

FOR FURTHER READING

Bibbero R. J., and D. M. Stern, *Microprocessor Systems Interfacing and Applications,* John Wiley & Sons, Inc., New York, 1980.

Lobelle, M., "VME Bus Interfacing: A Case Study," *Interfaces in Computing,* Vol. 1, No. 3, 1983.

MacWilliams P. D., et al., "Microcontroller Serial Bus Yields Distributed Multispeed Control," *Electronics,* Vol. 57, No. 3, 1984.

Motorola Schottky (LS)TTL Integrated Circuits, Motorola Semiconductor Products, Inc., 1980.

Stone, H. S., *Microcomputer Interfacing,* Addison-Wesley Publishing Company, Inc., Reading, Mass., 1982.

Thurber, K. J., E. D. Jensen, et al., "A Systematic Approach to the Design of Digital Busing Structures," *AFIPS, Proceedings of the 1972 FJCC,* Vol. 41, Part III, AFIPS Press, 1972.

4

COMMUNICATIONS AND DATA TRANSFER

Communications interfaces for computers and related equipment are becoming increasingly necessary with the advent of low-cost computational power and the sharing of information among many users. Information can be transmitted in one direction only (*simplex*), two directions nonsimultaneously (*half-duplex*), and two directions simultaneously (*full-duplex*).

4.1

SERIAL COMMUNICATIONS

Serial communications links are those that utilize a single data line to transmit information. The advantage of this method is the savings in wiring and interface circuitry while the disadvantage is the time taken to transmit a message in sequence, bit by bit. Alternatively, parallel transmission is faster but requires more wiring and interface circuitry. Figure 4.1 contrasts serial and parallel interfaces.

4.1.1 ASYNCHRONOUS TRANSMISSION

Asynchronous serial transmission produces sequences of data on a line with no specific clock timing governing the interval between each piece of data. In asynchronous transmission, each transmitted character is usually represented by 5–8 data bits (least significant bit first) bounded by start and stop timing bits. An optional parity bit can follow the data bits for error checking. Naturally, the more data bits utilized, the more characters that can be represented. Figure 4.2 illustrates the general format of asynchronous data transmission. The timing for each bit is determined by separate clocks in the transmitter and receiver. The high to low transition of the start bit initiates the timing at the receiver to search for the value of the following data bits and optional

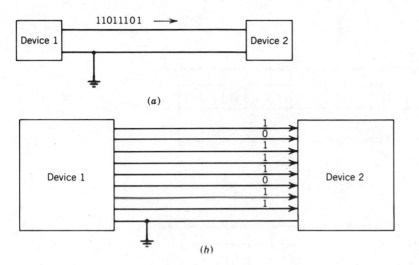

FIGURE 4.1 Serial (a) versus parallel (b) transmission.

parity bit. Common asynchronous serial transmission bit rates are 110, 300, 600, 1200, 2400, 4800, 9600, and 19,200 bits/s. If 11 bits (1 start bit, 8 data bits, and 2 stop bits) are chosen in the asynchronous format to represent one character, then the character transmission rate is given by the transmission bit rate divided by 11. For example, at 1200 bits/s transmission rate, the character transmission rate is 1200/11 or 109 characters per second.

The parity bit is optional in asynchronous serial transmission and is used as a check bit to detect some transmission errors. Typically, even-1 or odd-1 parity is used. In even-1 parity, an even number of 1s, including the parity bit, are always transmitted. Odd-1 parity dictates that an odd number of 1s, including the parity bit, are transmitted. Figure 4.3 illustrates an even-1 parity message and Figure 4.4 shows an odd-1 parity transmission of the 8-bit data 11011101.

The 5–8 data bits in the asynchronous format can be used to represent numbers, alphabetic characters, and other miscellaneous characters through appropriate coding. The common code is ASCII, the American Standard Code for Information Interchange. The ASCII code is given in Appendix A.

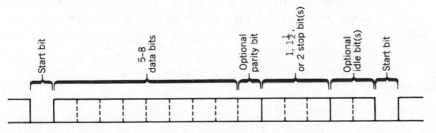

FIGURE 4.2 General asynchronous data transmission format.

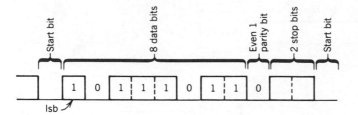

FIGURE 4.3 Even 1 parity example.

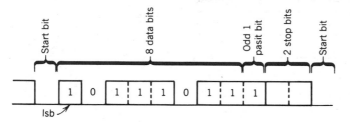

FIGURE 4.4 Odd 1 parity example.

Note that the 8-bit data 00110111 or 37 Hex represents the ASCII numeral 7. Similarly, 01001011 or 4B Hex is the ASCII code for the letter K.

Electrically, there are four popular conventions for transmitting asynchronous serial data. These conventions are the 20-mA current loop, EIA (Electronic Industries Association) standard RS-232-C, EIA standards RS-422, 423, 499, and EIA standard RS-485.

20 MILLIAMPERE CURRENT LOOP

The 20-mA current loop uses the presence of a current to represent a logic 1 or MARK while the absence of or minimal current denotes a logic 0 or SPACE. This type of data transmission has a high immunity to noise and can be run for relatively long distances, since current and not voltage represents the transmitted bits. Current loop transmitter and receiver interfaces with conventional LSTTL (Low Power Schottky TTL) logic using Hewlett-Packard HCPL-4100 transmitter and HCPL-4200 receiver IC's are given in Figures 4.5 and 4.6. The receiver also is compatible with CMOS logic. Optical coupling isolates the transmitter and receiver logic grounds and protects against equipment damage or logic errors.

RS-232-C[1]

The EIA standard RS-232-C defines the interface between data terminal equipment and data communication equipment employing serial binary data

---

[1] EIA Standard RS-232-C, Electronic Industries Asscociation, August 1969.

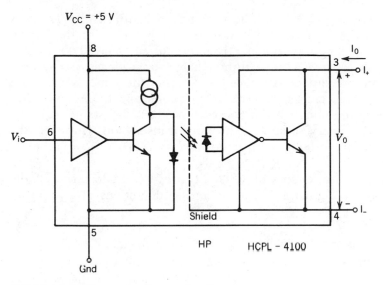

FIGURE 4.5 LSTTL to current loop transmitter.

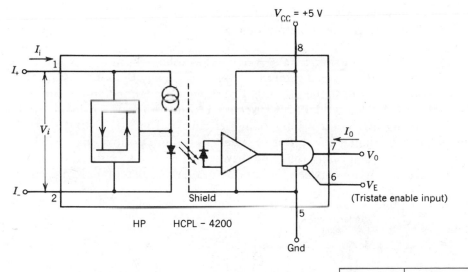

Logic 1 = MARK, ≥12 mA
Logic 0 = SPACE, ≤2 mA
Z = high impedance output state

I_i	V_E	V_0
H	H	Z
L	H	Z
H	L	Logic 1
L	L	Logic 0

FIGURE 4.6 Current loop to LSTTL and CMOS receivers.

interchange. Although EIA Standard RS-449, together with EIA RS-423 were designed to replace RS-232-C, RS-232-C is still widely used. The standard is defined for data signaling rates ranging from zero to a nominal 20,000 bps. In some instances, the term *baud* is used interchangeably with bits per second. Technically, there are situations where they are not the same, but in most instances, baud and bits per second are used synonomously. In this book, they can be considered the same. In this context, the RS-232-C standard is discussed in terms of binary serial asynchronous transmission. However, the standard is applicable to both synchronous and asynchronous serial binary data transmission.

The electrical characteristics of RS-232-C are defined in terms of the equivalent circuit of Figure 4.7. The following definitions also apply to Figure 4.7.

V_0 open-circuit driver voltage

R_0 driver internal dc resistance

C_0 total effective capacitance associated with the driver, measured at the interface point and including any cable to the interface point

V_1 voltage at the interface point

C_L total effective capacitance associated with the terminator, measured at the interface point and including any cable to the interface point

R_L terminator load dc resistance

E_L open-circuit terminator voltage (bias)

For data transmission, a MARK is defined as voltage V_1 more negative than -3 V with respect to signal ground. A SPACE is voltage V_1 more positive than $+3$ V with respect to signal ground. The signal is undefined and is said to be in the transition region if V_1 is between $+3$ V and -3 V.

In timing and control interchange circuits, a function is OFF when V_1 more negative than -3 V and is ON when V_1 is more positive than $+3$ V.

The standard specifies that "the driver shall be able to withstand an open circuit, a short circuit between the conductor carrying that interchange circuit

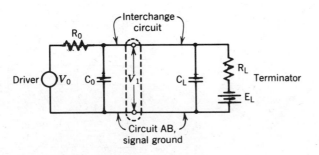

FIGURE 4.7 RS-232-C equivalent circuit.

in the interconnecting cable and any other conductor in that cable, or any passive noninductive load connected between that interchange circuit and any other interchange circuit including Circuit AB (Signal Ground), without sustaining damage to itself or its associated equipment."

The RS-232-C definitions and specifications that have been presented to this point along with additional parameters are summarized in Table 4.1.

TABLE 4.1 Some Important Parameters of EIA RS-232-C and CCITT V.24 Standards

Parameter	Symbol	Min	Max	Units	Comments
Driver output voltage with load resistance					
Logic 0	V_{OH}	5	15	Volts	
Logic 1	V_{OL}	−15	−5	Volts	$7\ K\Omega \geq R_L \geq 3\ K\Omega$
Driver output voltage open circuit					
Logic 0	V_{OH}		25	Volts	
Logic 1	V_{OL}	−25			
Receiver input threshold					
MARK output from driver		−3		Volts	
SPACE output from driver			+3	Volts	
Driver source resistance					
Power on	R_O		Not specified		
Power off	R_O		300	Ohms	$2\ V \geq V_{out} \geq -2\ V$
Driver output short-circuit current		−500	500	Milliamperes	
Driver output switching characteristics					
Slew rate			30	Volts/μs	
Rise and fall time			±14%	of pulse interval	
Bit rate		0	20K	Hertz	
Receiver input resistance	R_{IN}	3K	7K	Ohms	$25\ V \geq V_{in} \geq 3\ V$
Receiver input voltage		−25	25	Volts	

TABLE 4.2 RS-232-C Connector Pins and Signals

Signal	Pin
Protective ground (chassis)	1
Transmitted data (TD)	2
Received data (RD)	3
Request to send (RS)	4
Clear to send (CTS)	5
Data set ready (DSR)	6
Signal ground (common return)	7
Received line signal detector (CF)	8
(Reserved for data set testing)	9
(Reserved for data set testing)	10
Unassigned	11
Secondary received line signal detector (SCF)	12
Secondary clear to send (SCB)	13
Secondary transmitted data (SBA)	14
Transmitter signal element timing (TSET)	15
Secondary received data (SBB)	16
Receiver signal element timing (DD)	17
Unassigned	18
Secondary request to send (SCA)	19
Data terminal ready (DTR)	20
Signal quality detector (CG)	21
Ring detector (CE)	22
Data signal rate selector (CH/CI)	23
Transmit signal element timing (DA)	24
Unassigned	25

Mechanically, the standard recommends that the distance between the transmitter and receiver be less than 50 ft or 15 m. If the load capacitance, including the signal terminator, is not greater than 2500 pF, longer distances are possible. The connector used with RS-232-C is a standard 25-pin male connector such as the Cinch DC-255, Amp 205207-1, or the Amphenol 17-10250-1. The connector pins and signals are given in Table 4.2. Not all pins on the connector have to be used. The most commonly used pins are 1–7, 15, and 20.

RS-232-C SPECIFICATION LIMITS

Recall that the specification gives the maximum cable drive length as 50 ft and the maximum transmission rate as 20 K bps. Since the principal factor limiting the cable length and transmission rate is cable capacitance, longer cable lengths can be used if the transmission rate is reduced, and vice versa. These parameters relate back to the specification on the time required for the signal to pass through the -3 V to $+3$ V transition region. This time must not

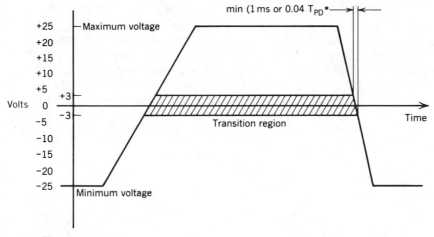

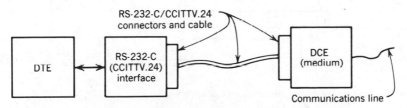

FIGURE 4.8 Transition region signal timing.

exceed 1 ms or 4% of the nominal duration of a signal element (T_{PD}) on the interface circuit, whichever is less. This timing is illustrated in Figure 4.8.

TYPICAL RS-232-C CONNECTION AND TIMING

The equipment used for information processing in communications is referred to as *data terminal equipment* (*DTE*); equipment that connects data terminal equipment to the communications line is called *data communications equipment* (*DCE*). In an RS-232-C application, an RS-232-C interface is connected between the DTE and DCE. Since the CCITT (International Telegraph and Telephone Consultative Committee) Standard V.24 is identical to RS-232-C, all discussions relative to RS-232-C apply to CCITT V.24. Figure 4.9 illustrates a typical connection of DTE to the communications line using RS-232-C.

The physical RS-232-C/CCITT V.24 connections for the DTE to the DCE through the cable, based on standard telephone company lead designations at the DCE, are given in Figure 4.10. Typical timing sequences on the interface

FIGURE 4.9 Typical RS-232-C/CCITTV.24 communications connection.

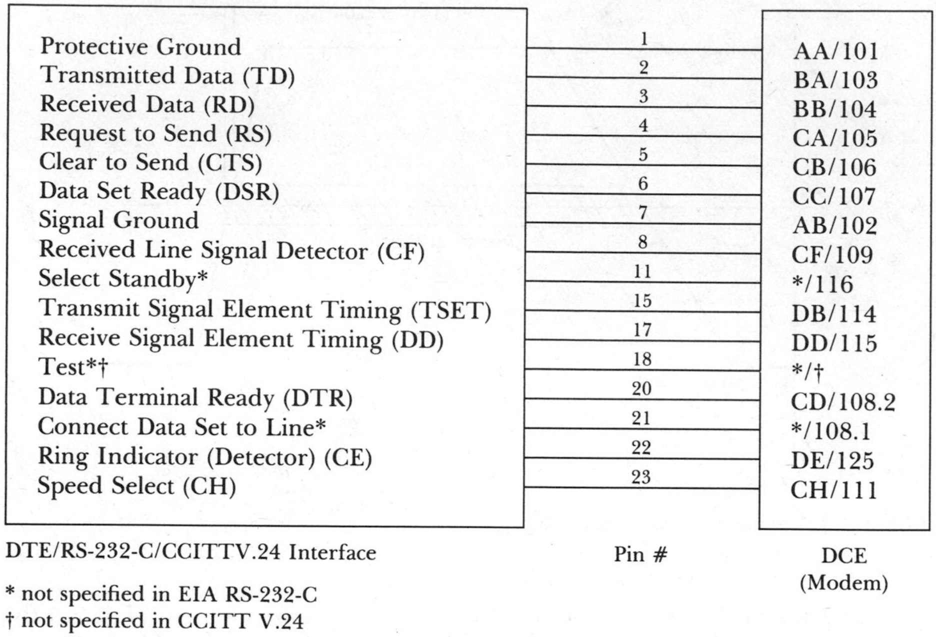

	Pin #	
Protective Ground	1	AA/101
Transmitted Data (TD)	2	BA/103
Received Data (RD)	3	BB/104
Request to Send (RS)	4	CA/105
Clear to Send (CTS)	5	CB/106
Data Set Ready (DSR)	6	CC/107
Signal Ground	7	AB/102
Received Line Signal Detector (CF)	8	CF/109
Select Standby*	11	*/116
Transmit Signal Element Timing (TSET)	15	DB/114
Receive Signal Element Timing (DD)	17	DD/115
Test*†	18	*/†
Data Terminal Ready (DTR)	20	CD/108.2
Connect Data Set to Line*	21	*/108.1
Ring Indicator (Detector) (CE)	22	DE/125
Speed Select (CH)	23	CH/111

DTE/RS-232-C/CCITTV.24 Interface   Pin #   DCE
(Modem)

* not specified in EIA RS-232-C
† not specified in CCITT V.24

FIGURE 4.10 Typical DTE to DCE RS-232-C/CCITTV.24 interface.

- - - indicates direct link timing that is different from dial-up timing.

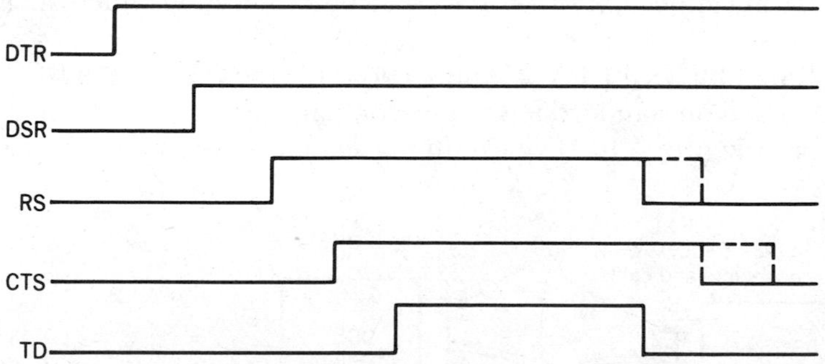

FIGURE 4.11 Dial-up and direct link timing sequences.

for dial-up (switched) communications links and direct connection links are given in Figure 4.11.

RS-422-A, RS-423-A, RS-449

EIA Standards RS-422-A, RS-423-A, and RS-449 are intended to gradually replace EIA standard RS-232-C. Standards RS-422-A and RS-423-A are the electrical interface standards for digital interface circuits while standard RS-449 specifies the functional and mechanical characteristics applicable to RS-422-A and RS-423-A. With a few modifications, equipment designed in conformance with Standard RS-232-C can operate with equipment designed to the newer standards. Some of the ways RS-422-A, RS-423-A, and RS-449 differ from RS-232-C are that signaling rates up to 2 million bits per second are included, a larger connector (37-pin) is specified, a separate 9-pin connector is specified to handle secondary channel interchange circuits, and new names and mnemonics have been defined.

RS-422-A[2]

This standard specifies the electrical characteristics of balanced voltage digital interface circuits. A balanced interface is advantageous for higher data transmission rates (up to 10-Mbs), long cable runs, and noisy environments, and where inversion of the signals may be necessary. The standard applies to the balanced circuit model of Figure 4.12. The generator circuit should have an output impedance of 100 Ω or less and produce a differential voltage between 2–6 V across the balanced interconnecting cable. Logically, a binary 1 (MARK or ON state) is defined as the A terminal of the generator being negative with respect to the B terminal. Conversely, a binary 0 (SPACE or OFF state) is given as the A terminal being positive with respect to the B terminal. Also, the voltage between either of the generator output terminals (A and B) and circuit ground should not exceed 6 V. The generator should be able to limit the current resulting from a direct connection of either terminal A or B to circuit ground to a maximum of 150 mA.

At the receiver, a differential input voltage of 200 mV should be the maximum required to develop the intended binary state. The complementary binary state is obtained by reversing the polarity of the receiver input voltage. These conditions should be valid over a *common-mode* voltage range of −7 to +7 V at the receiver input terminals (A' and B'). *Common-mode* voltage is defined as the algebraic mean of the two voltages at A' and B' with respect to circuit ground. The receiver should maintain operation for differential input signal voltage magnitudes from 200 mV to 6 V, and should continue to operate with a maximum differential input voltage of 12 V. Some of

[2] EIA RS-422-A: Electrical Characteristics of Balanced Voltage Digital Interface Circuits, Electronic Industries Association, December 1978.

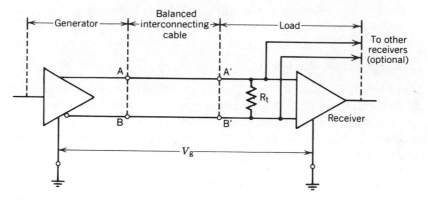

FIGURE 4.12 Balanced digital circuit model.

TABLE 4.3 Summary of RS-422-A

	Parameter	Conditions	Min	Max	Units
V_o	Driver unloaded output voltage			6	V
V_o				−6	V
V_T	Driver loaded output voltage	$R_T = 100\ \Omega$	2		V
V_T			−2		V
R_S	Driver output resistance	Per output		50	Ω
I_{os}	Driver output short-circuit current	$V_o = 0$ V		150	mA
	Driver output rise time			10	% unit interval
I_{ox}	Driver power OFF current	−0.25 V ≤ V_o ≤ 6 V		±100	μA
V_{TH}	Receiver sensitivity	$V_{CM} = \pm7$ V		200	mV
V_{CM}	Receiver common-mode voltage		−12	12	V
	Receiver input offset		±3		V

RS-422

the major electrical characteristics of the RS-422-A interface are given in Table 4.3.

RS-423-A[3]

RS-423-A covers the same areas as RS-422-A but for unbalanced (single-ended) voltage digital interface circuits. Unbalanced circuits can be used in

[3] EIA RS-423-A: Electrical Characteristics of Unbalanced Voltage Digital Interface Circuits, Electronic Industries Association, December 1978.

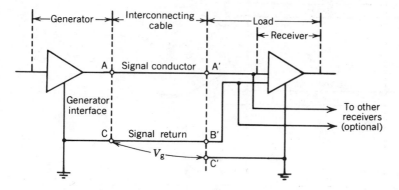

FIGURE 4.13 Unbalanced digital circuit model.

the same interconnection as balanced circuits. For example, balanced circuits could be used for high speed data transmission and unbalanced circuits could handle lower speed handshaking and control signals. The unbalanced digital specification states that the data rate may not exceed 100K bits/s. Also, the RS-423-A specifications should not be used in areas of high extraneous noise. The circuit model for RS-423-A is that of Figure 4.13.

A logic 1 (MARK state) at the generator output is defined as the A terminal in Figure 4.13 being negative with respect to terminal C. Logic 0 (SPACE state) is characterized by the voltage at terminal A being positive with respect to terminal C. Open circuit output (voltage magnitude) of the generator shall be in the range from 4–6 V. The binary state is determined by the polarity of the voltage. With generator output short-circuited to ground, the generator output current is to be limited to a maximum of 150 mA.

The receiver electrical specifications are identical to those of RS-422-A.

Signal returns in the interface can be connected in a number of ways. Two examples involving multiple signal paths are given in Figures 4.14 and 4.15. Table 4.4 summarizes some of the primary specifications of Standard RS-423-A.

RS-449[4]

RS-449 is aimed primarily at data applications on analog networks and defines signal characteristics (through RS-422-A and RS-423-A), interface mechanical characteristics, functional description of interchange circuits, and standard interfaces for selected communication system configurations.

There are two interface connectors specified in RS-449: One is a 37-position connector that includes connections to all circuits except secondary channels. These secondary channels are interfaced through a 9-position con-

[4] EIA RS-449: General Purpose 37 Position and 9 Position Interface for Data Terminal Equipment and Data Circuit Terminating Equipment Employing Serial Binary Data Interchange, Electronic Industries Association, November 1977.

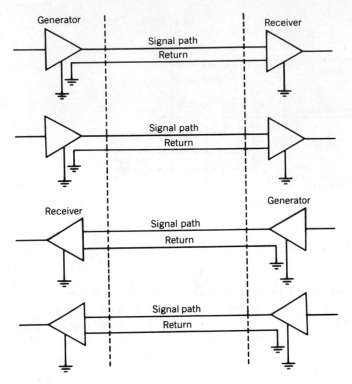

FIGURE 4.14 Multiple line RS-423-A interface.

TABLE 4.4 Summary of RS-423-A

	Parameter	Conditions	Min	Max	Units
V_o	Driver unloaded output voltage		4	6	V
V_o			−4	−6	V
V_T	Driver loaded output voltage	$R_L = 450\ \Omega$	3.6		V
V_T			−3.6		V
R_S	Driver output resistance			50	Ω
I_{OS}	Driver output short-circuit current	$V_o = 0$ V		±150	mA
	Driver output rise and fall time	Baud rate ≤ 1K baud		300	μs
		Baud rate ≥ 1K baud		30	% unit interval
I_{OX}	Driver power OFF current	$V_o = \pm6$ V		±100	μA
V_{TH}	Receiver sensitivity	$V_{CM} \le \pm7$ V		±200	mV
V_{CM}	Receiver common-mode range			±10	V
R_{IN}	Receiver input resistance		4000		Ω
	Receiver common-mode input offset			±3	V

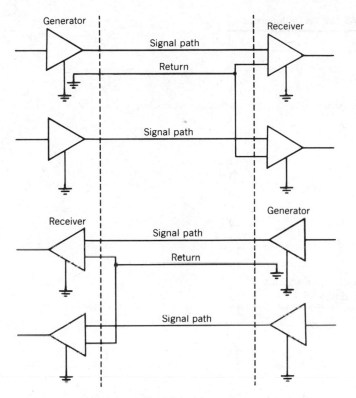

FIGURE 4.15 Multiple line RS-423 interface common return.

nector. Both connectors are to have dimensions detailed in the portions of MIL-C-24308 applicable to connectors. Pin numbers and corresponding signals for some principal RS-449 functions are given in Table 4.5 for the primary 37 position connector and in Table 4.6 for the secondary 9 position connector. These are given as examples of typical signals but are not complete, since connections depend on the type of generator used (balanced or unbalanced). Refer to the EIA specification document for further details.

RS-485

RS-485 is a balanced voltage interface standard similar to RS-422-A but differs in that it permits multiple drivers and receivers to operate on the same two-wire bus. Recall that the RS-422-A standard defines a single driver, multiple receiver architecture. In particular, RS-422-A defines 1 driver and up to 10 receivers, whereas RS-485 supports a maximum cable length of 4000 ft and a maximum data rate of 10 Mbs. RS-485 has additional advantages in that it can withstand higher common-mode voltages ($+12$ V to -7 V) than RS-422-A specifies ($+6$ V to -0.25 V) and also generates less noise, since voltage rise

TABLE 4.5 Some Primary RS-449 Pin Designations

Signal	Pin
Shield	1
Signaling rate indicator (SI)	2
Spare	3
Secondary receive data (SRD)	4
Send timing (ST)	5
Receive data (RD)	6
Request to send (RTS)	7
Receive timing (RT)	8
Clear to send (CTS)	9
Local loopback (LL)	10
Data mode (DM)	11
Terminal ready (TR)	12
Receiver ready (RR)	13
Remote loopback (RL)	14
Incoming call (IC)	15
Select frequency (SF) and signaling rate selector (SR) both share pin 16	16
Terminal timing (TT)	17
Test mode (TM)	18
Signal ground (SG)	19
Receive common (RC)	20
Spare	21
Terminal in service (IS)	28
Select standby (SS)	32
Signal quality (SQ)	33
New signal (NS)	34
Standby indicator (SB)	36
Send common (SC)	37

TABLE 4.6 Secondary RS-449 Pin Designations

Signal	Pin
Shield	1
Secondary receiver ready (SRR)	2
Secondary send data (SSD)	3
Secondary receive data (SRD)	4
Signal ground (SG)	5
Receive common (RC)	6
Secondary request to send (SRS)	7
Secondary clear to send (SCS)	8
Signal common (SC)	9

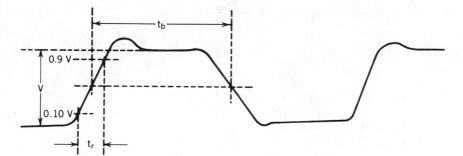

FIGURE 4.16 RS-485 timing.

times can be 3 times longer than RS-422-A. The latter specification is illustrated using Figure 4.16. In the figure,

V is the difference in steady state voltage levels.

t_r is the waveform rise time (transition time from 10% of V to 90% of V).

t_b is pulse width measured between successive 0.5-V levels.

Thus RS-485 defines $t_r \leq 0.3t_b$ while RS-422A specifies that t_r is approximately equal to $0.1t_b$. If t_r is known for a particular RS-485 driver, the maximum data transmission baud rate, f_t, can be derived as follows:

$$f_t = \frac{1}{t_b} = \frac{0.3}{t_r}$$

A typical RS-485 driver such as that contained in the TI SN74172B quad driver has a maximum value of $t_r = 75$ ns. Thus the worst case transmission rate of the SN74172B is $0.3/(75 \times 10^{-9})$ or 4 M baud.

A diagram of a general terminated RS-485 bus is shown in Figure 4.17. Since multiple drivers can be connected to the RS-485 bus, they must be internally protected so as not to be damaged when contention occurs. To cover this situation, the RS-485 standard specifies that a driver should not sustain damage if its output is connected to a voltage source ranging from -7 V to $+12$ V in any of its normal output states. Additionally, an RS-485 driver should be able to drive up to 32 unit loads and a total line termination resistance of 60 Ω. A unit load is defined as drawing 1 mA in the worst case under a maximum of 12 V common-mode voltage stress, not including any additional load such as a 60-Ω termination resistance. Table 4.7 provides a comparison between the RS-422A and RS-485 differential standards.

Typical drivers for RS-485 are the Texas Instrument SN75172B and SN75174B chips. Their associated receivers are, respectively, the SN75173A and the SN75175A devices.

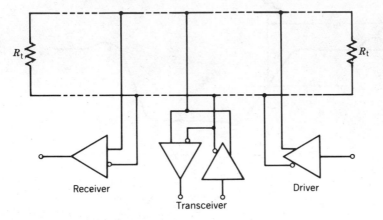

FIGURE 4.17 General RS-485 bus.

TABLE 4.7 RS-422A and RS-485 Comparison

Parameter	RS-422A	RS-485
Maximum cable length (ft)	4000	4000
Maximum data rate (bps)	10M	10M
Drive output (V)	±2 min	±1.5 min
Driver load (Ω)	100 min	60 min
Driver output short-circuit current limit	150 mA to ground	150 mA to ground, 250 mA to −8 V or +12 V
Driver output resistance (Ω), high impedance state		
Power on	N/A	120K
Power off	60K	120K
Maximum number of drivers and receivers on line	1 driver 10 receivers	32 drivers 32 receivers
Receiver sensitivity	±200 mV	±200 mV
Receiver input resistance (Ω)	4K	12K
Maximum common-mode voltage	+6 V, −0.25 V	+12 V, −7 V

4.1.2   SYNCHRONOUS TRANSMISSION

Synchronous transmission will be developed through the HDLC/SDLC protocols. This approach was chosen because these protocols are widely used and exemplify synchronous transmission principles.

HDLC/SDLC DATA LINK CONTROL

High Level Data Link Control (HDLC) and *Synchronous Data Link Control (SDLC)* are two similar protocols for managing information transfer on a data channel. HDLC was established by the *International Standards Organization*

(*ISO*) and is used in the ISO X.25 standard for packet switching networks. SDLC was developed by IBM under its *System Network Architecture* (*SNA*).

HDLC and SDLC are frameworks for information to be sent from one point to another. These protocols provide for the destination address, control commands, information, and error checking necessary in digital message transmission. The basic differences between SDLC and HDLC are that HDLC offers extended address and control fields and the character for indicating a transmission ABORT is seven contiguous 1's in HDLC as opposed to eight contiguous 1's in SDLC. Because of the similarity in the two protocols, the remainder of the discussion will concentrate on SDLC instead of referring to both SDLC and HDLC. The material presented draws heavily on that in a previous book by the author.[5] SDLC is becoming pervasive in many types of communications applications and is being supported for microprocessors by SDLC controller peripheral interface chips such as the Intel 8273. The 8273 will be discussed in Chapter 5 in an IBM PC application. As its name implies, SDLC is a synchronous data transmission protocol and, unlike the earlier IBM *Binary Synchronous Communication* (*BSC or BISYNC*) procedure, is bit oriented rather than character oriented.

In SDLC, one station acts as a primary node and the others as secondaries. Data transmission can be initiated by a secondary node as well as the primary, but transmissions are always to or from a primary node. SDLC can be used in half or full duplex mode.

The unit of SDLC transmission is called a *frame*. A standard SDLC frame consists of six fields as shown in Figure 4.18. These fields in the order of their transmission are the 8-bit FLAG (F), 8-bit ADDRESS (A), 8-bit CONTROL (C), variable bit length (in multiples of 8 bits) INFORMATION (I), 16-bit FRAME CHECK SEQUENCE (FCS), and the 8-bit FLAG (F). Figure 4.19 indicates the fields subject to 0-bit insertion (to be discussed shortly) and used in the FCS.

The SDLC frames fall into three categories: *Nonsequenced* (*management*) frames, *Supervisory* frames, and *Information* frames. A *Nonsequenced* frame conveys commands and responses. It is used for control of secondary nodes and indication of transmission errors. This type of frame is sent by the primary or by the secondary as a response to a primary poll.

A *Supervisory* frame is used for supervisory functions such as acknowledgments, busy indications, and calls for retransmission. The data or information to be transmitted are contained in the *Information* frame.

Each field in a frame serves a unique purpose, and the characteristics of each field are what make SDLC a flexible and effective protocol. The initial FLAG in the frame or *Open* FLAG, as it is sometimes called, signals the beginning of a new frame. It provides a reference for the following fields and signals that **all bits following the FLAG field,** with the exception of the *Close* FLAG and inserted zeros, should be included in error checking procedures. The FLAG pattern is a constant pattern of 01111110 and must not occur

[5] Krutz, *Microprocessors and Logic Design,* John Wiley & Sons, Inc., New York, 1980.

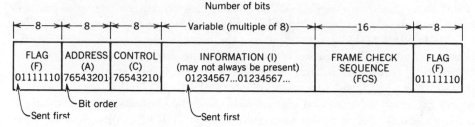

FIGURE 4.18 SDLC frame field designation.

anywhere else in the frame except in the last or *Close* FLAG at the end of the frame. If this pattern is present in any other fields of the frame to be transmitted, SDLC protocol requires that a zero be inserted after every group of five consecutive 1's before the data are entered into the I field. Consequently, these inserted 0's must be deleted by the receiver SDLC interface.

SDLC assumes that synchronism is provided by a modem or DTE (Data Terminal Equipment). The DTE is considered as the origin or destination of transmitted data. If timing for received data is not performed by a modem, *NRZI (Non-Return to Zero Inverting)* code is used. In this code, the signal level remains unchanged for the transmissionof a 1 while a transition to the opposite level represents a 0. The contrast between conventional Non-Return to Zero (NRZ) coding and NRZI is shown in Figure 4.20. Since the transitions of the signal level can be used for resynchronization between transmitter and receiver, the insertion of a 0 after five consecutive 1's guarantees transitions to opposite levels during long strings of 1's and, thus, eliminates relatively long periods without any level changes. Since the FLAG pattern is known by the receiver, this pattern is used for synchronization of the clocks between the transmitter and receiver. The *Address* field designates the destination or destinations of the frame sent by a primary station. A secondary station will synchronize only if its address is present in the address field. No address field is required when sending from a secondary station to the primary.

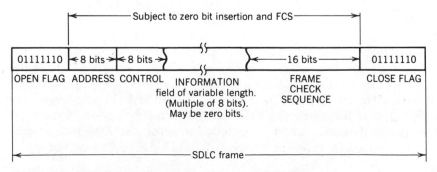

FIGURE 4.19 SDLC fields subject to zero bit insertion and FCS.

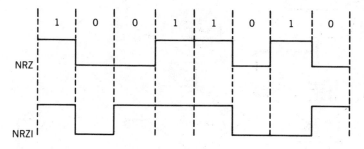

FIGURE 4.20 Difference between NRZ and NRZI codes.

The *Control* Field has three different formats, corresponding to either an Information, Supervisory, or Nonsequenced (Management) frame. If the leftmost bit (bit 7), of the Control field is 0, the Control field of an Information frame is identified. If bit 7 of the Control Field is a 1, the Control field of a Supervisory frame is specified by bit 6 = 0 and the Control field of a Nonsequenced frame is specified by bit 6 = 1. The Control field of an Information frame provides for one of the important features of SDLC, namely, **that multiple frames can be transmitted without necessarily waiting for an acknowledgment from the receiver following each frame transmission.** In fact, up to seven frames can be transmitted without acknowledgments.

Following the seventh frame, if no acknowledgment has been received, the transmitter will wait for an acknowledgment from the receiver in terms of the number of frames correctly received or request an acknowledgment. Actually, the acknowledgment sent back to the transmitter is the number, N_r, that corresponds to the next frame expected by the receiver. The number of frames transmitted is denoted by the symbol, N_s. For example, if frames 0, 1, and 2 have been transmitted and none immediately thereafter, N_s would equal 2, indicating frames 0, 1, and 2 have been transmitted, and N_r would be 3 if the **three frames were received without error.** Since the control field of each frame contains the values of N_s and N_r, known by a station, SDLC has the feature of a station's being able to verify and acknowledge another station's transmission while continuing to transmit. Of the three types of frames, Information frames are the only ones that are counted for N_s and N_r counts. This type of frame is also referred to as a *Sequenced frame*. Flowcharts illustrating a procedure for Information frame transmission, verification, and retransmission are given in Figure 4.21. Related examples are given in Figure 4.22. In these two figures, a central microcomputer, C, is the primary station and is transmitting data to a secondary terminal, T.

The Poll/Final (P/F) bit of the Control field, as shown in Figure 4.23 serves two purposes. If the frame is being sent from a primary station to a secondary station P/F = 1 requests (polls) the secondary to confirm the present N_s count. The P/F is also set to 1 in the final frame of transmission by a secondary station transmitting to a primary to indicate an End of Transmission (EOT).

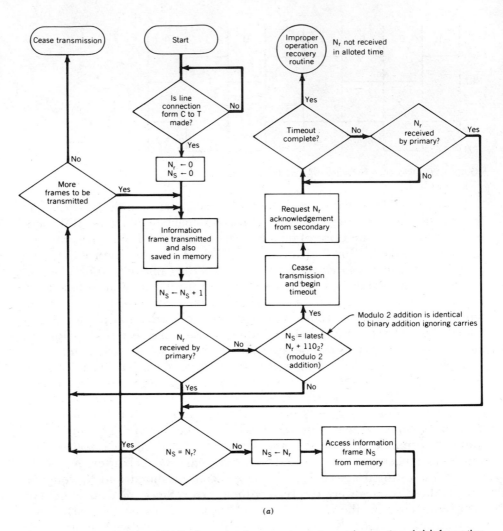

FIGURE 4.21 Flow charts of SDLC information frame transmission and reception: (*a*) Information frame transmission procedure flowchart for primary station; (*b*) information frame reception procedure flowchart for terminal T.

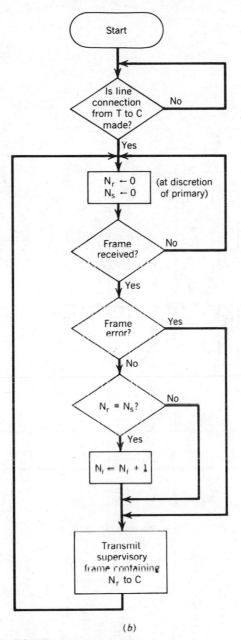

(b)

FIGURE 4.21 (Continued).

Primary (C)
Information
Frame transmission

| $N_S = 0$ | $N_S = 1$ | $N_S = 2$ | $N_S = 3$ | $N_S = 4$ | $N_S = 5$ | $N_S = 6$ | $N_S = 7$ | $N_S = 0$ |

Secondary (T)
Supervisory
Frame acknowledgment

$N_r = 3$ $N_r = 5$

Time ———→

(a) Error-free reception of information from C by T

Primary (C)
Information
Frame transmission

| $N_S = 0$ | $N_S = 1$ | $N_S = 2$ | $N_S = 3$ | $N_S = 2$ | $N_S = 3$ |

Secondary (T)
Supervisory
Frame acknowledgment

$N_r = 2$

Time ———→

(b) Retransmission of frame 2 by C as initiated
by $N_r = 2$ acknowledgment from T.

FIGURE 4.22 Examples of SDLC frame transmission from C to T.

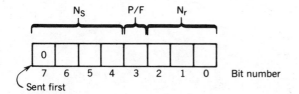

FIGURE 4.23 Control field of an information frame.

A Supervisory frame Control field has, as previously noted, bit 7 = 1 and bit 6 = 0. The remainder of the field consists of a P/F bit, two bits for Command/Response functions, and three N_r bits (Figure 4.24).

The Command/Response functions of the Supervisory frame and their Control field patterns are

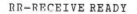

RR–RECEIVE READY

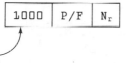

(Acknowledges receipt of frames $N_r - 1$ and states that station is ready to receive)

RNR–RECEIVE NOT READY

(Indicates temporary condition in which frames requiring buffer storage space cannot be received)

REJ–REJECT/RETRANSMIT REQUEST

(Acknowledges receipt of frames N_r 1 and requests transmission of N_r and succeeding frames.)

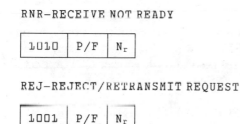

The Nonsequenced frame Control field contains five Command/Response bits, in addition to identification bits 7 and 6, and a P/F bit. This type of frame is sent only by a primary or as a response to a primary poll. **Nonsequenced and Supervisory frames are not counted as N_r and N_s transmissions** (Figure 4.25).

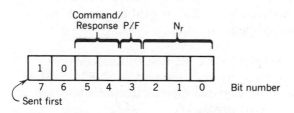

FIGURE 4.24 Control field of supervisory frame.

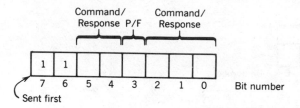

FIGURE 4.25 Control field of a nonsequenced frame.

The Command/Response functions of a Nonsequenced frame along with their Control field pattern are

NSI

| 1100 | P/F | 000 |

(Sent first)

NONSEQUENCED INFORMATION (Response to or command for nonsequenced information; NSI frames are not acknowledged)

SNRM

| 1100 | P | 001 |

SET NORMAL RESPONSE MODE (Primary station command designating secondary station; limits secondary to solicited transmissions only; Normal Response Mode condition changed by · DISC or SIM command; N_s and N_r counts of primary and secondary are reset to 0; expected response is NSA)

SIM

| 1110 | P | 000 |

SET INITIALIZATION MODE (Command from primary station resetting N_s and N_r counts of primary and secondary to 0; begins prespecified procedures at secondary; expected response is NSA)

DISC

| 1100 | P | 010 |

DISCONNECT (Command from primary station putting secondary station off line and terminating other modes; secondary station can be reconnected by SNRM or SIM command; expected response to DISC is NSA)

NSA

| 1100 | F | 110 |

NONSEQUENCED ACKNOWLEDGMENT (Confirming acknowledgment to SNRM, SIM or DISC commands; additional transmissions are at discretion of primary)

RQI

1110	F	000

REQUEST FOR INITIALIZATION (Secondary station request for SIM command; any response other than SIM command by primary will initiate another RQI by secondary)

CMDR

1110	F	001

COMMAND REJECT (Response by secondary station to nonvalid command; nonvalid commands include unassigned commands and, optionally, an I field that contains too many bits for receiving buffer storage)

ROL

1111	F	000

REQUEST ON LINE (Secondary station response indicating that it is presently off line)

XID

1111	P/F	101

EXCHANGE STATION IDENTIFICATION (Command to or response for the address of the receiving secondary station; an optional I field may be included in the frame to identify transmitting or responding station)

NSP

1100	1/0	100

NONSEQUENCED POLL (Transmission request by primary to secondary, this request is a command if P/F is a 1, but response is optional if P/F = 0; I field is not permitted)

TEST

1100	P/F	111

TEST (Command to or response for testing purposes; an optional I field can be included in the command that will then elicit a TEST Nonsequenced frame from the receiver that will contain the same I field)

The Information (I) field itself contains the data to be transmitted in multiples of 8 bits. For each individual 8-bit grouping, the high-order bit is transmitted **last** in time.

The Frame Check Sequence (FCS) field or *Block Check* (*BC*), contains a 16-bit pattern that is computed by the transmitter and that is a function of previous bits in the frame. To generate the FCS, the bits transmitted after the Open FLAG and before the Close FLAG (excluding inserted zeros) are considered to be coefficients of a polynomial in X. For n coefficients, the order of the polynomial is $n - 1$. The FCS field follows the Information field (if there is one) in Information and Nonsequenced frames or the Control field in all other situations.

Consider the 8-bit "message" 10010011. This message could be considered as the polynomial

$$M(X) = (1)X^7 + (0)X^6 + (0)X^5 + (1)X^4 + (0)X^3 + (0)X^2 + (1)X^1 + (1)X^0$$

or

$$M(X) = X^7 + X^4 + X + 1, \quad \text{where } n = 8$$

Now, for a general polynomial $M(X)$, a series of 16 check bits corresponding to the FCS is to be added to the message before transmission and used by the receiver to verify that a valid transmission has occurred. In order to obtain and prepare for the 16 check bits that are to be added to $M(X)$, $M(X)$ must effectively be shifted 16-bit positions to the left or, equivalently, multiplied by X^{16} while ignoring any carries during multiplication (modulo 2 multiplication). In SDLC, the vacant 16-bit positions to the right of $M(X)$ are set to 1's and the resulting bit string including $M(X)$ is divided (modulo 2, ignoring carries or borrows) by a polynomial, $G(X)$. Then $G(X)$ is known as the *generating polynomial*.

The generating polynomial must have an order equal to the number of check bits that, for SDLC, is 16. The generating polynomial is constant for a particular coding scheme and for SDLC is $X^{16} + X^{12} + X^5 + 1$. The quotient, $Q(X)$, of the division is ignored and the one's complement of the remainder, $R(X)$, becomes the FCS field. It is important to note that in carrying out the modulo 2 division, we perform modulo 2 subtraction. Furthermore, modulo 2 addition and subtraction are identical, since carries and borrows are ignored. Both operations are the EXCLUSIVE OR function of corresponding bits of the two words involved. The appending of the FCS field to the other fields involved in error checking will produce a polynomial to be transmitted, $T(X)$. Then $T(X)$ is divided (modulo 2) at the receiver by $G(X)$, where $T(X)$ includes the transmitted FCS field. The remainder obtained at the receiver is a constant and is characteristic of the divisor. For the $G(X)$ of SDLC, this remainder must always be 1111000010111000 for a correct transmission indication. If this constant is not obtained after division at the receiver, the frame is assumed to be in error and is not accepted. Then N_r is not incremented.

This procedure is summarized as

$M(X)$	message polynomial in X of order n
$G(X)$	generating polynomial in X of order 16
$X^{16}M(X)$	message polynomial shifted 16-bit positions to the left with vacant positions filled by 1's
$Q(X)$	quotient resulting from division of $X^{16}M(X)$ by $G(X)$
$R(X)$	16-bit series that is remainder of $X^{16}M(x)/G(x)$
$R(X)'$	one's complement of $R(X)$ and is FCS field

T(X)   transmitted word

T(X) $= X^{16}M(X) + R(X)$[1]

$\oplus$ modulo 2 addition

Dividing the message polynomial by G(X) gives

$$T(X) = \frac{X^{16}M(X)}{G(X)} = Q(X) \oplus \frac{R(X)}{G(X)}$$

or, since modulo two addition is the same as modulo two subtraction,

$$T(X) = X^{16}M(X) \oplus R(X) = Q(X)G(X)$$

The effect in SDLC of presetting dividend/remainder hardware registers to all 1's, transmitting the one's complement of R(X) as FCS, and similar operations at the receiver produces results equivalent to

$$X^{16}M(X) \oplus R(X)' \oplus \text{F0B8Hex} = Q(X)G(X)$$

Continuing, we obtain

$$X^{16}M(X) \oplus R(X)' = Q(X)G(X) \oplus \text{F0B8Hex}$$
$$\frac{X^{16}M(X) \oplus R(X)'}{G(X)} = \frac{Q(X)G(X)}{G(X)} \oplus \frac{\text{F0B8Hex}}{G(X)}$$
$$\frac{X^{16}M(X) \oplus R(X)'}{G(X)} = Q(X) \oplus \frac{\text{F0B8Hex}}{G(X)}$$

The last equation shows that dividing by G(X) at the receiver produces a quotient Q(X) that is discarded and a remainder F0B8Hex or 1111000010111000.

A code obtained in the manner just described is referred to as a *CRC* or a *Cyclic Redundancy Check.*

The modulo 2 division process is accomplished serially using shift registers and EXCLUSIVE OR gates. The shifting process performs the alignment of the highest-order nonzero divisor bit with the highest-order bit of the dividend. The first modulo 2 subtraction is implemented by an EXCLUSIVE OR function between each pair of bits of corresponding weight. This procedure is then accomplished repeatedly in the loading and shifting process of the shift register with the $M(X)X^{16} \oplus R(X)'$ sum exiting serially from the shift register.

A block diagram of a shift register CRC generator for the SDLC generating polynomial $X^{16} + X^{12} + X^5 + 1$ is given in Figure 4.26. The flip-flops comprising the shift register are D type with shift/preset control. The CRC generator can be used at both the transmitting and receiving stations.

At the transmitting station, the shift register is initialized to all 1's by

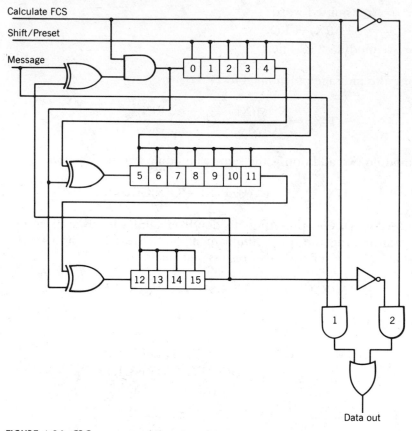

FIGURE 4.26 CRC generator shift register for
$$G(X) = X^{16} + X^{12} + X^5 + X$$

momentarily placing the SHIFT/PRESET line in the PRESET logic level. Then, with the SHIFT/PRESET line at the SHIFT logic level and the CALCULATE FCS line at a HIGH level, output AND gate 1 will be enabled and the Message consisting in order of the Address, Control, and Information (when desired and permitted) fields will appear at the Data out line in bit sequential form. These bits bypass the shift register and will be clocked out sequentially until completely transmitted. As this transmission is occurring, the Message bits are also being shifted into the shift register EXCLUSIVE OR string that was PRESET with 1's. This shifting process will perform the serial division of the message by $G(X) = X^{16} + X^{12} + X^5 + 1$. This division is accomplished by virtue of the recirculation of the output of flip-flop 15 (corresponding to X^{16}) through an EXCLUSIVE OR gate to the input (corresponding to X^0 or 1) and the placement of other EXCLUSIVE OR gates at the inputs of flip-flops X^5 and X^{12}. In the transmitting station, the FCS bits will appear following the last message bit at the Data out line as the CALCULATE FCS line is brought LOW following the last message bit output, thus switching out AND gate 1 and switching in AND gate 2. The 16-bit FCS character is formed by inverting the remainder, R(X), which emerges from the output of flip-flop X^{15} after the last message bit is transmitted.

At the receiving station, the circuit of Figure 4.26 is utilized in the same manner as at the transmitting station except that the FCS character is included in the division with the Message data. As with the transmitting station, inserted zeros are not included in the CRC generation. The remainder, R(X)′, of this operation must be the 16-bit sequence 1111000010111000 for the receiving station to assume that no error has occurred. In this sequence of bits, the leftmost bit corresponds to X^0 and the rightmost bit to X^{15}.

In order to keep a data link open and maintain an active line state, a series of contiguous FLAG characters can be sent by the transmitting station. A station sending in this manner is called an *Idle Station*. This action can be used to prevent the occurrence of timeouts at the other station on the link. For example, the primary station may transmit a frame with a Control field requesting a response from the secondary. If response is not received within a specified time interval or an unintelligible response is received, recovery procedures may be initiated. Some typical timeout intervals are 900 ms, or 3–20 s.

If it is desired to break a data link at any time, the transmitting station and only the transmitting station can transmit an ABORT signal by sending eight consecutive 1's. No FCS field or Close FLAG are produced after an ABORT.

4.2

IEEE STANDARD DIGITAL INTERFACE FOR PROGRAMMABLE INSTRUMENTATION, IEEE STD 488-1978,[6] PARALLEL/ASYNCHRONOUS

This standard is aimed at instruments and system components that are in relatively close proximity and provides a means to transfer digital data among these devices. The basic functional specifications of IEEE 488-1978 limit the number of devices that may be interconnected on a contiguous bus to 15, the maximum data rate on a signal line to 1 Mb/s, and the maximum transmission path length over the interconnecting cables to 20 m. Although the standard generally applies to laboratory instruments and production test situations, it is also used with storage devices, processors, and terminals in a wide variety of applications.

[6] IEEE Standard Digital Interface for Programmable Instrumentation, Standard 48801978, IEEE Standards, 345 E. 47th St., New York, N.Y., 10017.

4.2.1   TERMINOLOGY AND DEFINITIONS

The standard is concerned only with the interface characteristics of instrumentation systems and data transfer is performed in a byte-serial, bit parallel manner. There are three categories of system devices defined in the IEEE 488-1978 standard, namely *talker, listener,* and *controller.* A *talker* can send messages to another device connected to the interface system. These messages are termed *device dependent,* since they are carried by, but not used or processed by, the interface system directly. A listener receives device-dependent messages from another device on the interface system. A *controller* can send *interface* messages that are used to manage the interface system. For example, the controller can send messages to other devices instructing them to listen or talk. Interface messages are also available to perform serial poll, parallel poll, device clear and trigger, and remote/local functions. A *serial poll* is initiated when a device on the system with talker capability transmits a service request message to the controller. The controller then reads a status byte of all devices on the system in sequence to determine which device requested service. The *parallel poll* mode gives a number of devices the ability to request service, on the controller's demand, simultaneously with other devices on the system. The *device clear* and *device trigger* functions allow the controller to initialize or trigger one, many, or all devices on the system. The remote/local function allows a device to use either data provided on the bus or data provided locally, such as from the front panel of an instrument.

Capabilities to talk, listen, or control can be present individually or in any combination in the devices connected on the bus. It is important to note that the bus, itself, is passive and the circuitry associated with talking, listening, and controlling is contained within the instruments connected to the bus.

The IEEE 488-1978 bus is physically defined as 16 signal lines functionally grouped as an 8-line *data bus,* a 3-line *handshake* or *data byte transfer control bus,* and a 5-line *general interface management bus* as shown in Figure 4.27. The 8-line data bus, labeled DI01 through DI07, is a bidirectional bus that carries messages and data asynchronously in a byte-serial, bit-parallel fashion. The 3-line data byte transfer control bus is for handshaking to accomplish the transfer of data bytes on the data bus to one or more listeners. These data will have originated from a controller or talker. This bus is comprised of the DAV (data valid) line, the NRFD (not ready for data) line, and the NDAC (not data accepted) line. The DAV line indicates the availability and validity of data on the databus. The NRFD line is used by a device to indicate its state of readiness to receive and accept data. The NDAC line of the handshake bus shows the condition of acceptance of the data by the device.

The five lines of the general interface management bus are ATN (attention), IFC (interface clear), SRQ (service request), REN (remote enable), and EOI (end or identify). These lines are used to manage the flow of information on the interface system. A controller uses the ATN to specify how data on the data bus are to be interpreted and which devices must respond to the data.

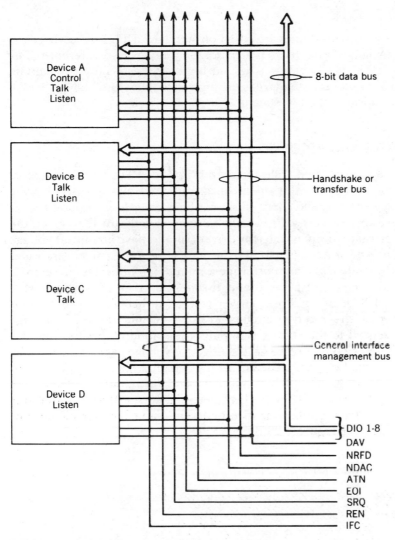

FIGURE 4.27 IEEE 488-1978 Bus.

The IFC line is used by the controller to put the interface system into a known quiescent state. Devices use the SRQ line to interrupt the controller and request attention. REN is used by the controller along with other messages to select between two different sources of device programming data. One example is to disable an instrument's front panel controls and enable remote control operation. A talker uses EOI to signal the end of transfer of multiple bytes. EOI is also used by a controller in conjunction with ATN to implement polling.

4.2.2   LOGIC LEVELS

The logic levels on the bus lines are based on standard TTL levels, where the power supply voltage is no greater than +5.25 V dc ground reference. The standard utilizes a low voltage = true and high voltage = false convention. Specifically, a logic 0 in the standard is the high state that electrically is > +2.0 V dc and the logic 1 is the low state that is < +0.8 V dc.

4.2.3   DRIVERS AND RECEIVERS

The standard specifies open collector drivers for the SRQ, NRFD, and NDAC signal lines. Tri-state drivers or open collector drivers may be used for the DAV, IFC, ATN, REN, EOI, and DIO 1–8 signals lines with the exception that in parallel polling applications, DIO 1–8 shall use open collector drivers. In the low state, the tri-state or open collector drivers should have an output voltage < +0.5 V dc at +48-mA sink current. The plus sign of the current indicates current into the node (sink current) while a minus sign indicates current out of the node (source current). For the high state, the tri-state output voltage must be > +2.4 V dc at −5.2 mA. Driver voltage and current specifications are a function of the receiver circuits, resistive termination, and voltage clamping. In regard to the voltage clamping, the standard specifies that if a signal line is connected to a receiver, the negative voltage excursions must be limited. This is usually accomplished by a diode clamp in the receiver circuits. For the situation where the driver, receiver, and terminating resistors are connected and the driver is in the high impedance state, each signal line interface should have the following direct current load characteristics and be in the shaded area of Figure 4.28.

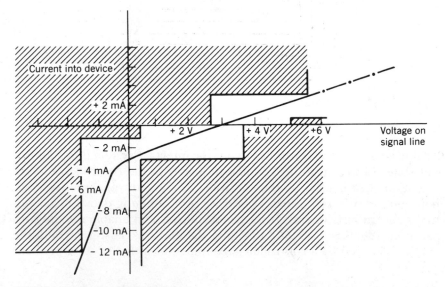

FIGURE 4.28   Dc load boundaries.

1. If $I \leq 0$ mA, V shall be <3.7 V.
2. If $I \geq 0$ mA, V shall be >2.5 V.
3. If $I \geq -12.0$ mA, V shall be > -1.5 V (only if receiver exists).
4. If $V \leq 0.4$ V, I shall be < -1.3 mA
5. If $V \geq 0.4$ V, I shall be > -3.2 mA
6. If $V \leq 5.5$ V, I shall be <2.5 mA
7. If $V \geq 5.0$ V, I shall be >0.7 mA or the small-signal Z shall be ≤ 2 kΩ at 1 MHz

4.2.4 EXAMPLE CIRCUITS

A typical driver receiver interface circuit is given in Figure 4.29. Characteristics of this circuit are

1. R_{L1} 3 kΩ $\pm5\%$ (to V_{cc})
2. R_{L2} 6.2 kΩ $+5\%$ (to ground)
3. Driver   Open collector driver output leakage current is $+0.25$ mA maximum at $V_0 = +5.25$ V. Tri-state driver output leakage current is ±40 μA maximum at $V_0 = +2.4$ V
4. Receiver   Input current is -1.16 mA maximum at $V_0 = +0.4$ V
Input leakage current is $+40$ μA maximum at $V_0 = +2.5$ V and $+1.0$ mA maximum at $V_0 = 5.25$ V
5. $V_{cc} = +5$ V $\pm$ 5%

Some actual interface circuits are given in Figures 4.30 through 4.33.

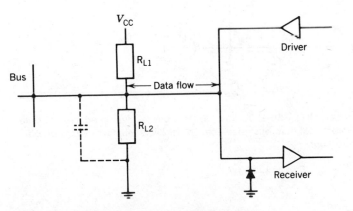

FIGURE 4.29 Typical driver receiver interface circuit.

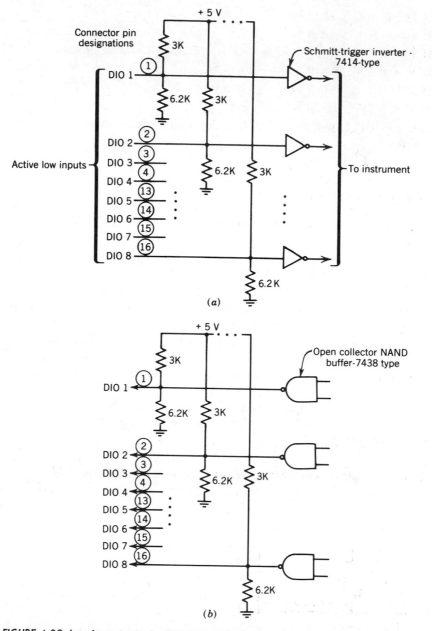

FIGURE 4.30 Interface circuits for IEEE 488-1978 Bus: (a) Data bus receiver interface; (b) date bus driver interface; (c) bidirectional data bus reference; (d) multiline interface; (e) listen address detection unlisten detection.

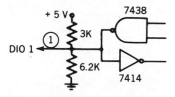

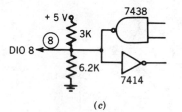

(c)

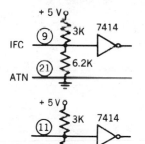

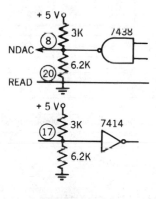

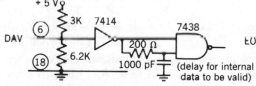

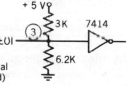

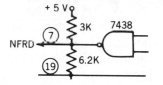

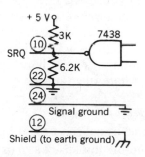

(d)

FIGURE 4.30 (Continued).

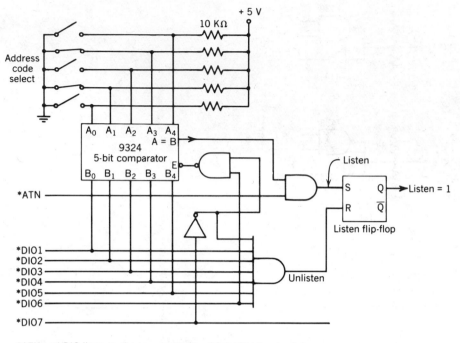

*ATN and DIO lines as shown are outputs of Schmitt-trigger gates.

(e)

FIGURE 4.30 (*Continued*).

4.2.5 HANDSHAKE TIMING

The handshaking process that is used to exchange data between a source and acceptor occurs on the 3-line data byte transfer control bus. This process can be described by means of a signal timing diagram, a flowchart, or a state diagram. The signal timing diagram is given in Figure 4.34. The dotted lines in Figure 4.34 represent multiple, superimposed waveforms indicating that two or more listeners are accepting the same byte of data at slightly different times. Figure 4.35 provides the same information in flowchart form. The number key to Figures 4.34 and 4.35 is given in the following list of events. A third representation of the handshaking process is shown in Figure 4.36 in the form of a state diagram.

List of Events for Handshake Process
1. Source initialized DAV to high (H) (data not valid).
2. Acceptors initialize NRFD to low (L) (none are ready for data), and set NDAC to low (L) (none have accepted the data).
3. $t - 2$. Source checks for error condition (both NRFD and NDAC high): then sets data byte on DIO lines.

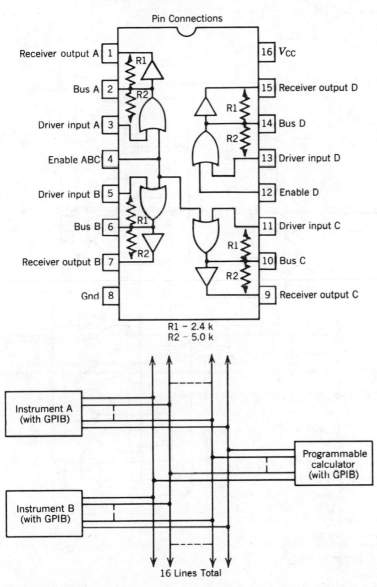

FIGURE 4.31 Motorola MC3446 Quad Interface Bus Transceiver. (Courtesy of Motorola Inc., Integrated Circuit Division).

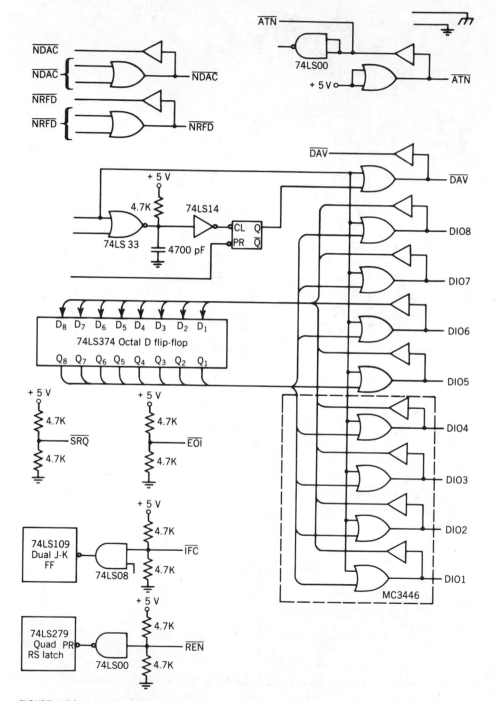

FIGURE 4.32 A typical portion of HP-1B interface using the MC3446.

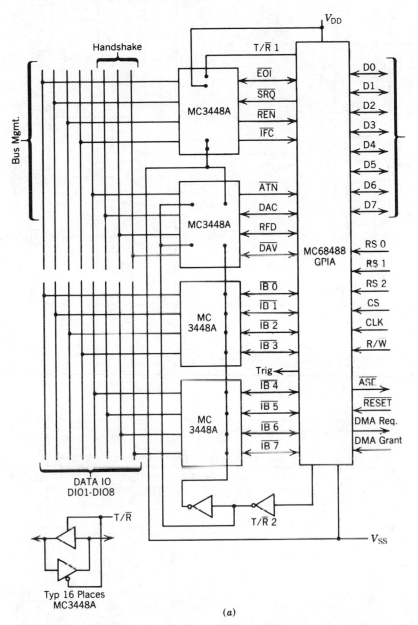

FIGURE 4.33 (a) MC3448A use with MC68488GPIA; (b) MC3447 bus receiver. (Courtesy of Motorola, Inc., Integrated Circuit Division).

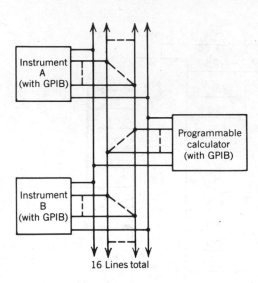

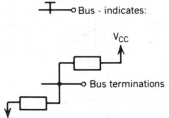

16 Lines total

Bus - indicates:

Bus terminations

V_{CC}

Gnd

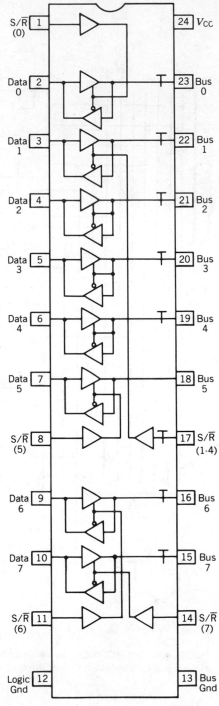

(b)

FIGURE 4.33 (Continued).

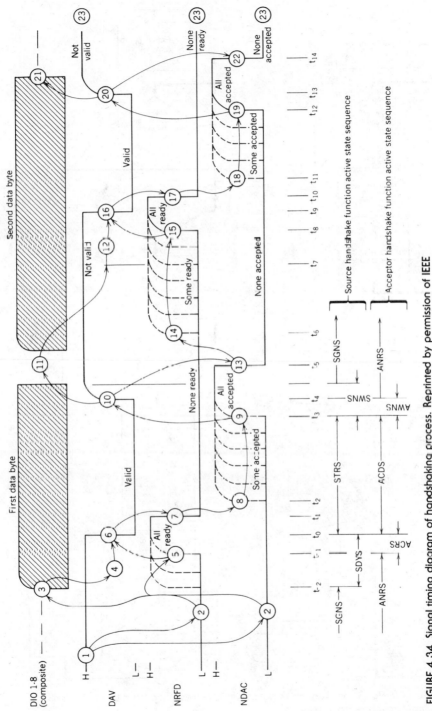

FIGURE 4.34 Signal timing diagram of handshaking process. Reprinted by permission of IEEE Standards, New York, NY.

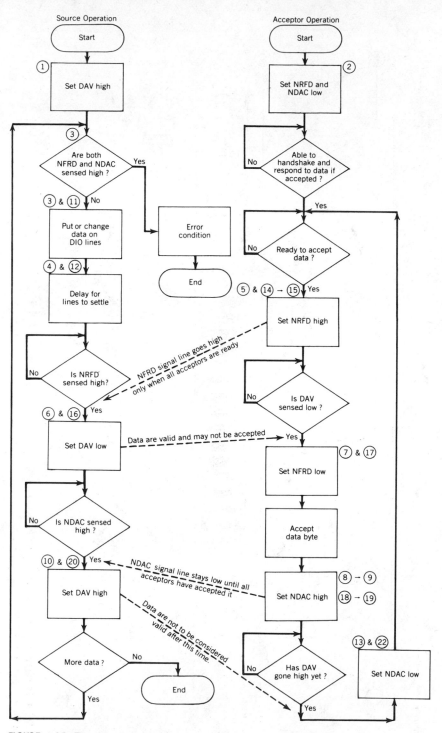

FIGURE 4.35 Flow chart diagram of handshaking process. Reprinted by permission of IEEE Standards, New York, NY.

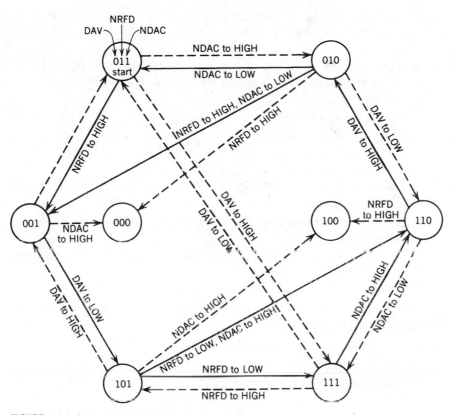

FIGURE 4.36 State diagram of handshaking process.

1. $t - 2 \rightarrow t_0$. Source delays to allow data to settle on DIO lines.
5. $t - 1$. Acceptors have all indicated readiness to accept first data byte; NRFD lines goes high.
6. t_0. Source, upon sensing NRFD high, sets DAV low to indicate that data on DIO lines are settled and valid.
7. t_1. First acceptor sets NRFD low to indicate that it is no longer ready, then accepts the data. Other acceptors follow at their own rates.
8. t_2. First acceptor sets NDAC high to indicate that it has accepted the data. (NDAC remains low due to other acceptors driving NDAC low.)
9. t_3. Last acceptor sets NDAC high to indicate that it has accepted the data; all have now accepted and the NDAC line goes high.
10. t_4. Source, having sensed that NDAC is high, sets DAV high. This indicates to the acceptors that data on the DIO lines must now be considered not valid.
11. $t_4 \rightarrow t_7$. Source changes data on the DIO lines.
12. $t_7 \rightarrow t_9$. Source delays to allow data to settle on DIO lines.

13. t_5. Acceptors, upon sensing DAV high (at 10) set NDAC low in preparation for the next cycle. NDAC line goes low as the first acceptor sets the line low.

14. t_6. First acceptor indicates that it is ready for the next data byte by setting NRFD high. (NRFD remains low due to other acceptors driving NRFD low.)

15. t_8. Last acceptor indicates that it is ready for the next data byte by setting NRFD high; NRFD signal line goes high.

16. t_9. Source, upon sensing NRFD high, sets DAV low to indicate that data on DIO lines are settled and valid.

17. t_{10}. First acceptor sets NRFD low to indicate that it is no longer ready, then accepts the data.

18. t_{11}. First acceptor sets NDAC high to indicate that it has accepted the data [as in 8].

19. t_{12}. Last acceptor sets NDAC high to indicate that it has accepted the data [as in 9].

20. t_{13}. Source, having sensed that NDAC is high, sets DAV high [as in 10].

21. Source removes data byte from DIO signal lines after setting DAV high.

22. t_{14}. Acceptors, upon sensing DAV high, set NDAC low in preparation for next cycle.

23. Note that all three handshake lines are at their initialized states [as in 1 and 2].

4.2.6   GENERAL INTERFACE MANAGEMENT BUS[7]

The general interface management bus is used primarily by the instrument acting as the controller. The controller transmits interface messages to other instruments on the bus to direct them in talking and/or listening. Messages can be sent by means of the eight data lines (multiline messages) or by means of one of the five interface management bus lines (uniline). The interface management lines are labeled as follows.

IFC Interface Clear

ATN Attention

SRQ Service Request

REN Remote Enable

EOI End or Identify

The *Interface Clear line* (IFC) is used by the controller to initialize all the instruments to an idle state in which they are neither talkers nor listeners but

---

[7] The material in this section borrows heavily from the author's earlier book, *Microprocessors and Logic Design,* John Wiley & Sons, Inc., New York, 1980.

simply monitors of the bus. All instruments must go into the idle state *relative to the bus* (they can still be functioning) within 100 μs following setting of IFC to a LOW. The *Attention* (ATN) *line* defines the eight data lines as being in either the data mode (when ATN = HIGH) or the command mode (when ATN = LOW). In the command mode, the data lines are used to transfer control information while the data mode is used to transfer data between talkers and listeners. At any time, the controller can set ATN to a LOW from a HIGH "to gain the attention" of the instruments on the bus. The instruments must then respond by setting NDAC and NRFD LOW within 200 ns of ATN going LOW, although NRFD is permitted to go HIGH immediately thereafter. A talker that may have been using the bus must release it. This operation would be employed if the controller had been interrupted by the SRQ (Service Request) line and had to initiate a serial poll to determine which instrument had requested service.

The *Service Request line* is an interrupt line to the controller that permits an instrument to request access to the bus. This line is activated by an instrument pulling it to a LOW value. The controller then determines which instrument or instruments initiated the interrupt request. The *Remote Enable line* (REN is pulled LOW by the controller if it is desired to disable the front panel control of an instrument and revert to remote control operation. The LOW state of REN permits an instrument that receives its listen address and is capable of IEEE-488 remote operation to enter the REMOTE State. When this remote control is terminated by setting REN to a HIGH, the instrument must return to the front panel mode within 100 μs.

The *End or Identify line* (EOI) along with the ATN line is used to implement a parallel poll of the instruments on the bus that have corresponding facilities for this type of operation. If ATN and EOI are pulled to a LOW simultaneously by the controller, an instrument can transmit 1 bit of information on a previously assigned line of the eight-line data bus. Thus, when the SRQ line is pulled LOW by an instrument to interrupt the controller, the controller can initiate a parallel poll of up to eight instruments by pulling ATN and EOI LOW simultaneously and sampling the individual lines of the data bus to determine which instrument(s) initiated the interrupts. The instrument(s) must respond by pulling their respective data line(s) LOW within 200 ns of ATN and EOI going LOW. The availability of one data line for each instrument when ATN and EOI are simultaneously set to a LOW can also be used to transmit status information between the controller and an instrument. The EOI = LOW can also be used to indicate an end of transmission when the data lines are in the data mode of operation.

When the controller desires to select a particular instrument as a talker, the data lines are put into the command mode (ATN = LOW), and a talk address is sent out on the data lines in the following talk command format.

DIO 8	DIO 7	DIO 6	DIO 5	DIO 4	DIO 3	DIO 2	DIO 1
X	1	0	A5	A4	A3	A2	A1

where DIO 1 through DIO 8 are data line designations, A_1 through A_5 are the *talk address* in binary form, and X can either be a 1 or 0. Any talk address code can be used except 11111, which is defined as an *untalk command*. The standard specifies that only one talker can be defined at any time. Thus any talker whose address does not match the talk address sent out when ATN = LOW is disabled from talking. Since a talker can have any address except 11111, sending out 11111 as a talk address disables all instruments from sending data and, therefore, accomplishes an *untalk* operation. A talk command does not affect the previously defined status of listeners. A *listen address* can be transmitted on the data lines when ATN = LOW and will select an instrument as a listener. Multiple listeners can be selected in sequence and a listen command will not affect previously selected listeners or disable listeners. The format of a listen command is

$$X01A_5A_4A_3A_2A_1$$

with the stipulation that address 11111 is reserved for the *unlisten* command. An unlisten command, similar to an untalk command, will disable all previously selected listeners.

A third type of command is the *universal* command, which affects all instruments connected to the bus (whether or not they have been addressed) provided that they are equipped to carry out the command. The three lines, IFC, ATN, and REN, of the interface management bus are uniline universal commands while the data lines in the command mode (ATN = LOW) can be used for multiline commands. The format of a multiline universal command is

$$X001A_4A_3A_2A_1$$

where A_1 through A_4 specify a particular universal command.

A fourth type of command is the *addressed* command, which is multilined and similar to a universal command except that only instruments that have been addressed as talkers and/or listeners are affected. The format of the addressed command is

$$X000A_4A_3A_2A_1$$

The last type of command is the *secondary command*, which is multilined and is used following an addressed, universal, or address (talk or listen) command to provide additional bits for expanding their respective codes.

4.3

MICROPROCESSOR PARALLEL COMMUNICATIONS

Parallel digital interfaces to computers are becoming more sophisticated and are endowed with increased computational power and programming capability. Typical LSI/VLSI parallel interfaces, even though designed for specific applications, usually have the common requirements of handshaking with the external world, handshaking with the computer, and accepting or transmitting data. In this section, the circuitry underlying these common requirements will be explored.

4.3.1 ADDRESSING

Controlling data input and output by a computer utilizing a peripheral interface chip is accomplished by either *isolated I/O* or *memory-mapped I/O*. In isolated I/O, the processor uses specific input and output instructions that are part of the instruction set. The execution of these specific input and output instructions activates a pin or pins (input and output) of the processor chip. The active voltage on these pin(s) can then be used to initiate I/O operations by the peripheral chip. Since these I/O pins are activated only when an I/O instruction is executed by the processor, they can be used to distinguish an I/O device address on the processor address bus from a memory address on the address bus. Thus the full address space is available for memory addressing and is not affected or reduced by I/O addressing.

In memory-mapped I/O there is no distinction between memory and I/O address space, and all instructions that reference memory can be used to access I/O devices. Thus a portion of the memory address space must be set aside for I/O addressing if I/O devices are to be used. Figure 4.37 illustrates isolated I/O and Figure 4.38 shows memory-mapped I/O. Both figures use the Intel 8255A Programmable Peripheral Interface (PPI) as the parallel I/O device. In Figure 4.38 all addresses above $32,768_{10}$ are considered I/O addresses, since I/O addressing is enabled when the highest order address bit, A_{15}, is 1.

If a large number of I/O chips is not required, an address line can be used to select an I/O device. Naturally, this reduces the address space available but eliminates the need for a decoder circuit. This method is known as *linear selection*.

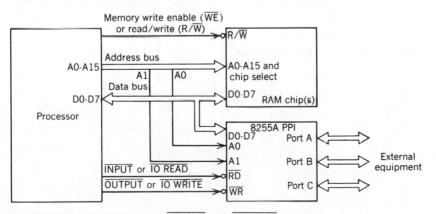

FIGURE 4.37 Isolated I/O using $\overline{\texttt{INPUT}}$ and $\overline{\texttt{OUTPUT}}$ lines.

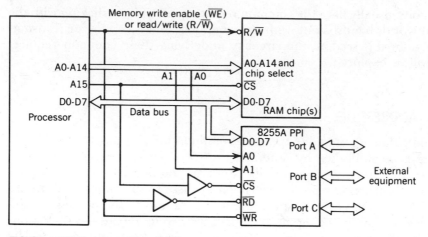

FIGURE 4.38 Memory-mapped I/O.

4.3.2   BASIC I/O CIRCUIT

The components of a basic I/O circuit are given in Figure 4.39. This circuit can accept data or transmit data, but only when the processor executes instructions to read from or write to the I/O port. Also, there is no handshaking with the external data source or recipient. Note that the input is through a nonlatched three-state buffer that ensures no load on the data bus when the I/O instructions are not being executed. The output data are latched and

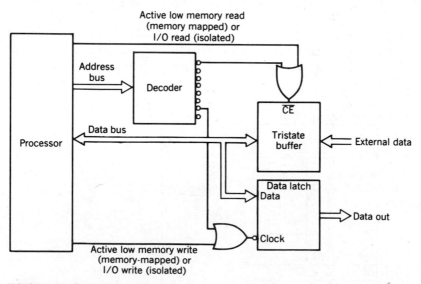

FIGURE 4.39 Basic I/O circuit.

retained present at the output terminals of the data latch until changed by another I/O write instruction.

4.3.3   STROBED I/O CIRCUITS

An improvement of the basic I/O circuit of Figure 4.39 is given in Figures 4.40 and 4.41. Figure 4.40 is the strobed input circuit and Figure 4.41 is the strobed output circuit. These circuits have handshaking capabilities that permit data exchanges at the initiation of external devices.

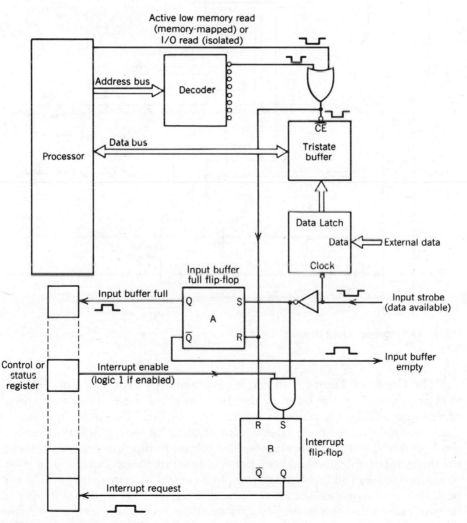

FIGURE 4.40 Strobed input circuit.

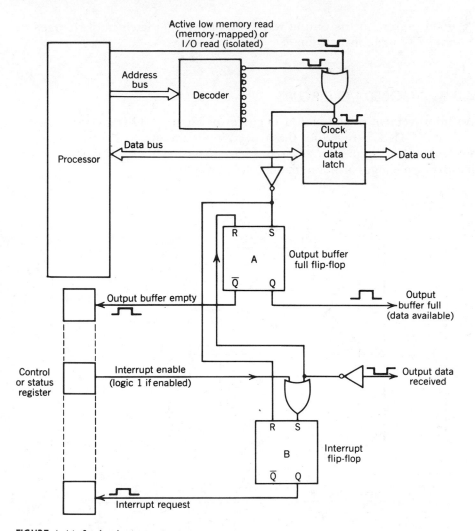

FIGURE 4.41 Strobed output circuit.

The circuit of Figure 4.40 can be explained by assuming that external data are present on the input of the data latch and the external equipment places a negative-going pulse on the input strobe line. This pulse indicates to the processor that data are available for reading by setting input buffer full flip-flop A and interrupt flip-flop B. The interrupt flip-flop can only be set if the interrupt enable line had been previously set to a logic 1 by the processor. The input buffer full flip-flop sets a bit in a control or status register that can be polled by the processor to determine if input data are available. It should be emphasized that this polling is a result of the processor executing instructions and checking for the input buffer full bit. If the processor is busy executing other code, the input buffer full status will not be known until the code polling the control or status register is executed. On the other hand, the interrupt request signal can interrupt the program presently being executed by the processor and request service immediately. In either case, when the processor executes an instruction to read the data, the tri-state buffer is enabled and the external data are transferred to the data bus and read into the processor. The pulse that enables the tri-state buffer also serves to reset the input buffer full and interrupt request flip-flops so that the read cycle can begin anew.

The circuit of Figure 4.41 operates in the opposite direction of that of Figure 4.40 in that data are transmitted or written out to an external device. When the processor executes an output instruction, the output data latch is clocked and data are transferred from the processor's data bus to the data latch and presented to the external equipment. Simultaneously, the pulse that clocked the data latch sets the output buffer full flip-flop and resets the interrupt flip-flop. The output buffer full flip-flop Q output presents a data available signal to the external equipment. Upon receipt of the data available signal, the external device can read the data presented on the output of the data latch. When the external device receives the data, it issues an output data received pulse that sets an output buffer empty bit in the control or status register or sends an interrupt request for new data to the processor if the interrupt flip-flop was enabled.

4.3.4   PSEUDOBIDIRECTIONAL I/O PORT

Another type of I/O circuit is the pseudobidirectional port of Figure 4.42. With this circuit, a line can be either an input or an output. As with the basic I/O circuit, the pseudobidirectional port of Figure 4.42 has a latched output and an unlatched input.

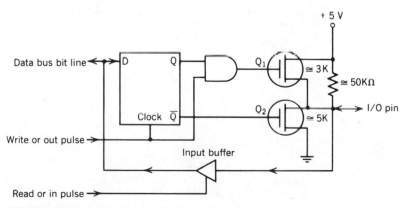

FIGURE 4.42 Pseudobidirectional I/O port.

This port is programmed to be an input port by first clocking a logic 1 into the flip-flop D input with the WRITE pulse, thereby setting the flip-flop Q output to a 1. Since the WRITE pulse is AND'ED with the Q output, Q1 will be momentarily turned on, providing a fast rise time as the I/O pin is pulled up to +5 V. Then, the I/O pin remains pulled up to +5 V through the approximately 50 kΩ resistance. Now the input can be read into the data bus bit line by the READ or IN pulse. A logic 1 or logic 0 on the I/O pin is transferred to the data bus when the READ pulse is momentarily applied. As an output port, the I/O line will reflect and retain the value read in to the D input of the flip-flop. The input/output line is TTL compatible.

FOR FURTHER READING

Deaton, G. A., Jr., "IBM Finally Gives U.S. Users Access to X.25 Networks," *Data Communications,* September 1981.

Hamming, R. W., *Coding and Information Theory,* Prentice-Hall, Inc., Englewood Cliffs, N.J., 1980.

Intel Peripheral Design Handbook, 1986.

Krutz, R. L., *Microprocessors and Logic Design,* John Wiley & Sons, Inc., New York, 1980.

Waitrowski, C. A., and C. House, *Logic Circuits and Microcomputer Systems,* McGraw-Hill Book Co., New York, 1980.

Yakobovitch, Sorin, "SDLC Interface Mates 6800 Peripheral to 8086," *Computer Design,* September 1981.

5

IBM PERSONAL COMPUTER AND 80386 PROCESSOR INTERFACING, ASPECTS OF IBM PERSONAL SYSTEM/2 FAMILY

Because of its popularity, the IBM personal computer (PC) will be used to apply and review some of the interfacing concepts covered in previous chapters. In addition, aspects of the IBM Personal System/2® family of personal computers will be introduced along with detailed interfacing techniques for the 80386, 32-bit microprocessor. The 80386 processor is utilized in the IBM Personal System/2, Model 80 along with an optional 80387 Math Co-processor.

5.1

IBM PC

The heart of the IBM PC is the system board. The microprocessor used on this board is the Intel 8088, which is an 8-bit external bus version of the 16-bit 8086 microprocessor. The 8 MHz 8086 microprocessor with optional 8087 Math Co-processor is the basis of the IBM Personal System/2, Model 30. The system board is comprised of the processor and its support elements, read/write memory and support, read only memory (ROM) and support, I/O adapters, and the I/O channel.

The 8088 processor has a 20-bit address bus (direct addressing of 1,048,578 bytes of memory), performs 16-bit operations and has an 8-bit external data bus. The 8088 in the PC operates at a 4.77-MHz clock rate with bus cycles of 840 ns (4 clock periods of 210 ns each). A coprocessor can be utilized with the 8088 as an added capability. Recall that interfacing examples involving the 8088 were developed in Chapter 3.

Associated chips on the PC processor board provide four DMA (Direct

Memory Address) channels, three of which are available for high speed I/O data transfer. The fourth channel is used on the system board for refreshing the dynamic RAM memory. The system board also has three programmable timer/counters (Intel 8253-5), and eight priority interrupts (Intel 8259A). The system board supports 48K × 8 EPROM or ROM. Of these 48K bytes, 40K bytes of ROM are delivered with the system and contain the cassette operating system, cassette BASIC interpreter, power-on self-test programs, I/O drivers, diskette bootstrap loader, and dot patterns for graphics characters. The system board can accommodate 16K × 9 to 64K × 9 bits of read/write memory (the extra bit in the word is for memory parity checking). Additional read/write memory can be added by using expansion slots provided in the PC housing.

5.1.1   SYSTEM BOARD/RAM INTERFACE

The 8088 microprocessor on the system board is configured in the maximum mode to permit large system implementation. Thus an 8288 bus controller receives status lines $\overline{S0}$, $\overline{S1}$, and $\overline{S2}$ as inputs and generates all bus control signals including Read and Write signals. (A discussion of a dynamic RAM design for the 8088 is given in Chapter 3.) The 8288 provides increased source and sink current capability for the bus control line and permits allocation of some 8088 microprocessor pins for large system implementation, including an auxiliary processor. A general block diagram of the IBM PC system board is given in Figure 5.1.

If the system board of the PC contains the maximum 64K dynamic RAM, additional RAM can be added using the system unit expansion slots or the expansion unit. The system unit Read/Write memory is organized as four banks of 16K × 9 bits. The memory word is one byte in width and incorporates an additional bit for memory parity check. The five I/O expansion slots are part of the basic PC system/unit whereas the expansion unit provides seven additional slots while utilizing one expansion slot in the system unit. There are 32K byte, 64K byte, and 64K/256K byte parity-check RAM expansion options. The 64K/256K byte option uses one to three 64K byte module kits to implement 64K byte to 256K byte total memory in the PC. As noted earlier, refreshing and addressing for dynamic RAM is generated on the system board. The 32K byte option utilizes 18, 16K × 1 dynamic RAM chips and the 64K byte option incorporates 18, 32K × 1 modules. The 64K/256K option utilizes up to three module kits. Each module kit is comprised of nine, 64K × 1 modules. Cycle and access times for the memory expansion options are given in Table 5.1. Pin designations for each module are listed in Table 5.2.

The memory expansion option utilizes control signals generated on the system card and available on the I/O expansion slots. Circuitry for the 32K byte expansion option is detailed in Figures 5.2, 5.3, and 5.4. In Figure 5.2, a

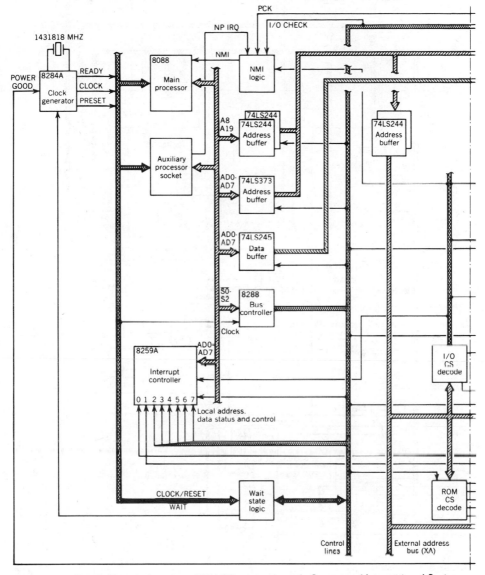

FIGURE 5.1 General block diagram of IBM PC system board. Courtesy of International Business Machines Corporation.

74S85 magnitude comparator (U22) is used to generate a card select signal, $\overline{CD}$ $\overline{SEL}$ $\overline{DEC}$, that enables the 32K byte memory expansion card. In the same figure, 74LS138 decoders develop the $\overline{CAS0}$, $\overline{CAS1}$, $\overline{RAS0}$, and $\overline{RAS1}$ signals for each of the two banks of 16K byte memory. Figure 5.3 illustrates the two memory banks and associated control signals. The addresses of the memory expansion option are set by DIP switches on the board.

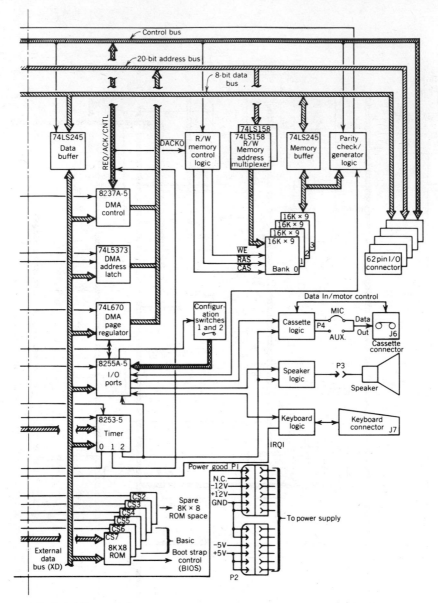

FIGURE 5.1 (*Continued*).

TABLE 5.1 Memory Expansion Cycle and Access Times

	Option		
	16K × 1	32K × 1	64K × 1
General cycle	410 ns	410 ns	345 ns
Absolute maximum access from $\overline{RAS}$	250 ns	250 ns	200 ns
Absolute maximum from $\overline{CAS}$	165 ns	165 ns	115 ns

TABLE 5.2 Memory Module Pin Designation (Courtesy of International Business Machines Corporation)

Pin	16K by 1 Bit Module (used on 32K option)	32K by 1 Bit Module (used on 64K option)	64K by 1 Bit Module (used on 64/256K option)
1	−5 Vdc	−5 Vdc	N/C
2	Data in[b]	Data in[b]	Data in[c]
3	Write	Write	Write
4	RAS	RAS 0	RAS
5	A0	RAS 1	A0
6	A2	A0	A2
7	A1	A2	A1
8	+12 Vdc	A1	+5 Vdc
9	+5 Vdc	+12 Vdc	A7
10	A5	+5 Vdc	A5
11	A4	A5	A4
12	A3	A4	A3
13	A6	A3	A6
14	Data out[b]	A6	Data out[c]
15	CAS	Data out[b]	CAS
16	GND	CAS 1	GND
17	[a]	CAS 0	[a]
18	[a]	GND	[a]

Note:
[a] 16K by 1 and 64K by 1 bit modules have 16 pins.
[b] Data in and Data out are tied together (three-state bus).
[c] Data in and Data out are tied together on Data bits 0-7 (three-state bus).

Similarly, Figures 5.5, 5.6, and 5.7 detail the 64K byte memory expansion option. The 74LS138 decoders shown in Figure 5.5 generate the $\overline{\text{CAS0}}$-$\overline{\text{CAS3}}$, and $\overline{\text{RAS0}}$-$\overline{\text{RAS3}}$ control signals for the 64K byte RAM option.

5.1.2 I/O CHANNEL

Interfacing to the IBM PC is most easily accomplished through the I/O channel. The I/O channel is essentially a buffered and demultiplexed version of the 8088 bus. I/O devices on the channel are normally isolated I/O as opposed to memory-mapped I/O and are allocated 9 bits of address (A0-A8). These 9 bits, then, allow 512 different I/O addresses, ranging from 000 Hex to 3FF Hex. The I/O channel has 62 signals that are provided on a 62-pin connector with 100-mil card tab spacing. These signals and pin designations are summarized in Table 5.3.

GENERAL INTERFACE

A general interface to the I/O channel that allows data I/O to be implemented will now be discussed. The interface should accept address, data, and

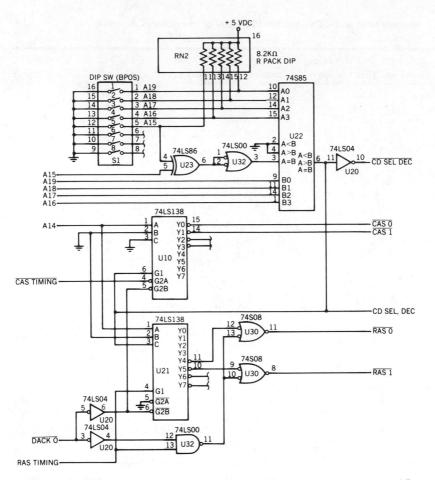

FIGURE 5.2 32-Kbyte expansion option control signals. Courtesy of International Business Machines Corporation.

control signals supplied on the I/O channel and transfer information to and from the processor by means of the data bus. A detailed diagram of the interface is given in Figure 5.8. This interface is analogous to the isolated I/O example of Figure 4.37. With this configuration, the 8088 microprocessor writes data from the data bus to an external destination or reads data onto the data bus from an external source. In either case, the data bus is buffered from the external device by a data bus transceiver such as the 74LS245 octal bus transceiver. This transceiver can transfer data bidirectionally to or from the data bus, and the direction of data transfer is determined by the logic signal applied to the DIR input of the transceiver. There is also an active Low Enable input, E, to the transceiver chip that, when high, isolates the data bus. The address and control signals are buffered through 74LS244 octal buffer/line drivers with three state outputs.

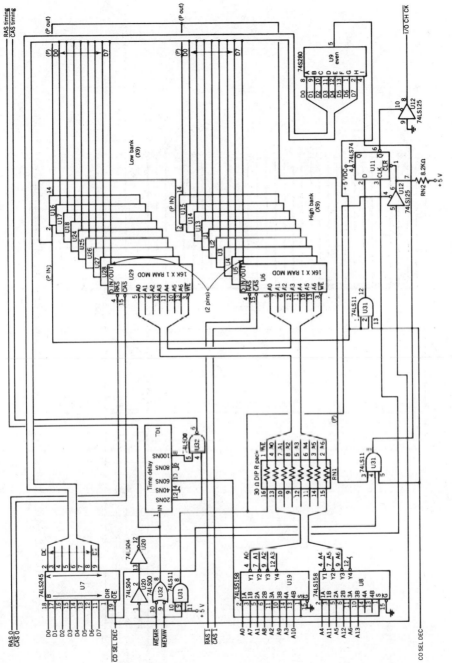

FIGURE 5.3 32 Kbyte memory banks. Courtesy of International Business Machines Corporation.

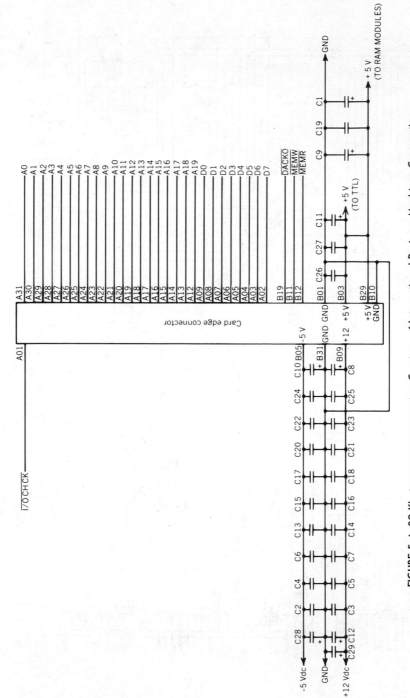

FIGURE 5.4 32 Kbyte memory connector. Courtesy of International Business Machines Corporation.

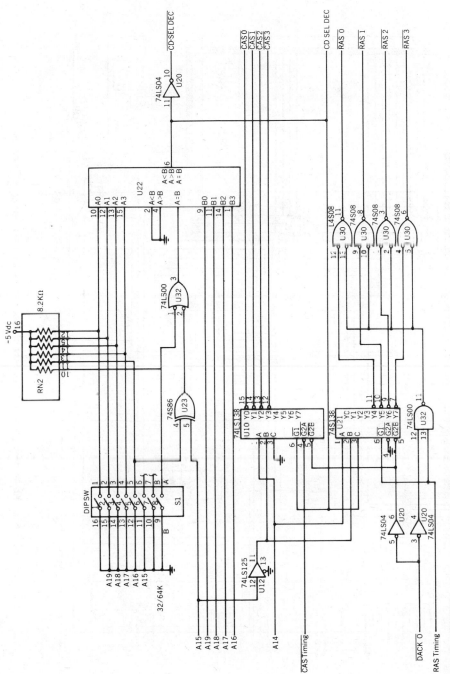

FIGURE 5.5 64 Kbyte expansion control signals. Courtesy of International Business Machines Corporation.

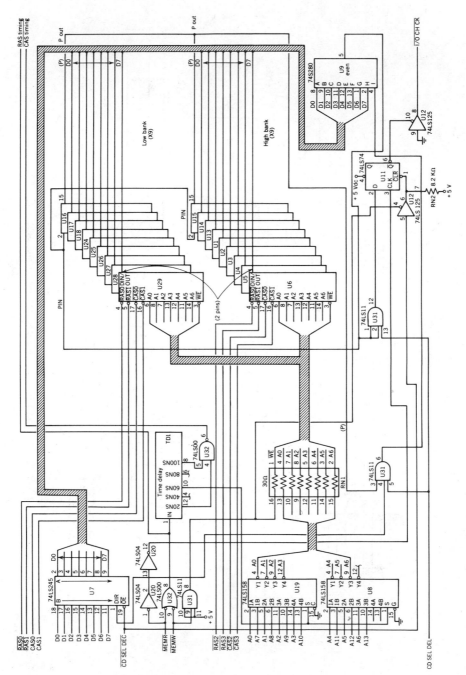

FIGURE 5.6 64 Kbyte memory banks. Courtesy of International Business Machines Corporation.

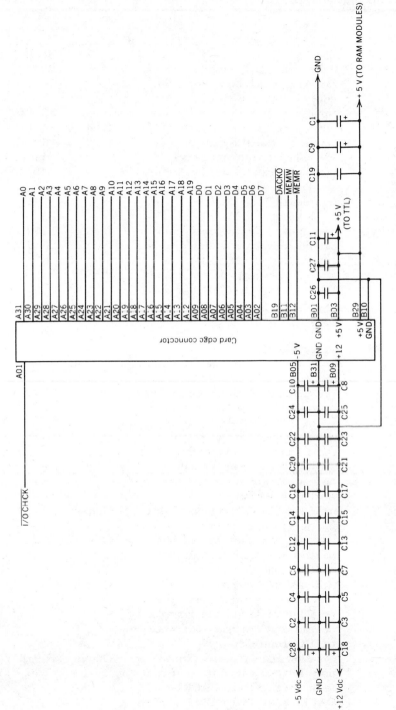

FIGURE 5.7 64 Kbyte memory connector. Courtesy of International Business Machines Corporation.

TABLE 5.3 I/O Channel Signals

Signal	Function	Direction	Pin
GND	Signal ground	—	B1, B31, B10
RESET DRV	Active high signal to reset system on power-up, synchronized to falling edge of clock	Output	B2
+5 V + 5%	Supply voltage	—	B3, B29
IRQ2–IRQ7	Interrupt request lines 2 through 7. Interrupt request is generated by a low-to-high transition that is held high until acknowledged by processor. Prioritized with decreasing priority from IRQ2 to IRQ7	Input	B4 (IRQ2), B21 (IRQ7) through B25 (IRQ3)
−5 V + 10%	Supply voltage	—	B5
DRQ1–DRQ3	Asynchronous DMA request lines prioritized with decreasing priority from DRQ1 to DRQ3. DMA requested by bringing line high and holding it high until acknowledged by DACK	Input	B18 (DRQ1), B6 (DRQ2), B16 (DRQ3)
−12 V + 10%	Supply voltage	—	B7
Reserved	—	—	B8
+12 V + 5%	Supply voltage	—	B9
MEMW	Active low memory write command that indicates to memory that data present on data bus are to store into memory. May originate with processor or DMA controller	Output	B11
MEMR	Active low memory read command that indicates to memory that it should place data on the data bus to be read. May originate with processor or DMA controller	Output	B12
IOW	Active low I/O write command that indicates to an output device that data are present on the data bus to be read. May originate with processor or DMA controller	Output	B13
IOR	Active low I/O read command that indicates to an output device that it should place data on the data bus to be read. May originate with processor or DMA controller	Output	B14
DACK0–DACK3	Active low DMA request acknowledge lines. DACK0 is used for system memory refresh and DACK1–DACK3 acknowledge DMA requests DRQ1–DRQ3, respectively	Output	B19, B17, B26, B15

TABLE 5.3 (Continued)

Signal	Function	Direction	Pin
CLOCK	4.77-MHz system clock that is derived by dividing oscillator output (pin 30) by 3. Clock period is 210 ns with 33% duty cycle	Output	B20
T/C	Active high terminal count line that will present an output pulse when a terminal count is reached on any DMA channel	Output	B27
ALE	Address Latch Enable line that is used by system board to latch valid addresses generated by the processor. When used with AEN signal, it can identify valid processor addresses. These addresses are latched by using the falling edge of ALE. ALE is generated by the 8288 bus controller	Output	B28
OSC	14.31818 MHz clock oscillator signal with 70-ns period and 50% duty cycle	Output	B30
I/O CH CK	Active low signal that indicates a parity error associated with data in memory or I/O devices	Input	A1
D0–D7	Active high data bits 0 7. D0 is lsb		
I/O CH RDY	I/O channel ready line that is normally high and is pulled low by memory or I/O devices to lengthen a memory or I/O cycle. Pulling this line low extends machine cycles by an integral number of 210-ns clock cycles. This feature allows slow devices to interface with the processor. I/O CH RDY should not be held low more than 10 system clock cycles	Input	A10
AEN	Address enable line that, when it is high, essentially isolates the processor and other devices from I/O channel. Thus DMA controller takes over address bus, data bus, IOR, IOW, MEMR, and MEMW lines to effect DMA transfers	Output	A11
A0–A19	Active high address bus lines; A0 is lsb and A19 is msb.	Output	A31–A12

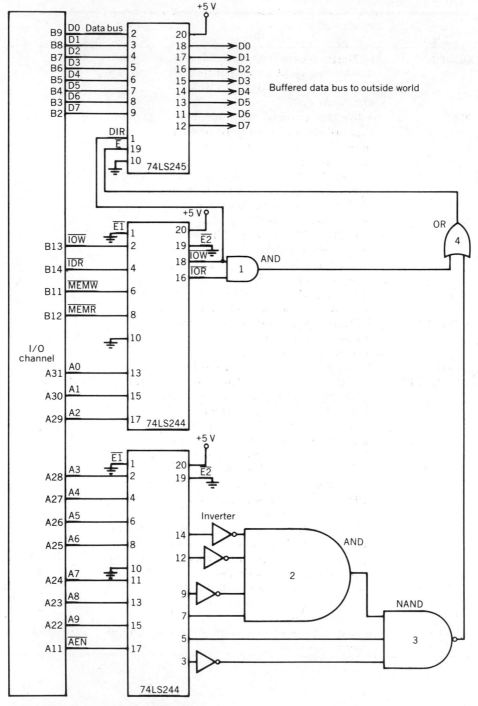

FIGURE 5.8 General I/O channel interface.

The IBM PC has allocated addresses 300 Hex through 31F Hex for such an interface, labeled the prototype card. The example in Figure 5.8 utilizes these same addresses. To decode addresses 300 Hex to 31F Hex, it can be seen that Address lines A0-A4 can be either 0's or 1's, lines A5-A7 must be 0's, and lines A8 and A9 must be 1's. Also, AEN, address enable, must be in its active low state. These signals are decoded by gates 2 and 3 and present an active low to gate 4 when addresses 300 Hex to 31F Hex are on the address lines and AEN is low. However, to prevent the interface from being activated when a memory address encompassing 300 Hex through 31F Hex is sent out, gate 4 must have an additional qualifying input that indicates that either an I/O read ($\overline{\text{IOR}}$) or I/O write ($\overline{\text{IOW}}$) is occurring. This qualification is accomplished by gate 1. Thus, when an $\overline{\text{IOR}}$ or $\overline{\text{IOW}}$ is indicated and the proper address is presented, gate 4 will produce a low output that enables data transfer through transceiver 74LS245. If an $\overline{\text{IOW}}$ is occurring, the DIR input to the transceiver is low and the transfer of data is set up by the transceiver from the microprocessor to the outside world. Conversely, if an $\overline{\text{IOR}}$ is being executed, the DIR line is high and data are transferred from the outside world to the microprocessor data bus through the transceiver.

With this interface, data can be transfered from the microprocessor data bus to an external device by executing an I/O write operation to any of the addresses 300 Hex through 31F Hex. Data can be read into the microprocessor through the data bus by executing an I/O read operation on the same addresses.

5.1.3   SDLC CONTROLLER (8273) AS USED ON SDLC COMMUNICATIONS ADAPTER

The IBM PC SDLC communications adapter board as programmed by communications software provides half-duplex, synchronous, serial transmission. It contains EIA drivers and receivers and connects to a standard EIA 25 pin, D shell, male connector. The maximum data transmission rate is 9600 bits per second. Key elements of the SDLC adapter are the 8255A-5 Programmable Peripheral Interface (PPI), the 8253-5 Programmable interval timer, the 8273 SDLC protocol controller, a data bus buffer, and address decode logic.

The heart of the SDLC adapter is the 8273 SDLC protocol controller. The controller operates from a single +5-V power supply, is HDLC/SDLC compatible, performs automatic zero bit insertion and deletion, and provides automatic frame check sequence (FCS) generation and checking. A block diagram of the 8273 is given in Figure 5.9.

In controlling the 8273, the processor accesses the seven registers in the 8273. The registers are selected by the chip select (CS) line; lines A0 and A1, which are the low-order bits of the adapter address byte; and the processor read (RD) and processor write (WR) lines that are obtained from the system control bus. Table 5.4 gives the codes for selection of the appropriate internal 8273 registers.

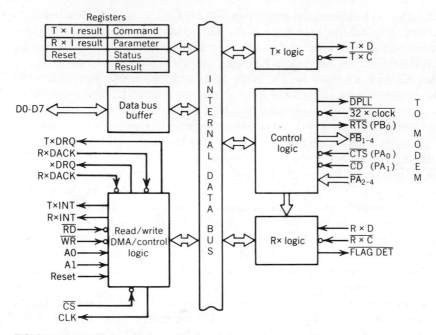

FIGURE 5.9 8273 protocol controllers.

These registers either provide information to external devices or cause actions to occur when particular patterns are written to them. The command register is used to initialize operations and is activated by writing a command byte to this register. As its name implies, the status register gives the status of the interface from the 8273 to the CPU. It also provides processor/adapter handshaking during operation of the 8273. The parameter register is used to accept additional information that is required by some commands. The result register receives a result byte as a consequence of a command execution. This result is read by the processor. The reset register provides software reset for

TABLE 5.4 8273 Register Select Codes

A1	A0	CS	RD	WR	Register
0	0	0	1	0	Command
0	0	0	0	1	Status
0	1	0	1	0	Parameter
0	1	0	0	1	Result
1	0	0	1	0	Reset
1	0	0	0	1	T × I result
1	1	0	0	1	R × I result

Reprinted by permission of Intel Corporation, Copyright, 1986.

the 8273. The T × I Result (Transmit Interrupt Result) register contains the result of a transmit operation. An interrupt to the processor is generated when the result is available. Similarly, the R × I Result (Receive Interrupt Result) register contains the result of a receive operation. An interrupt to the processor is also generated when the result is available. The R × Int and T × Int interrupt lines from the 8273 to the processor indicate that the receiver or transmitter result register should be read or data should be transferred. The interrupt lines also have corresponding bits in the status register, thus allowing for polled operation as opposed to interrupt-driven operation. In this mode, the processor can check the appropriate bits in the status register and, under program control, initiate the reading of the results register or transferring of data.

The 8273 supports DMA and initiates a DMA data transfer by raising the DMA request line (T × DRQ for transfer from memory to the 8273 or R × DRQ for transfer from the 8273 to memory). Granting of the DMA request (DRQ) is sensed by the 8273 through the appropriate T × DACK or R × DACK acknowledge signal and the WR (write to the 8273) or RD (read from the 8273) signal, all provided by the DMA controller. T × DACK and WR can also be used to transfer data to the 8273 in a non-DMA mode. Similarly, R × DACK and RD can also be used to read data from the 8273 in a non-DMA mode. This activation of the 8273 by the combination of T × DACK and WR, or R × DACK and RD is independent of the chip select (CS) of the 8273.

The DMA controller supplies timing and sequential addresses for the DMA transfer, beginning at an address specified by the processor.

The previous paragraphs described the interface of the 8273 to the processor and memory. The other major interface of the 8273 is to a modem. This interface is broken down into the modem control block and the serial data timing block.

As shown in Figure 5.9, Port A of the 8273 is a modem control input port. Port A bit assignments are as follows:

Bit 0 (PA0)   CTS (clear to send) input from modem. CTS must be active before a frame is transmitted. If a frame is being transmitted and CTS becomes inactive, the frame will be aborted and a CTS failure will be indicated in the corresponding interrupt result register.

Bit 1 (PA1)   CD (carrier detect) input from modem. CD must be active for a specified period before the arrival of a frame's address field. If a frame is being received and CD becomes inactive, a CD failure interrupt will be generated in the result register.

Bit 2 (PA2)   DSR (data set ready) input from modem.

Bit 3 (PA3)   Sense input bit to detect a change in CTS.

Bit 4 (PA4)   Sense input bit to detect a change in DSR.

Bit 5 (PA5)- ⎫
Bit 7 (PA7) ⎬ Not used and read as 1's when Port A is read.

(Bits PA2-PA4 are defined by IBM, whereas bits PA0 and PA1 are defined by the 8273.)

Port B is an output port for modem control. The bit assignments for Port B are

Bit 0 (PB0)	RTS (request to send) output to modem. If during normal operation the processor sets RTS to the active state, the 8273 will retain this active state. On the other hand, if the processor sets RTS inactive, RTS will be activated by the 8273 prior to each transmission and deactivated 1-byte time after each transmission.
Bit 1 (PB1)	Reserved.
Bit 3 (PB3)	Reserved.
Bit 4 (PB4)	Reserved.
Bit 2 (PB2)	DTR (data terminal ready) output to modem.
Bit 5 (PB5)	Flag detect output. Activated each time a flag sequence is detected.
Bit 6 (PB6)	Not used.
Bit 7 (PB7)	Not used.

(Bits PB1-PB4 are defined by IBM, whereas bits PB0 and PB5 are defined by 8273.)

The serial data timing is provided by the $\overline{T \times C}$ (user transmit) and $\overline{R \times C}$ (user receive) clocks. The serial timing block is made up of serial data logic and the digital phase-locked loop (DPLL). $\overline{T \times C}$, $T \times D$, $\overline{R \times C}$, and $R \times D$ are the signals associated with serial timing and their relationships are given in Figure 5.10.

Newly transmitted data out ($T \times D$) are generated on the active edge of $\overline{T \times C}$ while the active-to-inactive edge of $\overline{R \times C}$ is used to acquire the newly received data ($R \times D$).

The digital phase-locked loop in the IBM PC application is used in a diagnostic mode. In asynchronous applications the phase-locked loop is used to derive the clock from the receiver data stream.

5.1.4   PRINTER ADAPTER

The IBM PC Printer Adapter is a parallel interface for compatible printers. It can also be used as a general purpose I/O port if its control conventions are followed. Twelve TTL-compatible, buffered outputs are available from the

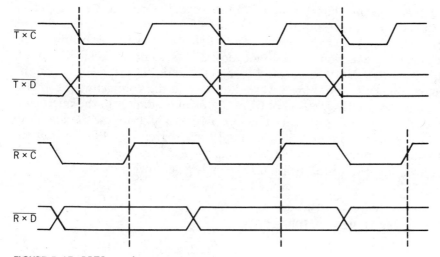

FIGURE 5.10 8273 serial timing signals.

adapter along with five steady state inputs. One of the five inputs can be used as a processor interrupt if needed.

For operation with a printer, 8 bits of data are written from the microprocessor data bus to the latched output port. A strobe signal then transfers the 8-bit data to the printer. The input ports can then be read to determine printer status or the interrupt line from the printer to the microprocessor could signal READY for the next character.

There are three input and two output instructions that utilize the printer adapter. The input instructions are implemented by the microprocessor IN instruction.

When the IN instruction reads from address 378Hex, the data should be identical to those last written to 378Hex. These data are present on pins 2 through 9 (8 bits) of the 25-pin D shell connector at the rear of the PC, with pin 2 the least significant bit. This command can be used to perform a loop self-test of the adapter. When the second IN instruction, input from location 379Hex is executed, the five binary values present on pins 15 through 11 of the 25-pin D shell connector are read by the microprocessor as bits 3 through 7, respectively. This command is used to obtain status information from the external device. The last input instruction utilized with the printer adapter is input from location 37AHex. Execution of this IN instruction causes the microprocessor to read the data on pins 1, 14, 16, and 17 as well as the IRQ interrupt bit. If no external data are applied to these pins, the data read will correspond to those last written to location 37AHex.

The two output instructions, implemented by the OUT command in the microprocessor, present the binary information on the data bus to the pins of the 25-pin, D-shell connector. The output instruction that writes to location

378Hex places latched bits 0 through 7 of the data bus on pins 2 through 9 of the D-shell connector, respectively. The drivers to these pins can sink 24 mA and source 2.6 mA and should not be pulled directly to the ground. The second output instruction, write to location 37AHex, presents latched bits 0 through 3 of the data bus to pins 1, 14, 16, and 17, respectively, of the connector. Bits 0, 1, and 3 are inverted as seen at the connector. Bit 4, also latched, enables the IRQ interrupt from the printer adapter to the microprocessor when it is set to a logical 1. The interrupt is a high-to-low signal on pin 10 of the connector. Pins 1, 14, 16, and 17 are driven by 7405 TTL open-collector drivers pulled up to +5 V through 4.7 kΩ resistors.

A block diagram of the printer adapter is given in Figure 5.11.

5.1.5   GENERAL RAM INTERFACE

The general RAM interface for the IBM PC is developed on the system board. The RAM is organized to have from 16K × 9 to 64K × 9 bits of dynamic

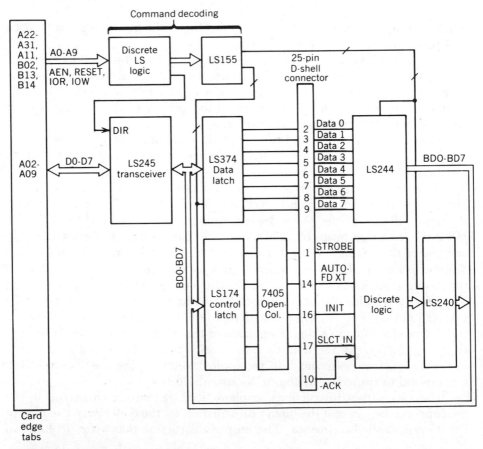

FIGURE 5.11 Block diagram of printer adapter.

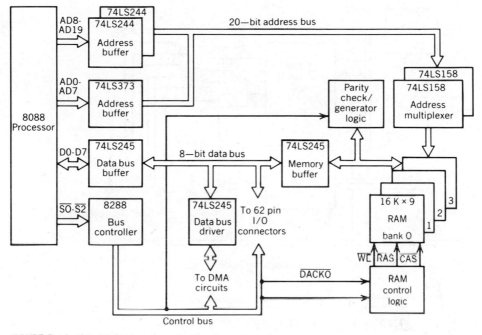

FIGURE 5.12 IBM PC RAM organization, system board.

memory. The extra bit in the word is used for parity checking. The chips used are 16K $\times$ 1 devices with access time of 250 ns and a 410-ns cycle time.

A block diagram of the RAM organization on the system board is given in Figure 5.12. A detailed schematic expansion of bank 2 of the system board RAM with appropriate signal designations is shown in Figure 5.13.

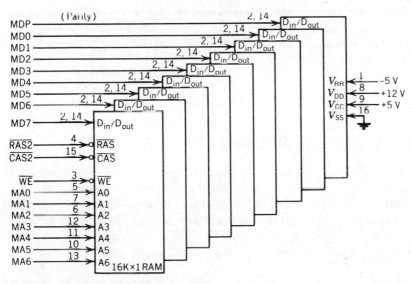

FIGURE 5.13 System board RAM, bank 2.

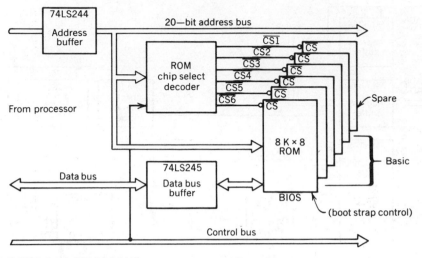

FIGURE 5.14 IBM PC ROM organization, system board.

5.1.6   GENERAL ROM INTERFACE

The system board has space for 48K × 8 ROM. The general interface from the processor to the ROM is given in Figure 5.14. Figure 5.15 is a more detailed schematic of the system board ROM circuitry.

5.1.7   IBM PC SUMMARY

The preceding material provides information on the common interfaces to the IBM PC through general I/O circuitry, the printer adapter, and RAM/ROM circuits. For additional documentation, refer to the IBM Personal Computer Technical Reference, Part Number 6025005.

5.2

IBM PERSONAL SYSTEM/2 SUMMARY

The IBM Personal System/2 family of computers provides a range of computing capability for a wide variety of users. The family consists of the Models 30, 50, 60, and 80. The system board of the Personal System/2 computers incorporates functions that occupied multiple adapter boards on the first generation IBM personal computers. This consolidation was accomplished through the use of VLSI, surface mount technology, and custom gate arrays.

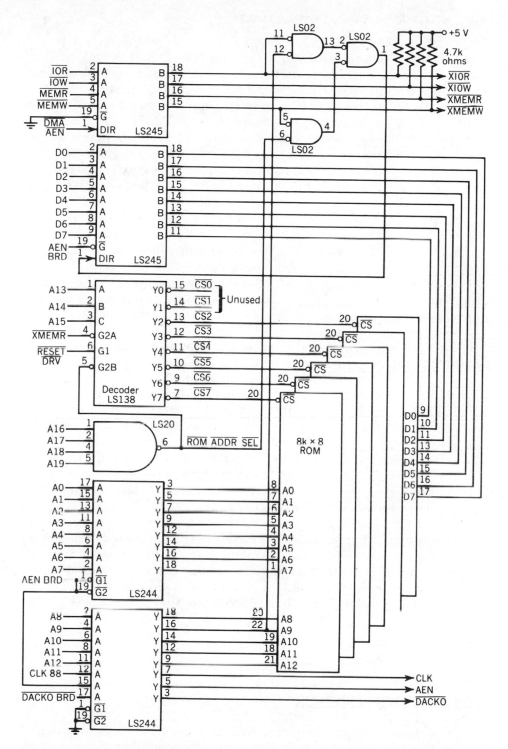

FIGURE 5.15 System board ROM.

The Model 30 is the successor to the IBM PC and IBM PC/XT, offering approximately 2.5 times the performance. Models 50, 60, and 80 are approximately 2 to 3.5 times faster than their IBM PC/AT predecessor.

The Model 30 uses the 16-bit 8086 microprocessor and supports up to 640Kb of RAM on the system board and 64Kb of ROM. There is also a serial port of up to 9.6Kbps and a bidirectional parallel port. Both ports are accessed through a standard 25-pin male connector. Three expansion slots that take the existing IBM PC attachment cards are also provided. Interfacing to a subset of the available cards was illustrated in the preceding sections on IBM PC interfacing.

Models 50, 60, and 80 are based on a Micro Channel architecture that offers features not available in the original PC architecture. The Micro Channel supports native multitasking, a 32-bit data path, concurrent multiple processors, and an analog channel for analog signals (including voice.)

The Personal System/2 Models 50 and 60 are designed with the 10 MHz 80286 microprocessor and an optional 80287 Math Coprocessor. Both computers provide 1Mb of RAM on the system board with expansion to 7Mb on the Model 50 and 15Mb on the Model 60. A parallel bidirectional port and a 19.2Kbps serial port are provided on both models. Three expansion slots are available on the Model 50 and seven on the Model 60.

The Models 80-041 and 80-071 use the 80386, 16 MHz microprocessor with the option for the 16 MHz 80387 Math Co-Processor while the Model 80 8580-111 incorporates the 20 MHz 80386. The system board on Model 80-041 provides 1Mb of RAM, expandable to 2Mb on the system board and to 16 Mb with adapter cards. Models 80-071 and 8580-111 support 2Mb of RAM on the systems board, expandable to 4Mb on the system board of the 8580-111. Expansion to 16Mb on both models is available through the addition of adapter cards. Four 16-bit expansion slots and three 16/32 bit expansion slots are available with the Model 80 computer.

At this writing, detailed design information was not available for the Personal System/2 family. However, methodology and design examples on interfacing with the 80386 microprocessor (used in the Personal System/2 Model 80) follow in the next section. Also, a large portion of the 8088 interfacing coverage in Chapter 3 and the IBM PC I/O interfacing material in this chapter are directly applicable to the 8086-based Model 30 machine.

5.3

80386 MICROPROCESSOR

Many advanced desktop computers, including the IBM Personal System/2 Model 80 incorporate the Intel 80386 microprocessor. This chip is a 32-bit device offering advanced functionality in both hardware and software.

5.3.1   80386 GENERAL DESCRIPTION

The Intel 80386 is a 32-bit microprocessor that incorporates a number of advanced features. These features include a pipelined architecture, multitasking support, and memory management. The pipelined architecture shortens the effective time of instruction execution by overlapping portions of the instruction fetch and execution cycles. Multitasking support permits the concurrent execution of tasks such as editing a document and printing out another. Memory management, including a protection mechanism, translates logical addresses to physical addresses in the address space and manages the task integrity and environment in multitasking applications. For example, the 80386 can perform all the functions involved in task switching such as saving the state (registers) of one task, loading the state of a new task, and resuming execution of the initial task in approximately 16 μs in a 16-MHz 80386.

5.3.2   80386 STRUCTURE

The 80386 is composed of six functional units. These units are the Bus Interface Unit, Code Prefetch Unit, Instruction Decode Unit, Execution Unit, Segmentation Unit, and Paging Unit.

The Bus Interface Unit (BIU) is the focus of interaction of the 80386 with external devices. It produces and/or processes the address, data, and control signals used for accessing external memory and I/O devices. The BIU also handles internal requests from the Code Prefetch Unit for accessing code from memory and from the Execution Unit for information transfers. Interfaces to coprocessors and outside bus masters are also provided by the BIU.

The Code Prefetch Unit, through the BIU, looks ahead and fetches instructions that will follow the one currently being executed. This reduces the overall execution time of the instructions by minimizing the time consumed waiting for the next instruction in a sequence to be fetched. The prefetched instructions are stored in a 16-byte code queue for processing by the Instruction Decode Unit.

The Instruction Decode Unit decodes the instructions from the 16-byte prefetch code queue and stores them in a three-deep FIFO called the Instruction Queue.

The Execution Unit takes the decoded instructions from the Instruction Queue and executes them.

The Segmentation Unit performs the translation of logical addresses to linear addresses. It also checks for bus-cycle segmentation violations. The linear addresses are then transmitted to the Paging Unit.

The Paging Unit, when enabled, translates the linear addresses provided by the Segmental or Code Prefetch units into physical addresses. The physical addresses are then sent to the BIU for accessing external devices or memory. If paging is not enabled, the linear address is equivalent to the physical address and no further translation is performed.

5.4

80386 BUS INTERFACE

The 80386 bus interface consists of four bus groupings. These groupings are the 32-bit data bus, the address bus consisting of 30 address bits and 4 byte-enable lines (BE0#–BE3#), 5 status lines, and a 12-bit control bus.

The data bus, bits labeled D0-D31, is bidirectional and can be allocated dynamically to transfer 32, 24, 16, or 8 bits of data as a unit. The address bus consists of address bits A2-A31 and four byte-enable lines, BE0#–BE3#. The 30-bit address lines are used to select one of 1,073,741,824 four-byte locations in memory while the four byte-enable lines select the active bytes in the four-byte location. This is equivalent to a 32-bit memory address that can access 2^{32} or 4,294,967,296 byte locations.

The five status output lines are Address Status (ADS#), Write/Read (W/R#), Memory/I/O (M/IO#), Data/Control (D/C#), and LOCK#. The ADS# output is active low and indicates that the addresses present on the address bus are valid. Memory/I/O or M/IO#, indicates a memory transfer if high and an I/O transfer if low. A high on Data/Control (D/C#) indicates data transfer involving memory, while a low indicates an instruction transfer involving memory. If LOCK# is low, a bus lock is in place. This locking means that another bus master cannot use the local 80386 bus and insures read-modify-write operations without interruption.

The control bus consists of 11 inputs and 1 output. The inputs are READY#, Next Address (NA#), Bus Size 16 (BS16#), CLK2, RESET, HOLD, Maskable Interrupt (INTR), Non-Maskable Interrupt (NMI), BUSY#, ERROR#, and Coprocessor Request (PEREQ). The output is Hold Acknowledge (HLDA). READY# is active low and, when active, terminates the current bus cycle. When a high input is placed on READY#, a wait state is added to the bus cycle. READY# is sampled in each additional wait state and adds another wait state each time until READY# is sampled low.

The Next Address (NA#) input is used to initiate and control pipelining (overlapping of bus cycles to reduce overall instruction execution time). NA# is active low and is generated by system logic external to the microprocessor when the address bus is no longer needed in the current cycle. In pipelining, the address of the next instruction or data location and associated status signals are placed on the address bus before the completion of the current bus cycle. Bus Size 16 (BS16#) is active low and is used to initiate a 16-bit data transaction, utilizing the lower 16 bits of the bidirectional data bus. Input CLK2 is for a double frequency input for microprocessor operation. For a 16-MHz 80386, CLK2 is a 32-MHz signal. This signal is divided by two internally for synchronous use by the microprocessor. A high signal activates the RESET line to the 80386 and places the microprocessor in a known state where it begins program execution at a RESET vector address. Another bus master can use the HOLD input to request that the 80386 release the external bus. The

80386, in turn, releases the bus and indicates so by generating an active high on the Hold Acknowledge (HLDA) output pin and placing output lines D0–D31, BE0#–BE3#, A2–A31, W/R#, D/C#, M/IO#, LOCK#, and ADS# in a high impedance (floating) state.

Non-Maskable Interrupt (NMI) and Interrupt (INTR) input lines are activated by a high input signal and transfer program control from the code presently being executed to an interrupt service routine. The INTR line is maskable in that it can be disabled and enabled by software to avoid interruption of a routine at a critical time. The NMI cannot be deactivated. The remaining three inputs, BUSY#, ERROR#, and Coprocessor Request (PEREQ) are used as interface signals to the 80287 or 80387 numeric coprocessors. These support chips provide hardware-based numeric processing capabilities to the 80386. Coprocessors provide operations such as high-precision integer and floating-point calculations. The 80387 performs 32-bit data transfers while the 80287 performs 16-bit transfers. The coprocessor is selected through I/O addresses produced by the 80386 through the execution of coprocessor instructions. An active low BUSY# input to the 80386 from the coprocessor indicates that the coprocessor is busy and cannot accept a new instruction at this time. A PEREQ input that is active high indicates to the 80386 that data must be transferred to or from memory by the coprocessor through the 80386. The active low ERROR# input to the 80386 indicates an error that has not been masked by the coprocessor control register has occurred following the execution of a mathematical instruction by the coprocessor.

5.4.1   BUS TIMING

The 80386 bus cycle consists of a minimum of two states, T1 and T2. The states are defined by the clock signals CLK2 and CLK. CLK2 is an externally applied synchronization signal for the 80386 that is divided by two internally by the microprocessor to produce the CLK signal. The bus states T1 and T2 and related timing are illustrated in Figure 5.16. Each state consists of two CLK2 phases as shown.

5.4.2   NON-PIPELINED READ CYCLE

A bus cycle to perform a non-pipelined memory read or write is initiated when, at T1, a valid address is placed on the address lines A2–A31 and BE0#–BE3#. Then ADS# is brought low to indicate that the output address bus levels are valid. Status line W/R#, M/IO#, D/C#, and LOCK# are also considered valid during T1 and are sampled when ADS# is low. During T2, devices external to the 80386 respond to signals initiated during T1. Recall that an active low READY signal terminates the present address cycle. To illustrate, the timing of a non-pipelined Read operation is given in Figure 5.17.

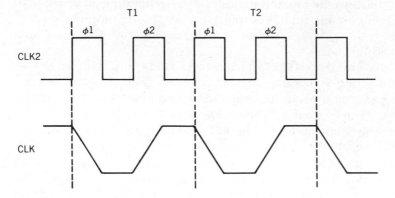

FIGURE 5.16 Basic 80386 bus cycle.

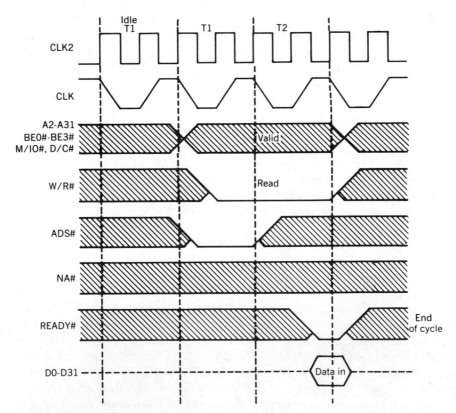

FIGURE 5.17 Non-pipelined read cycle.

To summarize, for a read cycle,

During T1
a. ADS# low.
b. A2–A31, BE0#–BE3# levels stable.
c. W/R# low.
d. M/IO# high for memory read, low for I/O read.
e. D/C# high if data to be read, low if instructions are to be read.
f. LOCK# low if bus cycle is locked.
g. Signals b–f above sampled while ADS# is low.

At end of T2
a. READY# is sampled. If READY# sampled low, data are read into micro-processor and read cycle ends. New cycle begins on next CLK.
b. If READY# is sampled high, a wait state (one cycle of CLK) is added to bus cycle to extend bus cycle before termination. Wait states are added until READY# is sampled low. This is used, for example, to allow for the incorporation of slow memories that cannot respond in the time allotted in a normal bus cycle.

5.4.3 NON-PIPELINED WRITE CYCLE

A non-pipelined write cycle is illustrated in Figure 5.18. The summary for the non-pipelined write cycle is as follows:

During T1
a. ADS# low.
b. A2–A31, BE0#–BE3# levels stable.
c. W/R# high.
d. M/IO# high for memory write, low for I/O write.
e. D/C# high.
f. LOCK# low if bus cycle is locked.
g. Signals b–f above sampled while ADS# is low.

At beginning of Ø2 in T1
a. Output data valid on data bus until Ø1 in next T1 state.

At end of T2
a. READY# is sampled. If READY# sampled low, write cycle ends. New cycle begins on next CLK.
b. If READY# is sampled high, WAIT states are added until READY# is sampled low.

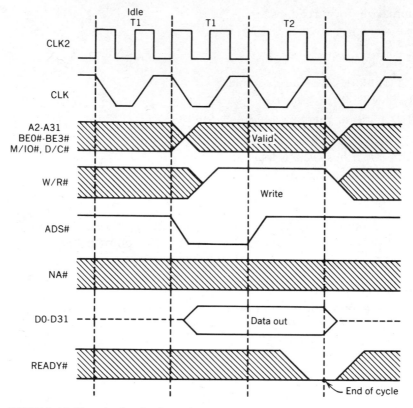

FIGURE 5.18 Non-pipelined write cycle.

5.4.4   PIPELINED CYCLES

A pipelined cycle is initiated by logic external to the microprocessor that indicates that the address bus is not needed at this time for a conventional memory or I/O access. An active low signal on the NA# input of the 80386 is used to initiate a pipelined address cycle. NA# is sampled on the leading (rising) edge of Phase 2 of each cycle of CLK in which ADS# is not active. Once NA# is sampled high, an additional T2 state (T2P) must be added to the present cycle to permit the output of address and status outputs for the next access before the end of the present address cycle. If NA# is sampled in its active state and additional pipeline cycles are desired following the one currently requested, NA# need not be active again until the next cycle of CLK following an active ADS#. Pipelining can continue for as many address cycles as desired. When pipelining is initiated, the address and status information for the following address cycle are presented during a state T2P that is added on to Phase T1 and T2 in the current address cycle. Once this pipelined instruction has occurred, address and status are presented in the second state of all following contiguous pipelined address cycles. The timing relationships among the various microprocessor signals for pipelined write initiation are given in Figure 5.19. The sequences for the execution of multiple, contiguous pipelined cycles are shown in Figure 5.20.

5.4.5 INTERRUPT ACKNOWLEDGE CYCLE

The 80386 can receive an interrupt through the Maskable Interrupt Request input, INTR. If the INTR input is not masked by software (interrupt flag of 80386 enabled), an active high on INTR will generate two consecutive interrupt acknowledge cycles. The result will be to cause the 80386 to temporarily cease execution of the present sequence of instructions and switch to the execution of an interrupt service routine located at a known series of locations in memory. Interrupts to the 80386 usually originate from the 8259A interrupt controller. The 8259A prioritizes up to eight interrupt requests and produces a single output to be connected to the 80386 INTR input. Upon

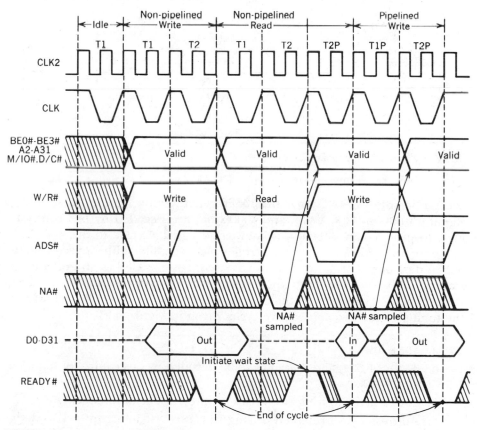

FIGURE 5.19 Initiation of pipelined read.

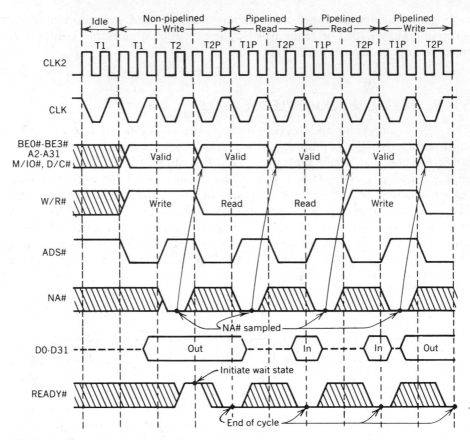

FIGURE 5.20 Multiple pipelined address cycles.

recognition of the INTR signal, the 80386 generates low logic levels on control signal outputs M/IO#, D/C#, and W/R# that are decoded by bus controller logic to produce an interrupt acknowledge signal, INTRA, to the 8259A. In response, the 8259A interrupt controller places an 8-bit address vector on the 80386 data bus. This address vector is used by the 80386 as an index to the starting address of the interrupt service routine corresponding to the interrupt request. A summary of the interrupt acknowledge bus cycle follows with the timing of the two consecutive cycles given in Figure 5.21.

a. ADS# low.
b. M/IO#, D/C#, and W/R# low to be decoded by bus controller to generate INTRA.
c. LOCK# active during two cycles of interrupt acknowledge.
d. Address of 4 presented during first cycle of interrupt acknowledge and address 0 presented during second cycle.

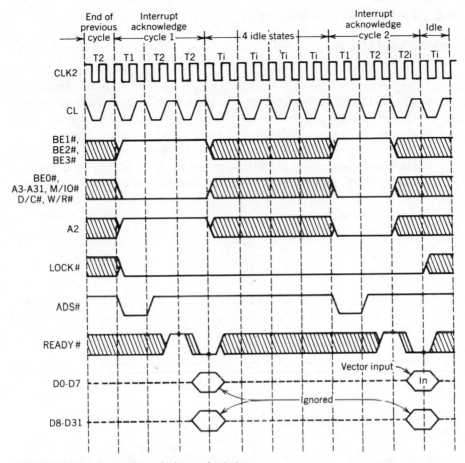

FIGURE 5.24 Interrupt acknowledge cycle timing.

e. BE1#-BE3# high for both cycles of interrupt acknowledge. A2 is high for first cycle and low for second cycle.

f. BE0#, A3-A31 low for both cycles of interrupt acknowledge.

g. D0-D31 floating for both cycles.

h. At end of second interrupt acknowledge cycle, interrupt vector presented by 8259A interrupt controller is read on data bus inputs D0-D7 to 80386.

i. READY# terminates each cycle with an active low level. (READY# must be delayed by system logic to allow for minimum pulse width requirements of 8259A. Also, four idle states, Ti, are inserted by the 80386 between the two interrupt acknowledge cycles to match the recovery time of the 8259A.)

5.4.6   TIMING CONSIDERATIONS

READY# TIMING

The READY# signal is used to end a bus cycle. When an active low READY# input is sampled at the end of a T2 state, the current bus cycle is terminated. The time elapsed from the presentation by the 80386 of the valid addresses to the active low transition of the READY# signal is the amount of time available to external system logic for the generation of the READY# signal. This time can be calculated by subtracting the address line settling time delay and the READY# setup time required by the 80386 from the time for four CLK2 cycles. For a 16-MHz 80386 with no pipelining, the calculation is

Four CLK2 cycles	125 ns
Address line delay (max)	−40 ns
READY# set up time (min)	−20 ns
Time to generate READY#	65 ns

For a pipelined address cycle in a 16-MHz 80386, an additional 62.5 ns are available for generation of READY#, allowing 65 ns + 62.5 ns, or 127.5 ns total. For any *n* Wait states that are added to the cycle, a further n × 62.5 ns will also be available.

If a cache subsystem is used, the READY# signal should be generated utilizing logic from the cache comparator to ensure rapid incorporation of cache hits.

READ TIMING

The memory access time (time from availability of valid address on the address bus to reading the data bus) for a nonpipelined cycle is calculated as follows:

Four CLK2 cycles	125 ns
Address line delay (max)	−40 ns
Data bus input setup time (min)	−10 ns
Memory access time	75 ns

For a pipelined address cycle, an additional 62.5 ns is added to the non-pipelined access time. Also, each additional WAIT state adds 62.5 ns to the access time total.

WRITE TIMING

A critical timing portion of a write cycle is the time available in the cycle for external logic to decode and latch a valid address. This is calculated as:

Four CLK2 cycles	125 ns
Address line delay (max)	−40 ns
Time to decode and latch address	85 ns

As in the READ cycle, a pipelined write cycle adds 62.5 ns to this time and each additional WAIT state adds 62.5 ns.

A second critical timing sequence is the time available to an external device to read data written on the data bus. There are setup and hold-time requirements that must be met for the external device. Note that data are available and valid on the data bus after the end of the bus cycle. This hold time is calculated by

One CLK2 cycle	31.25 ns
Data bus hold time (min)	1.00 ns
Data hold time	~32.25 ns

This time is not affected by WAIT states.

The setup time is calculated as:

Three CLK2 cycles	93.75 ns
Data bus output delay (max)	−50.00 ns
Setup Time	~43.47 ns

Additional WAIT states each add 62.5 ns to this time.

5.4.7   LOCK CYCLE

A LOCK cycle is activated by the LOCK prefix added to specific instructions of the 80386 instruction set. It is used to ensure that desired bus cycles will be executed consecutively and not interrupted by another bus master. In many cases, especially when setting or reading flags, the insertion of a bus cycle from another processor that is also accessing the same flags could result in erroneous data being interpreted by one or more microprocessors. The 80386 also activates LOCK# during interrupt acknowledge cycles, the execution of an XCHG instruction, a page table update, or a descriptor update.

The LOCK# output is activated (low) on the CLK2 edge at the beginning of the LOCKED bus cycle. LOCK# returns to an inactive (high) state at the end of the last LOCKED cycle when READY# is sampled low.

5.4.8   RESET TIMING

The discussion of timing of individual control signals will conclude with RESET signal timing. A low-to-high transition on the RESET input terminates the program execution in the 80386. RESET must remain high for a minimum of 15 CLK2 periods for proper initialization of the 80386. The 80386 also provides for self-test at reset, but RESET must remain high for a minimum of 80 CLK2 periods for self-test to be completed. A low level following the high level of RESET initializes the 80386 to a predetermined known state and then causes the 80386 to begin instruction execution from a specific RESET memory location.

The RESET signal is generated from the 82384 clock generator chip from the RES# input to the 82384. A high-to-low transition on the Schmitt-trigger RES# input initiates the RESET signal generation in the 82384. The CLK output of the 82384 is then initialized by the low-to-high initial transition of RESET. On the high-to-low transition of RESET, the 80386 adjusts the high-to-low transition of its internal CLK signal to be synchronized with the first CLK2 cycle following the high-to-low transition of RESET. These timing relationships are illustrated in Figure 5.22.

Also, on the high-to-low transition of RESET, BUSY# input is sampled by the 80386. If BUSY# is at a low level, the 80386 conducts a self-test prior to initiating instruction execution. The 80386 will continue the initialization sequence following the self-test, regardless of the outcome of the test. The RESET address at which the 80386 starts execution of instructions following a RESET operation is 0FFFFFF0 Hexadecimal.

A typical circuit used to generate the input to RES# of the 82384 clock generator is shown in Figure 5.23. If V_{cc} is changing prior to or during a reset, RESET must be held at a high level for a minimum of 1 ms after V_{cc} and CLK2 are stable.

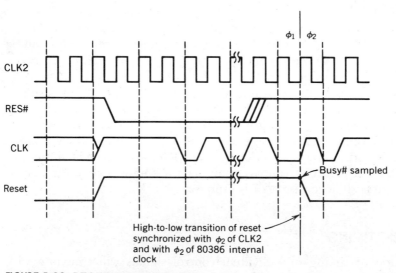

FIGURE 5.22 RESET timing relationships.

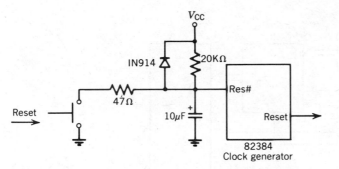

FIGURE 5.23 RESET input circuit. Reprinted by permission of Intel Corporation, Copyright 1986.

The execution of a reset causes the address lines LOCK#, D/C#, and ADS# to be at a high level while BE0#–BE3#, HLDA, W/R#, and M/IO# are placed in a low state. The data bus lines enter the three-state mode.

5.5

INTERFACING TO MEMORY

High speed memory components are necessary to take advantage of the performance capability of the 80386 microprocessor. Combinations of high speed and relatively lower speed memory can be used to achieve a compromise between speed and cost. One approach is to use a high speed cache memory subsystem and lower speed main memory. This approach is discussed under the Cache section of this chapter.

The basic memory interface consists of the components shown in the block diagram of Figure 5.24.

In the diagram of Figure 5.24, the Address Latch holds the address for the total bus cycle. In pipelining, the latches are required, since the address outputs of the 80386 are changing to the address of the next bus cycle before the current cycle has terminated. Typical latches used for this application are TTL devices 74F373 or 74AS373 and are controlled by the ALE# (Address Latch Enable) output from the bus controller.

The address decoder is used to generate chip select signals from the address lines. These chip selects activate a particular set of memory chips. They are produced by decode logic and connected to the memory subsystem through the bus controller.

On the data bus, bidirectional transceivers are required to permit data to flow in both directions at different times. The transceivers provide current drive capability for the bus that is not capable of being met by the direct data outputs of the 80386. Typical transceivers that can be used with the 80386 are the 74F245 and 74AS245 8-bit transceivers.

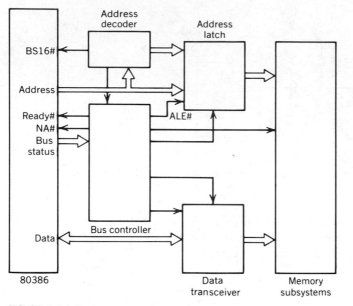

FIGURE 5.24 Basic memory system.

The latched W/R# output of the 80386 is used to enable the transceivers for writing to, or reading from, memory. The latched W/R# signal, known as DT/R#, is set low for reading and high for writing. The transceiver outputs are enabled or disabled by a separate signal, DEN#, that is produced by the bus controller.

5.5.1 DYNAMIC RAM SUBSYSTEM DESIGN

A design of a dynamic RAM subsystem will be implemented for illustration purposes. In this example, four Programmed Array Logic (PAL) elements are initialized. These are a DRAM State PAL, a DRAM Control PAL, a Refresh Address Counter PAL, and a Refresh Internal Counter PAL. They are programmed with combinatorial and sequential logic equations to perform their respective functions. The State PAL monitors the banks of RAM to determine the ones needing time for precharging (charging cycle required by RAM cells following an access), initiates refresh cycles upon receipt of refresh requests and reads 80386 DRAM chip-select logic outputs. The Control PAL produces most DRAM control outputs. The Refresh Address Counter and Refresh Interval Counter PALs provide the next refresh address and generate the refresh request signal, respectively.

The example will be of a 3-CLK design that refers to the number of CLK pulses required for a memory read. An alternative, more timing critical approach is a 2-CLK design.

The memory for this example is a 32-bit wide by 256K system. It is made up of 32, 64K × 4-bit chips. A block diagram of the memory system is given in Figure 5.25. In Figure 5.25, each rectangle in the DRAM array is the equivalent of a 64K × 8 DRAM composed of 2, 64K × 4 packages.

5.5.2   DYNAMIC MEMORY REFRESH

The required refreshing of dynamic RAMs can be accomplished by periodic accessing of all the rows in memory. This is accomplished by activation of the RAS (Row Address Strobe) signals without the need to activate the CAS (Column Address Strobe) signals. This approach is called a RAS-only refresh. A typical refresh sequence would be

- Address multiplexer placed in high impedance state.
- Refresh address counter PAL selected to place address of next row to be refreshed on the address bus.
- The state PAL activates row selects for all dynamic RAM banks and, thus, refreshes the addressed row in all banks simultaneously.
- Refresh address counter PAL increments so that another row is refreshed during the next refresh cycle.

Refreshing of the DRAMs used in the example must be accomplished at least every 4 ms. This interval is common for most larger DRAMs. As discussed in the preceding paragraph, only row addresses are needed for a refresh cycle. Therefore, the refresh address counter PAL must be programmed to present only 8 bits on the address bus for refreshing.

Refreshing can be accomplished at intervals throughout the 4-ms period (distributed) or at one time ("burst") during the period. The distributed approach has the advantage that program execution is not delayed for any relatively long period by a refresh cycle. It also is easier to implement.

In our example, there are 256 rows per bank that are to be refreshed. For an even distribution over the 4-ms refresh period, 4 ms/256 or approximately 15 μs per refresh cycle are required. The refresh interval counter PAL is, therefore, programmed to request a single refresh cycle every 15 μs. Based on 16-MHz CLK cycles, this would amount to 16 MHz × 15 μs or 240 CLK cycles. This number must be corrected, however, by the time required for the state PAL to respond to an interrupt request. The maximum latency is 4 CLKs for either the 2–CLK or 3–CLK design. Since these 4 CLKs are spread over the 256 refresh accesses, the latency per refresh access is $\frac{4}{256}$ CLKs or 0.015625 CLKs. This number must be subtracted from 240, yielding approximately 239 CLKs. This count is used to program the refresh interval counter to initiate a refresh request.

Following a power-up condition, dynamic RAMs may require warm-up

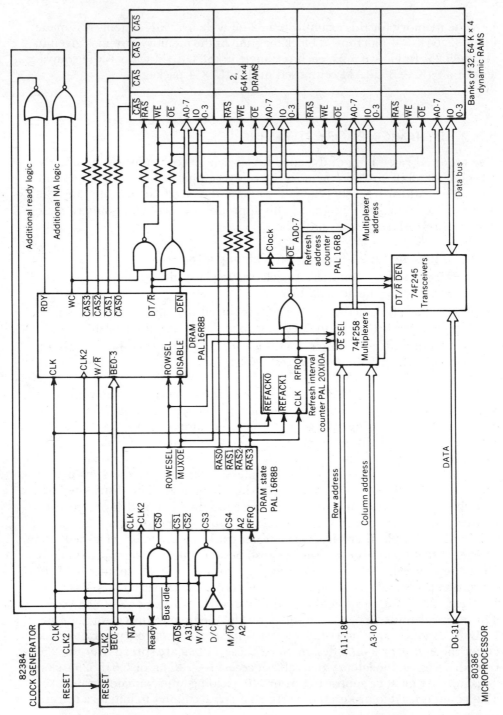

FIGURE 5.25 Dynamic RAM subsystem.

cycles before operation. This warm-up can be accomplished by externally activating refresh requests for a period of time or performing a number of memory accesses.

5.6

80386 CACHE MEMORY INTERFACING

Microprocessors are available that can operate at clock frequencies of 16 MHz and will eventually reach 32 MHz. For these two rates, the bus cycle times are on the order of 62.5 and 31.25 ns, respectively. If it is assumed that a minimum of two clock periods are required for a memory transfer, then minimum bus cycle times of 125 and 62.5 ns are required. Static random access memories are available with times in these ranges that are suitable for such an application. However, these static RAMs are expensive and the implementation of the entire memory with such RAMs would be very costly. Dynamic RAMs offer lower cost than equivalent size static RAMs but cannot meet the access time requirements. A useful compromise is to employ a relatively small amount of static RAM with a large dynamic RAM memory system. The goal is to make the entire memory appear as high speed static RAM by periodically transferring the anticipated next group of instructions and data that are to be accessed from the dynamic RAM to the static RAM. In this mode, the static RAM is known as cache memory. If the desired information is in static RAM, a cache "hit" has occurred. Otherwise, a "miss" has resulted and the required data and/or instructions must be loaded into cache from the main dynamic RAM system.

A typical cache memory subsystem is shown in Figure 5.26.

As implied by the preceding discussion, it is often possible to anticipate the location of instructions or data that will be accessed, since most programs access locations near to those recently addressed. In other words, instructions from a group of instructions are usually accessed. This is fortunate, since cache memories would not be practical if memory accesses were totally random. Where jumps to subroutines and references to buffers and lists occur, a

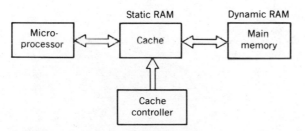

FIGURE 5.26 Cache memory subsystem.

discontinuity from the present group of instructions is introduced. Usually, sequential execution then takes place within the new group of instructions. There are exceptions, naturally, but accesses to memory are sequential or near-sequential enough times to make cache memories viable.

5.6.1   ORGANIZATION OF CACHE MEMORY

In order to closely match the general pattern of memory accessing that occurs in real systems, caches are usually organized into *blocks* of instructions. A block is a group of instructions that are located in sequential memory address locations. Blocks typically range in size from 2–16 bytes.

   If the blocks of instructions and data that were to be executed in the program could be predicted and were of "reasonable" number, they could be stored in a cache regardless of their respective locations in main memory. This concept, known as *fully associative cache,* is illustrated in Figure 5.27 for a 4-byte block, 2-block cache.

   In order to identify which blocks are stored in cache, additional *tag* memory is required. Tag static RAM contains the main memory address associated with a particular block in cache as shown in Figure 5.27. Thus a 32-bit processor address would typically consist of the fields as also illustrated in Figure

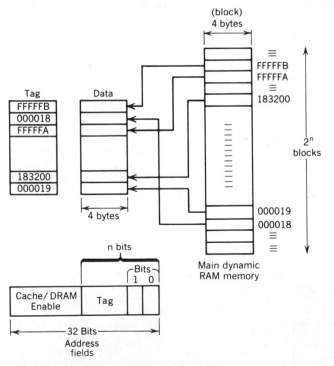

FIGURE 5.27 Fully associative cache concept.

5.27. An obvious drawback with a fully associative cache is that at each memory access all tags in the tag RAM must be checked to determine if the desired instruction or data are contained in cache.

Another approach that requires only one tag comparison is the *direct mapped cache*. In direct mapped cache, a block in main memory can be stored in only one possible location in cache. This is accomplished by looking at the processor address as a combination of an index and tag field. The tag field appears as a memory page. For example, three main memory addresses in Hex could be FFF800, FFF801, and FFF803. The two Hex characters on the right of each address can be considered tags or pages. Thus the subfield FFF8 specifies one of a number of addresses that can be stored in only one location of the tag RAM. Which FFF8 is stored there is determined by the rightmost tag characters. Thus, if address FFF801 is sent out by the processor, the cache control logic would use FFF8 as an index to the tag RAM to see if the hex characters 01 were stored there. If so, a "hit" has occurred. If not, a "miss" has occurred. In either event, only one comparison was required to determine if the contents of the desired address are contained in cache. Direct mapped cache organization is given in Figure 5.28.

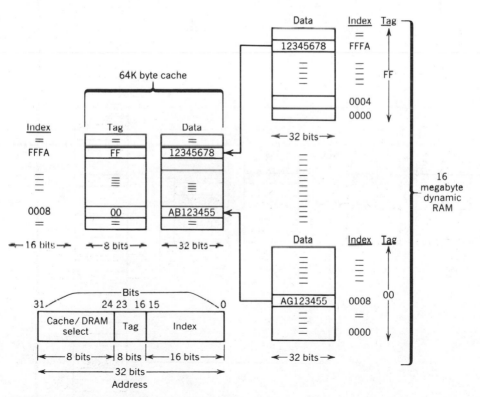

FIGURE 5.28 Direct-mapped cache organization.

A more complex approach is the *set associative cache* that combines the features of the fully associative and direct mapped caches. Essentially, a set associative cache is organized as parallel blocks or sets of direct mapped cache. Thus a particular index field can refer to more than one cache and identify a block of instructions or data in each of the cache sets. The tag field identifies the block that is desired. If four parallel sets of cache were employed, there would be four locations in cache in which each block could be stored. Four comparisons would, therefore, have to be made to determine if the addressed location is in cache. A two-way set associative cache is shown in Figure 5.29.

The set associative cache increases the probability of a "hit," since more than one block with the same index is available in the cache at any given time. Thus fewer main memory accesses are required and "hit" efficiency is increased. Some other observations relative to the set associative cache are that

1. The controller must make *n* comparisons, where *n* is the number of sets or groups, to determine if the requested instruction or data is in the cache.

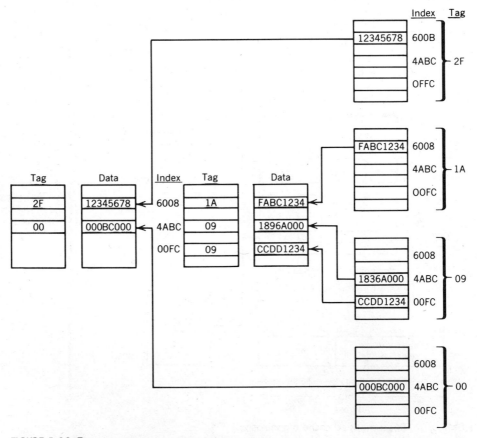

FIGURE 5.29 Two-way set associative cache example.

2. When information is to be transferred to cache from main memory, the controller must decide which set is to receive the information.

3. When a block overwrite in the cache is required, the controller must decide which block must be overwritten.

5.6.2 CACHE DATA INTEGRITY

When data or instructions are changed in cache, the corresponding information in main memory must also be changed lest there be a discrepancy in the contents of the two memories. This updating of main memory is usually accomplished by *write-through, buffered write-through,* or *write-back.* In write-through, data or instructions are copied into main memory immediately after being written to cache. This additional write in the access cycle decreases performance somewhat but is simple and relatively easy to implement. Buffered write-through is similar to write-through, in that data are written to main memory after a code access, but the data are first placed in an intermediate buffer. This allows the processor to begin a new cycle before the access to main memory is completed. With the write-back method, an additional bit, called the altered bit, is included with tag field to indicate if the block has been altered with new information. Then, before an overwrite operation, the cache controller polls the altered bit of the tag field to determine if new data have been written into cache. If so, the new data in the cache block are written to main memory before the overwrite operation is executed. This approach is more complex than the write-through method but reduces the number of write accesses to main memory.

5.6.3 TYPICAL CACHE PERFORMANCE

Cache performance is a function of many parameters, such as organization, up-date methods, cache size, and type of memory employed. However, some typical performance parameters can be listed for informational purposes and are presented in Table 5.5.

5.6.4 EXAMPLE DESIGN

A design example for a cache memory system will serve to summarize the main principles of cache design. The following specifications will serve as a basis for the design:

- Cache will be of direct access type, since costs of static RAM have decreased and, thus, the hit rate can be increased by storing more information in a larger cache rather than utilizing a smaller cache with more complex control logic.

TABLE 5.5 Typical Cache Performance

| | Cache Characteristics | | | |
Asso-ciative Organi-zation	Capacity	Block Size (bytes)	Hit Rate (%)	Percent of Performance of Dynamic RAM Without Cache
Direct	1K	4	42	91
Direct	16K	4	80	135
Direct	32K	4	85	137
Two-way	32K	4	88	140
Direct	64K	4	89	140
Two-way	64K	4	90	141
Direct	64K	8	92	142
Two-way	64K	8	93	143
Direct	128K	4	90	140
Two-way	128K	4	89	140
Direct	128K	8	93	142

Reprinted by permission of Intel Corporation, Copyright, 1986.

- Interface will be to the 80386 microprocessor.
- Block size will be 4 bytes to accommodate the 32-bit data bus of 80386.
- Both data and instructions will be stored in the cache.
- For simplicity, the write-through technique will be used for main memory updating. This technique is very effective if high speed dynamic RAM is used for main memory.
- Main memory is 8 megabytes.
- Cache will be 32K bytes.
- Higher-order bits of microprocessor address field (9 bits in this example) will be used for decoding to select this particular cache subsystem from other memory systems addressed by this CPU. A general layout of the example cache system is given in Figure 5.30.

As stated in the specifications, the cache subsystem will be used in conjunction with an 80386 32-bit microprocessor. One characteristic of the 80386 is that its address bus consists of 30 bits, A2–A31, and 4 byte enable lines, BE0#–BE3#. The byte-enable lines are used to select one of 4 bytes in a 4-byte location identified by the combination of the 30 address lines. This number, then, 4×2^{30} is the 2^{32} address space of the 80386.

The hardware implementation of the 80386 cache subsystem is shown in Figure 5.31.

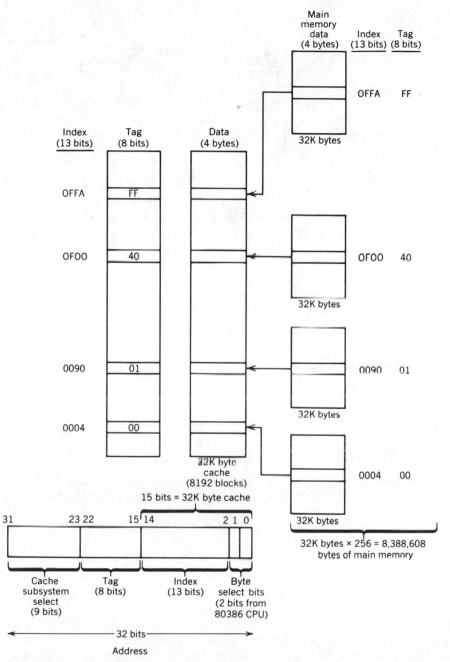

FIGURE 5.30 Cache subsystem example layout.

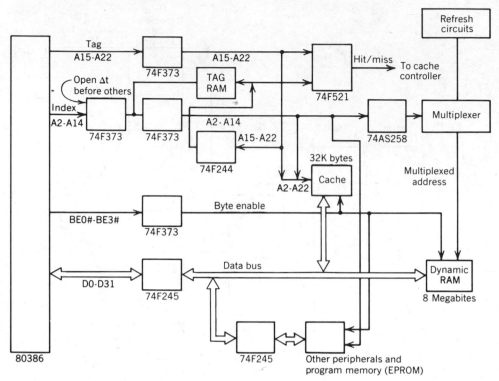

FIGURE 5.31 Cache subsystem hardware implementation.

5.7   I/O INTERFACING

I/O interfacing in some respects is similar to memory interfacing. Addresses of I/O devices are placed on the address bus and data are transferred on the data bus. If *memory-mapped* I/O is utilized, I/O devices are indistinguishable from memory from the processor's point of view. Any instructions that invoke memory can be used for I/O operations. The I/O devices are assigned a portion of the microprocessor's memory space and thus reduce the amount of space available for memory. The *I/O-mapped* addressing, however, utilizes additional, physical signals available on output lines of the microprocessor to indicate that a particular instruction being executed is an I/O instruction. These output pins are activated by specific I/O instructions when they are executed by the processor. These instructions are the IN, OUT, INS, and OUTS instructions of the 80386. Transfer of information in and out of the microprocessor is accomplished through the AL, AX, and EAX registers of the 386 with direct, 8-bit addressing. Using indirect addressing through the DX register, all 64-kilobyte I/O device addresses can be accessed.

5.7.1 GENERAL I/O INTERFACE

A general input/output interface that can be used with a variety of I/O devices is shown in Figure 5.32. In the general interface, the address decoder chip-select output signals are fed to the wait-state generator to determine the number of required wait states for a particular device to be accessed. Then the proper control signals are sent from the wait state generator to the bus control logic.

The bus control logic decodes 80386 status outputs W/R#, D/C#, and M/IO# and determines the command signals corresponding to the bus cycle to be executed. The bus cycles are memory data and code read, memory write, I/O read and write, and interrupt acknowledge. The memory data read and

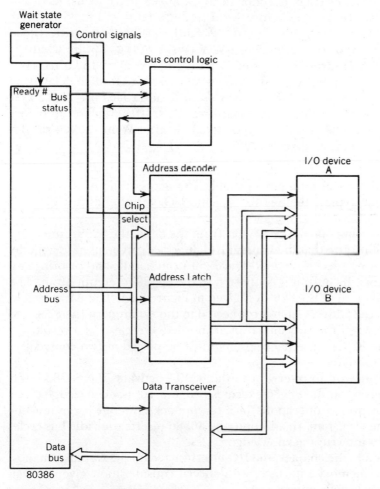

FIGURE 5.32 General I/O interface.

memory code read cycles generate the MRDC# command that causes the selected memory device to provide data. The I/O read and I/O write cycles generate the IORC# and IOWC# commands, respectively. These commands provide data to the 80386 (IORC#) or cause data to be read by the external device from the 80386 (IOWC#). The memory write cycle produces the MWTC# signal that initiates a transfer of data from the microprocessor to the external device. The INTA# signal is generated during the interrupt-acknowledge cycle and is sent to the 8259A interrupt controller (not shown in Figure 5.32).

A detailed implementation of the general I/O interface circuit is given in Figure 5.33. The maximum timings for the various I/O-related signals for a CLK2 of 32MHz are as follows:

SC1WS = 38.75 ns (max decoder delay), CS0WS = 31.75 ns, (activates NA#), tAR (address stable before Read [(IORC# fall)] = 132.75 ns, tAW (address stable before Write [(IOWC# fall)]) = 132.75 ns, tRR (Read (IORC#) pulse width) = 269.25 ns, tWW (Write (IOWC#) pulse width) = 300.5 ns, tRA (address hold after Read (IORC# rise) = 180.5 ns, tWA (address hold after Write (IOWC# rise) = 211.75 ns, tAD (data delay from address) = 335.5 ns, tRD (data delay from Read (IORC#)) = 253.25 ns, tDF (Read (IORC# rise) to Data Float) = 241 ns, tDW (data setup before Write (IOWC# rise)) = 289.5 ns, tWD (data hold after Write (IOWC# rise)) = 52.5 ns, tRV (command recovery time) = 331.75 ns.

These data indicate the minimum possible I/O access time of a particular I/O device. Some typical timings for specific I/O peripherals are given in Table 5.6.

An example using one of the devices from the table, the 8259A interrupt controller, will illustrate that the general I/O interface can be used as a model when implementing specific circuits. The 8259A can handle and prioritize up to eight interrupts. Its output is connected to the 80386 interrupt input. The input/output diagram of the 8259A is given in Figure 5.34. The A2 output of the 80386 from the address latch is connected to the A0 input of the 8259A to discriminate between the two interrupt acknowledge cycles. As a result, the addresses of the 8259A internal registers must be placed at two consecutive double word boundaries.

The occurrence of an interrupt to the 8259A activates the 80386 interrupt input, whereupon the 80386 executes two interrupt-acknowledge cycles in sequence. For proper operation, the interrupt-acknowledge cycles must be extended by one wait state. In addition, the 80386 inserts four idle bus cycles between the two interrupt-acknowledge cycles.

In conclusion to the chapter, the I/O interface technique is similar in most instances to the memory interface, with special consideration given to wait state insertions, where required.

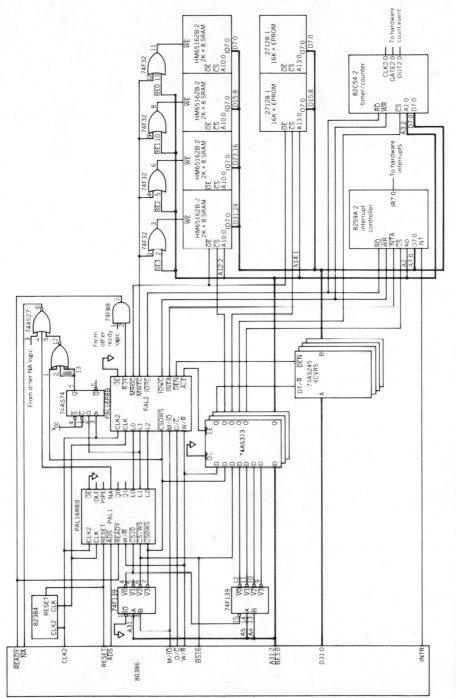

FIGURE 5.33 Detailed implementation of general I/O interface. (From Intel 80386 Hardware Reference Manual, No. 231732-001, 1986.) Reprinted by permission of Intel Corporation, Copyright 1986.

TABLE 5.6 Typical Timings for Some I/O Peripherals (from Intel 80386 Hardware Reference Manual, #231732-001, 1986)

	tAR	tAW	tRR	tWW	tRA	tWA	tAD	tRD	tDF	tDW	tDWF	tWD	tRV
Bus controller	39	39	269	300	180	211	335	253	241	289	—	52	331
8259-2	0	0	160	190	0	0	200	120	85	160	—	0	190
8254-2	30	0	95	95	0	0	185	85	65	85	—	0	165
82C54-2	0	0	95	95	0	0	185	85	65	85	—	0	165
82C55-2	0	0	150	100	0	20	—	120	75	100	—	30	300
8272	0	0	250	250	0	0	—	200	100	150	—	5	—
82064	0	0	200	200	0	0	—	70	200	160	—	0	300
8041	0	0	250	250	0	0	225	225	100	150	—	0	2500
8042	0	0	160	160	0	0	130	130	85	130	—	0	1120
8251	0	0	250	250	0	0	—	200	100	150	—	20	—
8273-4	0	0	250	250	0	0	300	200	100	150	—	0	1920
8274	0	0	250	250	0	0	200	200	120	150	—	0	300
8291	0	0	140	170	0	0	250	100	60	130	—	0	—
8292	0	0	250	250	0	0	225	225	100	150	—	0	—

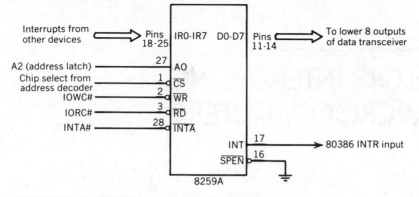

FIGURE 5.34 The 8259A interrupt controller interface.

FOR FURTHER READING

Intel 80386 Hardware Reference Manual, NO. 231732-001, Intel Corporation, 1986.

6

ANALOG INTERFACING TO MICROCOMPUTERS

This chapter will cover analog-to-digital and digital-to-analog converters and their interface with microcomputers. In many cases, analog design and evaluation techniques are not developed in as much detail as digital design. Interfacing to the analog world, however, requires skills that are different from those required for digital interfacing. Since most real-time signals originate as analog voltages or currents, it is extremely important to have an in-depth understanding of the analog/digital conversion process.

6.1

ANALOG-TO-DIGITAL CONVERTERS

Since most real-world occurrences are analog or continuous in nature, that is, voltage, current, velocity, temperature, illumination, and so on, a conversion to the digital or discrete domain is required for computer processing of these data. This analog-to-digital or *A/D* conversion can be accomplished in a variety of ways, depending upon the conversion speed, accuracy, cost, and noise immunity desired.

6.1.1 APPROACH

A basic approach to A/D conversion is the comparison of a series of "trial" analog signals to the analog signal to be measured until the two signals match within a specific tolerance. Since most analog signals can be represented in the form of a voltage, this discussion will be in terms of A/D conversion of voltages.

Comparison of two analog voltages is accomplished by means of an *analog comparator* circuit as shown in Figure 6.1. This circuit can be viewed as an analog amplifier with very large gain that amplifies the difference between the

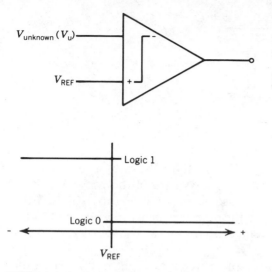

FIGURE 6.1 Analog comparator circuit.

input voltages. V_{REF}, the reference voltage, is the basis of comparison, and V_u, the unknown voltage, is the voltage to be determined. Since V_u is applied to the negative or inverting input of the amplifier, the difference between V_u and V_{REF} will be inverted, amplified, and presented at the output. The gain of the amplifier is set to be so large that only two states are possible at the output. When V_u differs from V_{REF} by a small threshold, the output is either a maximum (logic 1) or a minimum (logic 0). A logic 1 output is produced when V_u is negative with respect to V_{REF}; a logic 0 is produced when V_u is positive with respect to V_{REF}. In implementing an analog to digital converter, then, most schemes differ in the method used to generate the reference or "trial" voltage, V_{REF}.

6.1.2   ERRORS AND ACCURACY

If an A/D converter with binary output is viewed as a "black box," it would appear as in Figure 6.2. Note that there are n output lines that can represent the input voltage from zero to full scale in 2^n discrete steps. For a 4-bit converter (n = 4) with a binary output, the relationship between the analog input voltage and the output digital code is given in Figure 6.3. Since the output of the A/D can be only 1 of 16 possible codes to represent a continuous analog input voltage, there will be a "dead space" or interval between output codes, where analog input voltage changes are not reflected in the output. This interval is termed the *quantization* interval and is equal to the full-scale voltage, V_{FS}, divided by 2^n. This value is also equal to one least significant bit (lsb). In Figure 6.3, for example, the output code remains at 0000 for an

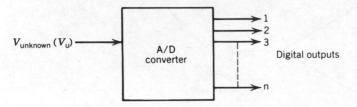

FIGURE 6.2 A/D converter black box.

analog input ranging from 0 V to $V_{FS}/32$. Thus, for the output code of 0001, the analog input voltage can have a value anywhere in the range $V_{FS}/32$ or $0.03125V_{FS}$ to $3V_{FS}/32$ or $0.09375V_{FS}$. Also, as shown in Figure 6.3, there are $2^n - 1$ or 15 threshold points at which the output switches to a new state as the input voltage increases. These threshold points are at $V_{FS}/32$, $3V_{FS}/32$, $5V_{FS}/32$, and so on, to $29V_{FS}/32$.

For an ideal A/D converter, the error between the input voltage value and the digital output value ranges over the quantization interval. When the input voltage is at the center of the quantization interval, the output code exactly

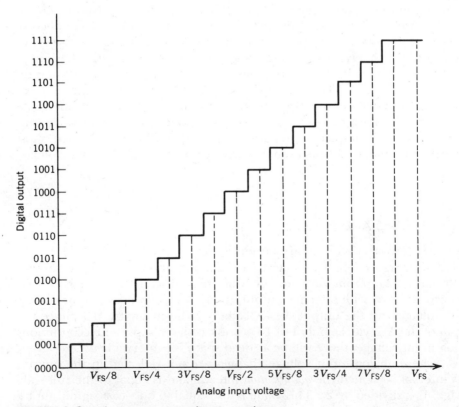

FIGURE 6.3 Digital output versus analog input voltage.

represents the input and the *quantization error* is 0. On either side of the center of the quantization interval, the maximum error is equal to one-half the quantization interval or $(V_{FS}/2^n)/2$ or $V_{FS}/2^{n+1}$. If the quantization error is plotted as a function of the analog input voltage for an ideal 4-bit converter, the sawtooth waveform of Figure 6.4 results. In the figure, Q is $V_{FS}/2^n$ or $V_{FS}/16$, the quantization interval, and E is the quantization error.

Up to this point, only the characteristics of ideal A/D converters have been discussed. Analyzing an actual A/D converter with less than ideal quantization characteristics is a valuable exercise and provides the basis for defining more terms associated with A/D conversion. Figure 6.5 characterizes an actual A/D converter by means of its input/output characteristics and quantization error.

From Figure 6.5 the deviation from the ideal input/output characteristic and quantization error plots can be observed. Recall that in the ideal situation, the step width of the input/output characteristic is equal to $V_{FS}/2^n$ or $V_{FS}/16$ in our example. In Figure 6.5a, output codes 0010 and 1100 remain, even though the input voltages have increased beyond the threshold points where the output codes should have changed to 0011 and 1101, respectively. The output codes remain at 0010 and 1100 over a voltage range that is greater than the quantization interval of an ideal converter by $V_{FS}/32$ or 0.5 lsb. Thus the A/D converter is said to have a differential linearity error of 0.5 lsb at these two points. Similarly, at code 1000 of Figure 6.5a, there is a differential linearity error of 1 lsb. Strictly speaking, *differential linearity error* is defined as the amount of deviation of any step in the input/output characteristic of the A/D converter from its ideal size of $V_{FS}/2^n$. The differential linearity error can also have a negative value such as for codes 0011, 1001, and 1101 that have values of -0.5, -1, and -0.5, respectively. Note that if the differential linearity is greater than 1 lsb, a code, 1001, is skipped between 1000 and 1010. This code is called a missing code. If an A/D converter has a differential linearity

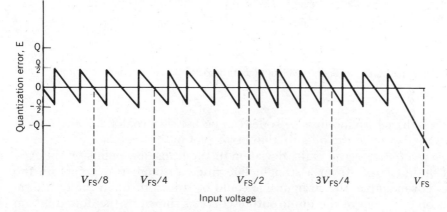

FIGURE 6.4 Quantization error for an ideal, 4-bit converter.

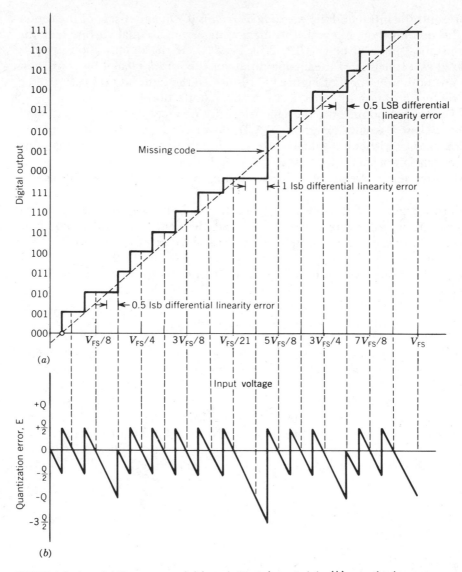

FIGURE 6.5 Actual A/D converter: (a) Input/output characteristic; (b) quantization error.

error less than 1 lsb, no missing codes can occur. Differential linearity errors are also reflected in the quantization error plot in Figure 6.5b.

Integral linearity error is the deviation of the transition points on the A/D input/output curve from a straight line fitted through two points on the curve. For example, the straight line could be drawn through the transition points on each end of the input/output curve as shown by the dotted line in

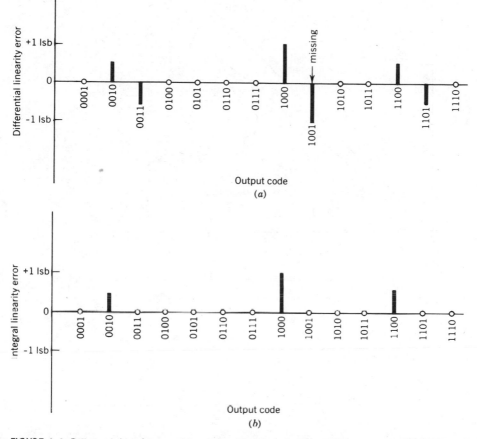

Output code

(a)

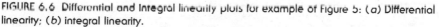

Output code

(b)

FIGURE 6.6 Differential and Integral linearity plots for example of Figure 5: (a) Differential linearity; (b) integral linearity.

Figure 6.5a. Figure 6.6 summarizes the differential linearity and integral linearity errors for the input/output curve of Figure 6.5a.

Another characteristic an A/D converter should display is *monotonicity*. An A/D converter is *monotonic* if the output is an increasing function of the input, that is, at no time does the output decrease as the input increases.

Two other errors associated with A/D converters are *gain error* and *offset error* as shown in Figure 6.7. The gain error is the difference between the slope of the ideal input/output curve of the A/D converter and the actual input/output characteristic. The offset error is the margin by which the input/output curve of the A/D converter does not pass through the origin.

Figure 6.8 shows the input/output characteristic of a 3-bit A/D converter with gain, linearity, and offset errors.

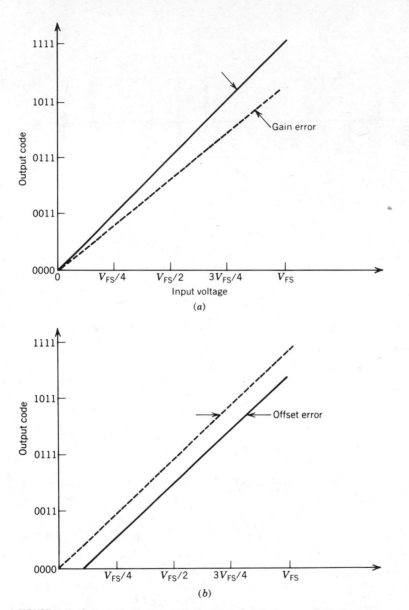

FIGURE 6.7 Gain and offset error illustrations: (a) Gain error; (b) offset error.

6.1.3   TEMPERATURE EFFECTS

Integral linearity, differential linearity, offset, and gain errors of an A/D converter are temperature-dependent. The changes in these errors are usually specified over a defined temperature range. The three temperature ranges used in specifying A/D converters and semiconductor devices in general are

Military range	$-55°C$ to $+125°C$
Industrial range	$-25°C$ to $+85°C$
Commercial range	$0°C$ to $+70°C$

Of principal concern in evaluating an A/D converter specification is that there are no missing codes and that the converter is monotonic over the temperature range. Since an A/D converter becomes nonmonotonic when the differential linearity error is greater than 1 lsb, the temperature change required to cause nonmonotonicity can be calculated from the temperature coefficient of differential linearity. This coefficient is usually supplied by the manufacturer in units of parts per million of full-scale range per degree Celsius. Assuming an initial value of $\frac{1}{2}$ lsb differential linearity error in the converter, the temperature change, δT, required to gain an additional $\frac{1}{2}$ lsb can be expressed in the following relationship:

$$\delta T \frac{(DTC)}{10^6} \times 2^n = \frac{1}{2}$$

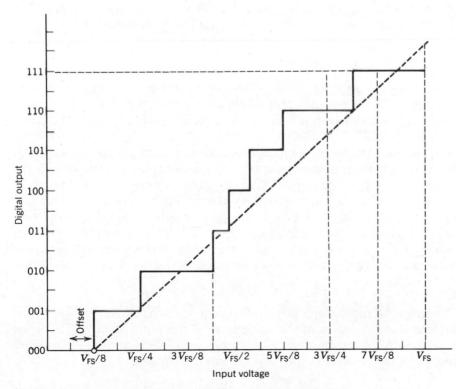

FIGURE 6.8 A/D converter with gain, linearity, and offset errors.

where

δT is the temperature change in degrees Celsius

DTC is the differential linearity error in parts per million of full-scale range

n is the number of output bits (resolution) of the converter

Solving for δT,

$$\delta T = \frac{10^6}{\text{DTC} \times 2^{n+1}}$$

δT is the maximum temperature change in degrees Celsius that can occur and still retain monotonicity.

6.1.4 METHODS OF A/D CONVERSION

One popular method of A/D conversion is generating an analog signal having a digital representation and comparing this signal to the unknown analog signal until a match within tolerances is achieved. The more accurate and rapid the desired conversion, the more expensive is the solution.

COUNTER CONVERTER

This counter converter serves as the basis of other more complex devices but can be used in many, nondemanding applications. The block diagram of the counter converter is shown in Figure 6.9. In the circuit, the A/D conversion process begins by resetting the flip-flop to allow the clock pulses from the clock generator to enter the n-bit counter and start the counting process. Each digital representation of the count (the counter output) is presented to the digital to analog (D/A) converter as a trial voltage, where it is converted to its analog equivalent and compared to the unknown analog voltage in the analog comparator. Since the comparator changes its output only when the trial voltage exceeds the unknown voltage, the A/D converter output represents the smallest digital value that is greater than the unknown voltage within the limits of the inherent errors in the converter, as discussed in the preceding section.

Timing. In the counter converter, the time of the conversion process is variable and is a function of the magnitude of the input voltage. The worst case maximum conversion time, T_{max}, can be calculated knowing the number of bits, n, of the counter output and the clock frequency, f_c.

$$T_{max} = \frac{2^n}{f_c}$$

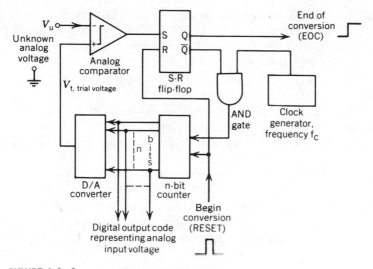

FIGURE 6.9 Counter A/D converter block diagram.

For example, an 8-bit counter converter with a clock frequency of 2 MHz requires 0.256 ms maximum for each conversion, which is equivalent to 3900 conversions per second. Also, these calculations assume that the input voltage, V_u, is held constant during the conversion process. If the voltage is varying, a *sample and hold* circuit can be used to acquire a sample of the unknown voltage and present a steady value to the A/D converter.

SUCCESSIVE APPROXIMATION A/D CONVERTER

A *successive approximation* A/D converter is a more sophisticated version of the counting converter that has a fixed conversion time, independent of the input voltage. In the successive approximation technique, digital representations of trial voltages are presented to the D/A by the *successive approximation logic* (SAL) of the A/D converter in a *binary search* pattern. Upon initiation of a conversion, the *most significant bit* (msb) of the SAL is presented to the D/A converter. This value normally represents $\frac{1}{2}$ of the full-scale voltage, V_{FS}. To obtain the proper input/output transfer function, an offset is added to the D/A converter, as was shown in Figures 6.3 and 6.5. This offset is equal to $\frac{1}{2^{n+1}}$ for an n-bit A/D converter. Thus, when the msb of the n-bit SAL output is set to a "1" and the other $n - 1$ bits are "0," a value of $\left[\left(\frac{1}{2}\right) - \left(\frac{1}{2^{n+1}}\right)\right] V_{FS}$ is seen by the D/A. This digital code is converted to an equivalent analog voltage by the D/A and presented as a trial voltage to the analog comparator. If the trial voltage is greater than the unknown voltage, V_u, as sensed by the SAL by means of the comparator output, the digital value to the D/A is decremented by an amount equal to $V_{FS}/4$. Conversely, if the trial voltage is less than V_u the digital value to the D/A is incremented by $V_{FS}/4$. The new analog voltage out of the D/A and V_u are again compared and the output of the SAL to the D/A is incremented or decremented by $V_{FS}/8$. This procedure continues at every clock cycle with the last increment/decrement occurring after n clock cycles when the factor is equal to $V_{FS}/2^n$. Note that this is in contrast to 2^n clock cycles for the counter A/D. For a 12-bit A/D conversion, then, the successive approximation converter can convert $2^{12}/12$ or 341 times faster than the counter A/D. Figure 6.10 gives a block diagram of a successive approximation A/D and Figure 6.11 is a diagram of the convergence of voltages of a 4 bit successive approximation A/D converter. The convergence follows the darkened path in Figure 6.12, which shows all combinations of the binary search for a 4-bit successive approximation A/D converter. Again, the assumption for these conversions is that the unknown analog input voltage remains constant or is fed through a sample and hold circuit.

FLASH CONVERTER

Even though the successive approximation A/D has improved speed over the counter converter, there are always instances where a yet higher conversion rate is required. A method of achieving this higher speed is through the flash or parallel converter. With this converter, the search to determine if each bit of the trial digital value is a 1 or 0 is accomplished in parallel and the time of conversion is determined by the delay through the circuit logic elements. The comparisons are made as shown by the 3-bit converter logic diagram in Figure 6.13. In this circuit, assuming high impedance (no loading)

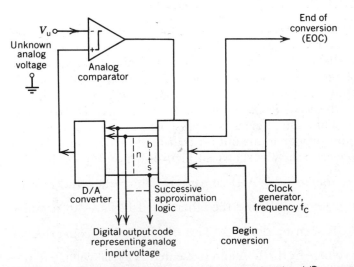

FIGURE 6.10 Block diagram of a successive approximation A/D converter.

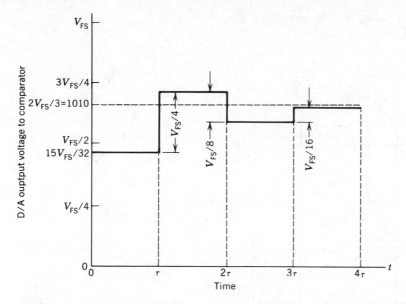

FIGURE 6.11 Convergence of 4-bit successive approximation A/D converter to 1010 using binary search.

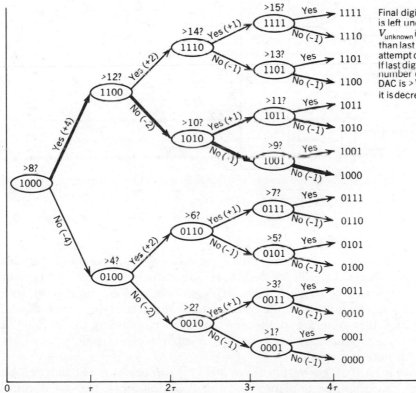

FIGURE 6.12 All combinations of binary search for 4-bit successive approximation A/D. Final digital output is left unchanged if $V_{unknown}$ is larger than the last digital attempt out of DAC. If the last digital number out of DAC is greater than $V_{unknown}$, it is decreased by 1.

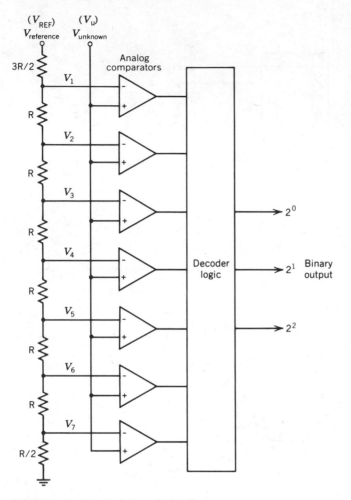

FIGURE 6.13 3-bit flash (parallel) A/D converter.

inputs into the comparators, the unknown voltage, V_u, is compared against seven trial voltages, $V_1 - V_7$, to determine the value of V_u. With the resistor values as shown, $V_1 = (\frac{13}{16})V_{REF}$, $V_2 = (\frac{11}{16})V_{REF}$, $V_3 = (\frac{9}{16})V_{REF}$, and so on, with $V_7 = (\frac{1}{16})V_{REF}$. The outputs of the comparators are fed into the combinational decoder logic that generates the binary output corresponding to the value of V_u. The resistor values are such that the comparators' transition points are spaced 1 lsb apart.

For an n-bit flash converter, there are 2^{n-1} comparators needed. A 4-bit converter can, then, be implemented with 8 comparators, but the requirements worsen as n increases. For example, an 8-bit converter requires 128 comparators and corresponding reference voltages. An alternative is to use

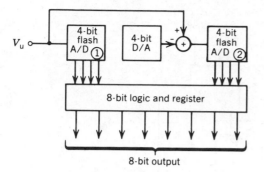

FIGURE 6.14 8-bit A/D using 4-bit flash A/D's.

two, 4-bit flash converters to develop an 8-bit A/D converter as shown in Figure 6.14.

In Figure 6.14, V_u is converted with 4-bit resolution in A/D ① and this result is converted back to analog form by the D/A. The analog output of the D/A is subtracted from V_u and the remainder is converted by A/D converter ②. The binary outputs of both A/D's are presented to the 8-bit logic and register where they are presented as an 8-bit output representing the analog input voltage.

Flash converters are hardware intensive but have the ability to convert at rates around 10^8 conversions per second.

INTEGRATING A/D CONVERTERS

Another method of conversion integrates the unknown input voltage and converts its value into a time period that is measured by a counter. There are three principal types of integrating A/D converters: the *single-slope, dual-slope,* and *triple-slope.* These types are also known as *single-ramp, dual-ramp,* and *triple-ramp,* respectively.

Single-Slope (Single-Ramp). The single-slope converter uses a varying reference signal of fixed slope that begins at some negative value, crosses zero volts and increases until it matches the unknown input voltage. The time from the zero volt crossing of the ramp signal to its matching the unknown input voltage is proportional to the input voltage. A digital representation of this time and, hence, the input voltage is obtained by starting a fixed-rate counter when the ramp voltage is zero and stopping the counter when the ramp voltage matches the unknown input. The value of the count is the digital output of the A/D. Figure 6.15 shows the block diagram and timing diagram of a single-slope A/D converter. The count is started by comparator ① when the constant slope ramp crosses zero volts and is stopped by comparator ② when the ramp voltage matches the unknown voltage. The accumulated count on the n-bit counter during this interval is proportional to V_u.

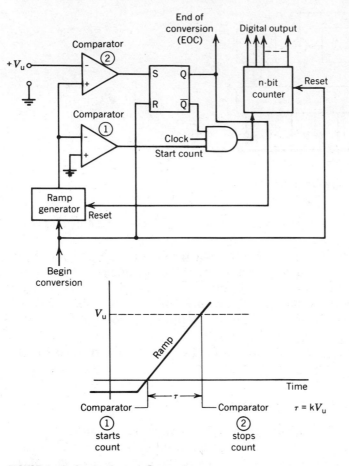

FIGURE 6.15 Single slope A/D converter.

The time of conversion of the single-slope converter is variable and is a function of the input voltage. Also, the ramp is usually generated by an RC network that is sensitive to temperature variations. The single-ramp converter has essentially been replaced by the dual-ramp converter that overcomes some of the drawbacks of the single ramp method.

Dual-Ramp (Dual-Slope) Converter. The dual-slope converter eliminates the temperature sensitivity problems of the single-slope converter by making the unknown voltage determination independent of the clock stability and long-term variations of the value of the integrating capacitor. This characteristic is achieved by making relative measurements instead of absolute measurements. A typical dual-slope integrating A/D converter circuit is shown in Figure 6.16. In the dual-slope technique, there are two periods involved in the conversion. In the first period, the unknown voltage charges

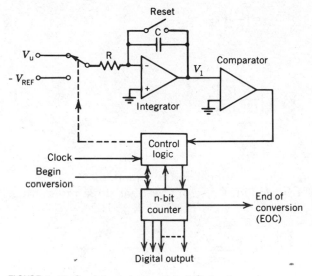

FIGURE 6.16 Dual-slope integrating A/D circuit.

the capacitor, C, for a fixed interval of time, t_1. Time t_1 is the interval from begin conversion to overflow of the n-bit counter. Then the control logic switches the integrator input from the unknown voltage, V_u, to the negative reference voltage, V_{REF}. The capacitor is then discharged while the counter starts again from zero until the output of the integrator crosses the zero voltage point. This crossing is detected by the comparator which, in turn, ends the integration and stops the counting. This second time period is t_2. The counter digital output is, then, a function of the input voltage.

The circuit defines that the integrals of the voltages of the two time periods must be equal as shown in Figure 6.17. Thus

$$\frac{1}{RC} \int_0^{t_1} V_u \, dt = \frac{1}{RC} \int_{t_1}^{t_1+t_2} V_{REF} \, dt$$

If the clock frequency is f_c, then for the n-bit counter,

$$t_1 = \frac{2^n}{f_c}$$

Similarly, if the accumulated count at the end of interval t_2 is X, then

$$t_2 = \frac{X}{f_c}$$

The average value of V_u is proportional to the ratio of t_2 to t_1, or

$$\overline{V}_u = \left(\frac{t_2}{t_1}\right) V_{REF}$$

Substituting for t_2 and t_1 gives

$$\overline{V}_u = \left[\frac{(X/f_c)}{(2^n/f_c)}\right] V_{REF}$$

$$= \left(\frac{X}{2^n}\right) V_{REF}$$

Note the values of R, C, and f_c do not appear in the equation.

The maximum time of conversion, t_{max}, for the dual slope integrating A/D converter is equal to $t_1 + t_2$ or $(2^n + X)/f_c$. Since X_{max} is 2^n,

$$t_{max} = \frac{2^n + 2^n}{f_c}$$

$$= \frac{2^{n+1}}{f_c}$$

For a 10-bit A/D converter with a 2-MHz clock,

$$t_{max} = \frac{2^{11}}{2 \times 10^6}$$

$$= \frac{2048}{2 \times 10^6}$$

$$= 1.024 \text{ ms}$$

This time yields a minimum conversion rate of 976.6 conversions per second.

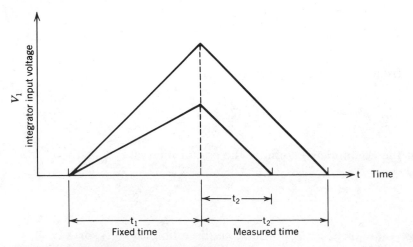

FIGURE 6.17 Dual-slope A/D timing diagram.

TABLE 6.1 A/D Error Calculations

Error Source (From Data Sheet)	Percent Error
Quantization error ($+ \frac{1}{2}$ LSB)	0.012
Gain error (35 ppm/°C at 25°C $\times$ 45°C)	0.1575
Offset error (10 ppm/°C at 25°C $\times$ 45°C)	0.0450
Differential linearity error ($\pm \frac{1}{2}$) LSB)	0.012
Differential linearity error with temperature	
(5 ppm/°C at 25°C $\times$ 45°C)	0.0225
Power supply voltage sensitivity (0.005%)	0.005
Total worst case error	0.254

The dual-slope integrating A/D converter provides a high degree of accuracy, since it averages the input over a fixed period of time, t_1. If t_1 is selected to be a multiple of the period of a signal of frequency f_t, then the converter will be essentially immune to interference from that signal, since its average over time t_1 will be zero. Thus, if $t_1 = \frac{1}{60}$ Hz or 0.0166 s, the converter will have immunity to power line frequency interference. Because of their accuracy, dual-slope integrating converters are widely used in panel meters.

Other refinements to the dual-slope method, such as triple and quad slope, are used to eliminate offset errors and reduce conversion time.

6.1.5 EXAMPLE ERROR CALCULATIONS

In using A/D converters, not only resolution error, but also the cumulative errors must be considered in estimating overall conversion accuracy. The example in Table 6.1 using specifications for 25°C from a data sheet of a 12-bit A/D converter operating to 70°C calculates the worst case error ($\delta T = 45°C$).

If statistical addition is used, assuming that errors are not all additive in the same direction, we have the error as:

$$[(0.012)^2 + (0.1575)^2 + (0.0450)^2 + (0.012)^2 + (0.0225)^2 + (0.005)^2]^{\frac{1}{2}}$$

or 0.16628%. If resolution alone (1 lsb) were used to estimate the error, the result would have been 0.024%. The statistical and worst case errors are approximately 7 and 10 times the resolution error, respectively.

6.2

DIGITAL-TO-ANALOG (D/A) CONVERTERS

Digital-to-analog converters perform the transformation from the digital domain to the analog domain. They take digital inputs and generate an analog

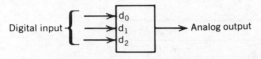

FIGURE 6.18 Block diagram of 3-bit D/A converter.

voltage or current as an output. Figure 6.18 shows a general block diagram of a 3-bit D/A converter; Figure 6.19 illustrates its ideal output characteristic.

The D/A converters can be implemented in a number of ways, depending upon the accuracy and temperature stability required. Two popular methods are the weighted current source and the R-2R resistor network.

6.2.1   WEIGHTED CURRENT SOURCE

The weighted current source can be easily visualized using a voltage switching approach and then applying the same techniques to current switching. Figure 6.20 depicts a simple 3-bit D/A using weighted voltage switching.

In Figure 6.20, the switches represent digital inputs to the D/A converter. When all switches are open, representing an input of 000 to the converter, the output analog voltage is zero. If bit 0 is closed, representing a digital input of 001, the output analog voltage is $(R/4R)V_{REF}$ or $V_{REF/4}$. If bit 0 and bit 1 are closed, the digital input is 011 and the analog output voltage is $\left(\dfrac{1}{4R} + \dfrac{1}{2R}\right) RV_{REF}$ or $3V_{REF}/4$. Table 6.2 gives the analog output voltages corresponding to the eight possible digital inputs to the 3-bit D/A converter.

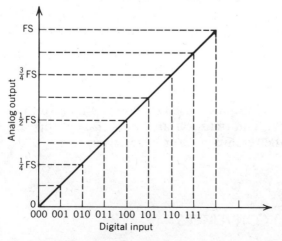

FS = full scale

FIGURE 6.19 D/A converter ideal input/output characteristic.

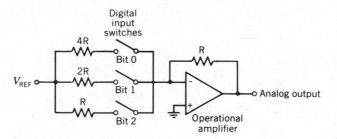

FIGURE 6.20 Simple 3-bit D/A converter.

Most D/A converters switch currents instead of voltages, since currents can be controlled and switched more accurately. Then, a current to voltage converter is used in the last stage to produce a voltage output. A weighted current source converter corresponding to the previous weighted voltage source example is shown in Figure 6.21.

In the weighted current source D/A converter, the bipolar transistors are current sources biased at constant voltage V_b. The currents through the current source transistors are held constant by the current reference circuit, which requires a stable reference voltage, V_{REF}. If a logic input is a positive voltage (logic 1), the current flows through the transistor to the current summing junction. If a logic input is zero volts (logic 0), the current is diverted or switched through the diode to ground and is not summed. The currents are summed at the operational amplifier summing junction and converted by the operational amplifier to an output voltage.

6.2.2   R-2R NETWORK D/A CONVERTER

The weighted current source approach runs into problems when higher resolution D/A conversion is required. For each additional input bit, a higher

TABLE 6.2 Analog Output Voltages for 3-bit D/A Converter

Digital Input Bit			Analog Output Voltage
2	1	0	
0	0	0	0
0	0	1	$V_{REF}/4$
0	1	0	$V_{REF}/2$
0	1	1	$3V_{REF}/4$
1	0	0	V_{REF}
1	0	1	$5V_{REF}/4$
1	1	0	$3V_{REF}/2$
1	1	1	$7V_{REF}/4$

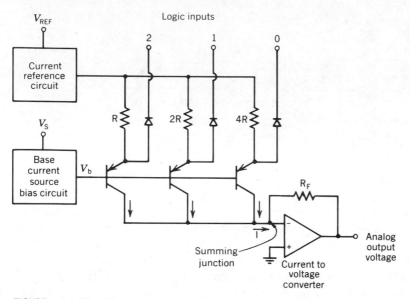

FIGURE 6.21 Weighted current source 3-bit D/A converter.

value resistor is needed in the network. For an 8-bit D/A, the resistor values would range from R to 128 R. This wide range of values causes difficulties in monolithic converter designs, limits switching speed, and is more sensitive to temperature variations. An alternative design is the R-2R approach that requires only two values of resistance for all practical bit resolutions. Figure 6.22 shows a general R-2R ladder D/A converter. It can be seen that the resistance to the right of any R, 2R resistor junction ($\textcircled{1} - \textcircled{n}$) is 2R. Thus the current entering any junction splits equally in a binary fashion down the ladder network as shown in Figure 6.22. If the transistor switches steer the current into the summing junction of the operational amplifier, the current is added to

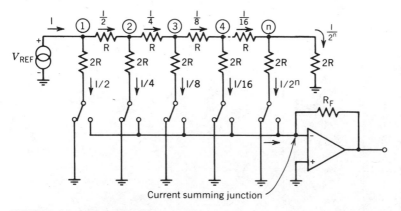

FIGURE 6.22 R-2R ladder D/A converter.

produce the analog voltage out. Since the resistance at node ① is 2R in parallel with 2R, or R, I is equal to V_{REF}/R. For all switches connected to the summing junction (all inputs = logic 1), the current at the summing junction is

$$\left(\frac{V_{REF}}{R}\right)\left(\frac{1}{2} + \frac{1}{4} + \frac{1}{3} + \frac{1}{16} + \cdots + \frac{1}{2^n}\right)$$

6.3

INTERFACING A/D AND D/A CONVERTERS

Now that we have discussed A/D and D/A converters, the question is, "How can these devices be connected to a microcomputer?" A good approach is to utilize an input/output chip, usually referred to as an I/O adapter, to interface peripheral devices, such as A/D and D/A converters to the microprocessor. The I/O adapter is selected by sending out its address on the address bus and then using control signals to either write to, or read from, the peripheral device. Figure 6.23 is a general diagram of this interface for an A/D converter.

A begin conversion signal is generated by the I/O adapter when the correct address (determined by the address decoder) transmitted by the microprocessor activates the I/O adapter. After conversion, the end of conversion signal sets a flag in the I/O adapter that indicates the A/D digital output is available for reading. This flag can then be read by the microprocessor when it polls its various I/O adapters in a sequential polling procedure. In addition, the I/O adapter can be programmed to send an interrupt signal to the microprocessor on one of the control lines, requesting immediate service and not waiting for the microprocessor to poll.

6.3.1 DATA ACQUISITION SYSTEMS

The A/D converters are used with other components to form data acquisition systems. A typical data acquisition system is given in Figure 6.24. The analog

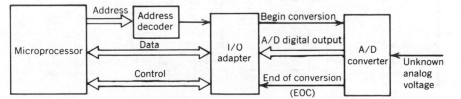

FIGURE 6.23 A/D converter interface to microprocessors.

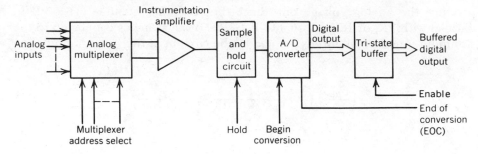

FIGURE 6.24 Typical data acquisition system.

multiplexer selects one of a number of analog inputs to be sent to the A/D converter through the instrumentation amplifier and sample and hold circuit. The address select on the multiplexer determines which of the analog inputs is to be sampled and converted. The instrumentation amplifier reduces common-mode noise that may be imposed upon the unknown analog voltage and provides amplification of the analog voltage. The sample and hold circuit samples the unknown analog voltage and holds the sampled value so that the A/D converter sees a stable, essentially nonchanging input while it is converting. Tristate buffers provide the means to connect a number of data acquisition systems to the same bus without causing circuit damage or contention problems. In addition to having conventional 1 and 0 outputs, the tristate buffer has a third, high impedance state. In this state, the buffered digital outputs are effectively disconnected from the outside world. The buffer outputs are connected to the external world when the enable line is activated. Thus, in a multiple data acquisition system application, only one enable line is permitted to be active at a time. The desired enable line is also selected by means of address decoding. A typical configuration is shown in Figure 6.25.

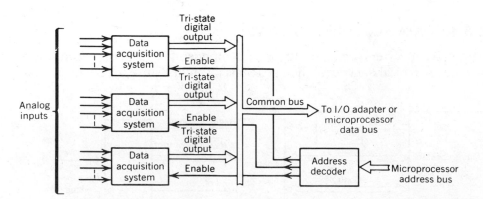

FIGURE 6.25 Multiple data acquisition system interfacing using tri-state buffers.

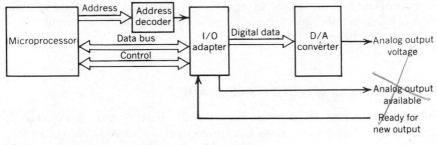

FIGURE 6.26 Microprocessor D/A converter interface.

6.3.2 D/A CONVERTER INTERFACE

In the D/A interface of Figure 6.26, the address and digital data are presented to the address and data bus, respectively, by a microprocessor OUTPUT, LOAD, or MOVE type of instruction. The address decoder selects the I/O adapter corresponding to the D/A. The I/O adapter then presents the digital data to the D/A converter, which converts the digital input to the analog output. The I/O adapter can also communicate with the outside world and

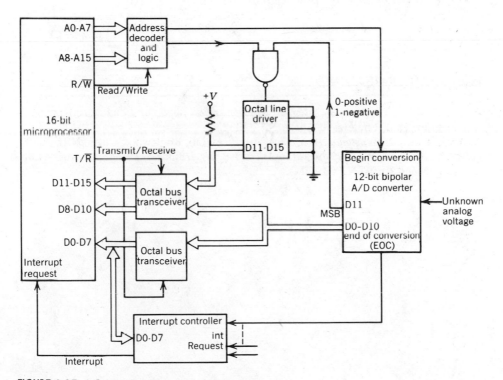

FIGURE 6.27 A/D interface with 16-bit microprocessor.

the microprocessor control bus to perform "handshaking" functions. The "analog output available" and "ready for new output" signals are examples of handshaking functions.

6.3.3   ADDITIONAL A/D INTERFACE EXAMPLE

Another A/D interface is given in Figure 6.27. In this case, a 12-bit A/D converter is interfaced to a 16-bit microprocessor. The A/D converter can read both positive and negative voltages with negative voltages represented in two's complement form. Thus, when the analog voltage is positive, indicated by the A/D msb D11 = 0, the "filler" bits D11–D15 presented to the microprocessor data bus should be all 0s. Conversely, if the analog voltage is negative, bits D11–D15 should all be 1s. This setting of bits D11–D15 is accomplished by the octal line driver that uses msb D11 from the A/D to generate all 0s or all 1s on data lines D11–D15. The direction of the data bus transceiver is determined by the transmit/receive bit generated by the microprocessor that indicates whether data are being written to, or read from, the data bus lines, D0–D15. By using an interrupt controller chip, the end of conversion indication from the A/D converter can be used to generate an interrupt to the microprocessor to read the A/D output on data bus inputs D0–D15.

FOR FURTHER READING

Andrews M. A., *Programming Microprocessor Interfaces for Control and Instrumentation*, Prentice-Hall, Englewood Cliffs, N.J., 1982.

Artwick B. A., *Microcomputer Interfacing*, Prentice-Hall, Englewood Cliffs, N.J., 1980.

Draft, G. D., and W. N. Toy, *Mini/Microcomputer Hardware Design*, Prentice-Hall, Englewood Cliffs, N.J., 1979.

7

LOCAL AREA NETWORKS

Local area networks, *LANs,* are becoming increasingly important as a means of connecting a wide variety of devices such as computers, printers, intelligent terminals, and so on. Local area generally refers to offices, buildings, and groups of buildings.

As microcomputers proliferate in office equipment, more and more data will originate and be transferred within the local area. In fact, surveys have shown that, in general, 80% of the information utilized in a local environment is generated within the environment.

It is important to recognize that protocols such as HDLC/SDLC discussed in Chapter 4 and LAN technology are part of the larger picture of system interconnection. In other words, how can computers, terminals, disks, robots, and so on, produced by different manufacturers talk to each other? At the present time, most of them cannot talk to each other. Also, strictly speaking, a LAN is different from a network formed by local private branch telephone exchanges (PBXs), modems, and telephone lines. In general, communication on a LAN is serial in nature and the transmission medium (wire, coaxial cable, optical fiber, etc.) is shared by the network devices. The transmission data rate is higher in a LAN, typically 10M bits/s as opposed to typically 9.2K bits/s in a conventional network. In addition, typical data exchange in a LAN is peer-to-peer, wherein any element on the network can communicate directly with any other element on the network. This communication does not have to go through a central switching device as in a PBX-based network.

It is obvious that some standards for the different aspects of LANs would be useful in allowing products of different manufacturers to communicate directly in a network. This issue will be discussed in detail in Section 7.3.1. First, definitions and establishment of terminology are in order.

7.1

TOPOLOGY

A network can be categorized by its topology. Points on the network are referred to as *nodes* or *stations.* A node or station is a basic unit of information processing that has a specific address. A printer, word processing intelligent terminal, and a computer are examples of nodes or stations. Nodes are connected by communication paths called *links*. There are three fundamental methods of interconnection among nodes. These methods are the *star, ring,* and *bus* configurations.

7.1.1   STAR

The star configuration, as shown in Figure 7.1, has all nodes joined at a central location. The control of communications among the nodes in a star network can reside at the common node, at one of the other nodes, or at each node in a distributed fashion.

7.1.2   RING

In the ring topology, the communication nodes and links form a closed path as shown in Figure 7.2. Messages travel around the ring addressed to a specific node. A characteristic of the ring topology is that a node must compare its

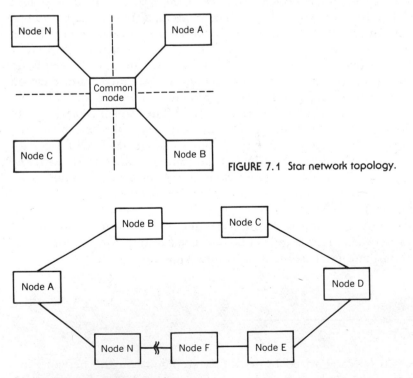

FIGURE 7.1 Star network topology.

FIGURE 7.2 Ring network topology.

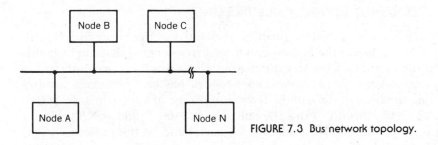

FIGURE 7.3 Bus network topology.

address with the address of each message traveling around the ring to determine if it is the destination of the message. If a node is not the destination, it must pass the message along to the next node on the ring. Control in a ring network may be distributed or exercised by a central control node. A ring with central control is also called a *loop*. An advantage of the ring is that message routing is simplified, since a node merely passes a message on to its neighbor. Disadvantages are that a failure in a node can interrupt the message traffic and the addition of a new node requires "breaking" the ring and making two connections to the new node. Bypassing techniques can be used to circumvent a failed node.

7.1.3 BUS

The *bus* topology can be characterized as a *broadcast* approach. There is a single physical medium that is shared by the nodes through taps into the medium. Figure 7.3 shows a bus network. Nodes can be added easily to a bus network, thus making this topology popular for use in LANs. Distributed control is the most sensible approach to the use of bus networks.

7.2

RELATED COMMUNICATIONS TERMINOLOGY

Information can be transmitted on the physical medium at maximum rates determined by the medium. Analog signal transmission rates are specified in terms of frequency (hertz, Hz) and digital transmission rates in bits per second. For example, coaxial cables can provide useful transmission of frequencies up to 300 megahertz (MHz), twisted pair up to 2 MHz, and optical fibers up into the gigahertz (GHz) region.

In order to obtain efficient utilization of the transmission medium, ways of sharing its message transmission capacity have been developed. The three main methods of accomplishing this sharing are *frequency division multiplexing, (FDM), time division multiplexing, (TDM),* and *space division multiplexing (SDM).*

7.2.1 FREQUENCY DIVISION MULTIPLEXING

The *bandwidth* of a transmission medium is defined as the range of frequencies (difference between the highest and lowest frequencies) that can be transmitted on the medium. Consider the twisted pair wires used presently in the existing voice telephone network. This network has been designed around transmitting the human voice in the frequency range of approximately 200 to 3400 Hz. The bandwidth of this transmission is $3400 - 200 = 3200$ Hz. In practice, a range of 4000 Hz is allotted for each band on the public-switched telephone network. If we assume that there is a way to allocate this 4000-Hz band across the 2-MHz bandwidth of the twisted pair wires, $2 \times 10^6/4000$ or 500 simultaneous phone conversations could take place. This allocation is possible by raising the voice frequencies to higher frequency ranges by *modulating* higher frequency carriers with the voice frequencies. By this means, a number of frequency *bands* are transmitted across the medium simultaneously. Conversely, upon reception, the frequencies are *demodulated* and converted back to the original 200 to 3400-Hz voice frequencies. This technique is defined as *frequency division multiplexing (FDM),* and it divides the existing frequency capacity of a transmission medium into separate frequency bands for simultaneous transmission. The equipment that modulates the voice frequencies on the transmitting end and demodulates these frequencies on the receiving end is called a *modem* (from *modulator–demodulator*). In FDM there are three basic modulating techniques: amplitude, frequency, and phase. Illustrations of each technique are given in Figure 7.4.

7.2.2 TIME DIVISION MULTIPLEXING

Another way to share the transmission medium is *time division multiplexing (TDM).* As the name implies, transmitting stations timeshare their transmissions based on some type of allocation basis. In TDM, portions of messages from each station are transmitted in time sequence following each other and must be sorted at the receiving end. Figure 7.5 illustrates the concept.

7.2.3 SPACE DIVISION MULTIPLEXING

A third and obvious approach to providing for multiple, simultaneous transmissions is to allot a separate set of wires to each station. Thus each station could transmit or receive independently. Since space for wires is being allocated, this approach is referred to as *space division multiplexing, (SDM).* As we noted earlier, twisted pair wires have a bandwidth of 1–2 MHz and a voice band only requires 4000 Hz. Therefore, SDM is not an efficient means of utilizing wires for communication but is used by necessity in homes and offices.

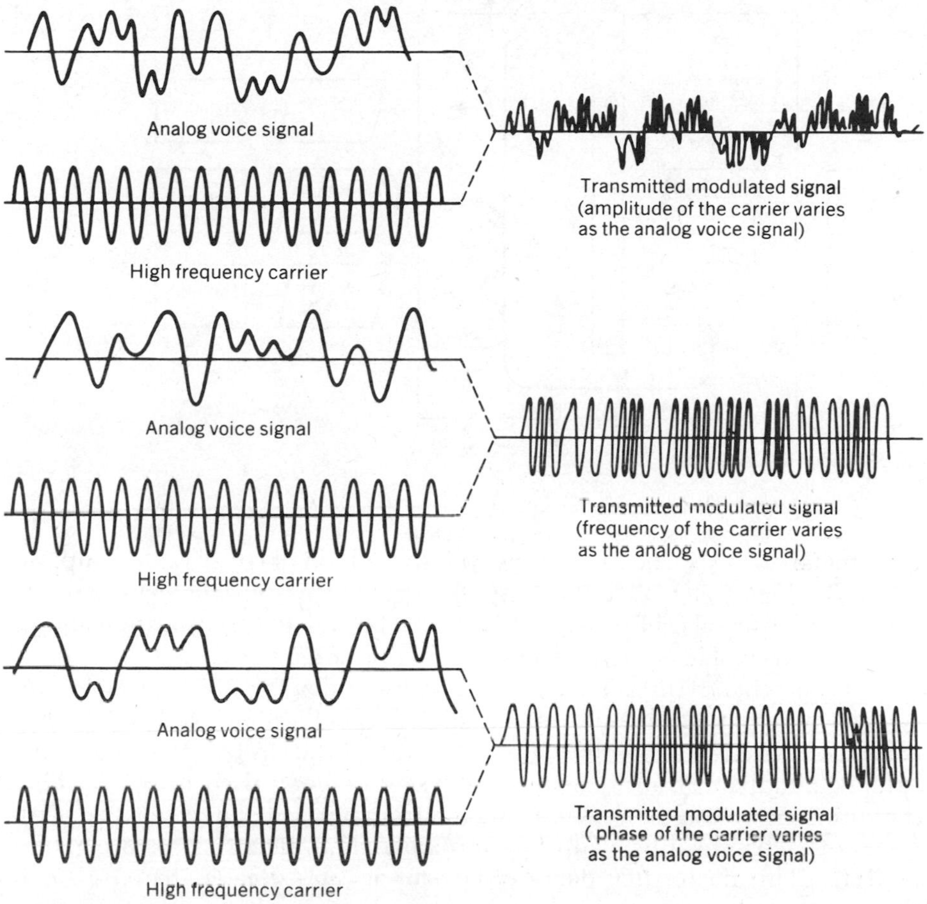

Analog voice signal

High frequency carrier

Transmitted modulated signal
(amplitude of the carrier varies
as the analog voice signal)

Analog voice signal

High frequency carrier

Transmitted modulated signal
(frequency of the carrier varies
as the analog voice signal)

Analog voice signal

High frequency carrier

Transmitted modulated signal
(phase of the carrier varies
as the analog voice signal)

FIGURE 7.4 Three basic methods of modulation.

7.2.4   BROADBAND AND BASEBAND APPROACHES

The terms *broadband* and *baseband* are closely associated with the bandwidth or transmission capacity of the medium but are really transmission approaches or philosophies. In the baseband mode the frequencies representing the information to be transmitted are placed on the medium without being shifted to a higher frequency range by modulating a carrier frequency. Thus only a single valid transmission can take place on the medium at any given time. The broadband approach supports multiple, simultaneous transmissions on the medium by shifting the messages to different frequency bands by modulation techniques, thereby dividing the bandwidth of the median into separate "slices." The medium commonly used for baseband and broadband transmissions is coaxial cable. Because of its high volume use in telephony and the Community Antenna TV (*CATV*) industry, coaxial cable is readily available at

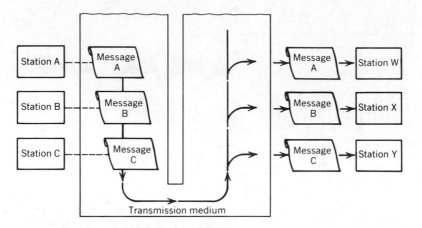

FIGURE 7.5 TDM example.

reasonable cost. Also, because of the widespread use of coaxial cable, supporting technologies such as couplers, splitters, and taps are well developed. The
$\frac{1}{2}$-in. CATV coaxial cable is typically used for broadband transmission and the
0.4-in. coaxial cable is used for baseband application.

The baseband transmission rate is approximately 10 M/bits/s for one
signal, whereas multiple simultaneous digital signals can operate in the broadband mode over the 300–400 MHz bandwidth of the CATV cable. To further
extend broadband transmission capacity, cables comprised of optical fibers
that transmit messages by conducting light can be used. These glass or plastic
fibers offer bandwidths of up to approximately 3 billion Hz or 3 gigahertz
(3 GHz). This means that data can be sent at rates greater than 1 gigabit/s
baseband on fiber optics and tens of thousands of simultaneous messages can
be transmitted by broadband. In addition, since optical fibers transmit light,
they are immune to conventional electrical noise and do not generate electromagnetic fields.

7.3

ARCHITECTURE

Since LANs are a communication network, they should be viewed in terms of
a general, standard *network architecture*. This network architecture can be pic-
tured as a group of layers, ranging from physical interconnections to operating system-like functions. These layers contain *modules* that perform specifically defined functions that implement the layer's tasks. In a particular layer,
the set of rules that allows cooperation with the same layer in another system
is called a *protocol*. The relationship between modules in the same layer or

different layers on a network is defined as an *interface*. The relationships are shown in Figure 7.6.

7.3.1 INTERNATIONAL STANDARDS ORGANIZATION OSI MODEL

One widely accepted network architecture model is the *International Standards Organization (ISO) Open Systems Interconnection (OSI)* model. This model, given in Figure 7.7, is a seven-layer structure in which messages originating in an upper layer are transmitted down through the lower layers to the physical transmission medium. The reverse procedure occurs at the receiving device. The message may originate and terminate at different levels of the architecture.

The OSI model does not give the specific details for implementation of each layer but provides a framework for standards to be developed for each layer. A catalyst that assisted in the development of the OSI model was the

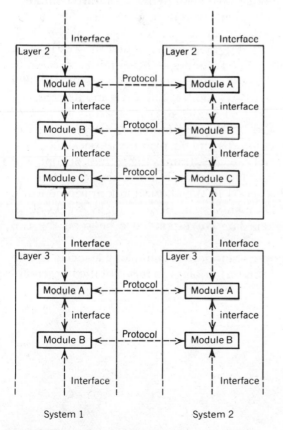

FIGURE 7.6 Layer, module, interface, and protocol relationships.

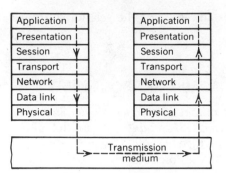

FIGURE 7.7 ISO open system interconnection network architecture model.

U.S. National Bureau of Standards (NBS) Institute for Computer Sciences and Technology (ICST). The ICST also developed test methods for determining if an implementation of a part of the OSI framework is correct and also for measuring performance of prototype implementations. Standards for some of the lower layers in the model have been defined and are complete, whereas standards for some of the higher levels do not yet exist.

To focus on possible LAN standards, the Computer Society of the IEEE (Institute of Electrical and Electronic Engineers) established the IEEE Standards Project 802 in February 1980. The resulting 802 Committee decided to work within the ISO OSI layered architecture model. As a result, the 802 standard has identified a three-level LAN model. The LAN model is associated with the two bottom levels in the OSI model, the *data link* and the *physical levels*. The physical layer makes and breaks the connection between devices involved and transmits the information, 1's and 0's, over the physical medium. The data link layer allocates channel capacity, frames the *packets* (portions) of the message to be sent, determines which station is addressed to receive the message, detects errors, and gains access to the physical link in the network. Thus the data link layer in the ISO Open System Model actually breaks down into two layers in the LAN model. The correspondence between the two models is given in Figure 7.8.

The 802 Committee also adopted standard terminology associated with the LAN model and has described this terminology in terms of the implemen-

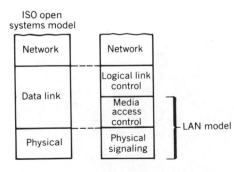

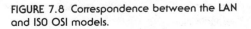

FIGURE 7.8 Correspondence between the LAN and ISO OSI models.

tation reference model of Figure 7.9. The Committee further proposed and adopted standardization on two medium access schemes: *token passing* and *carrier-sense multiple-access with collision detect (CSMA/CD)*. With token passing, each station gains access to the medium on a temporary exclusive basis by means of a single token or message that is passed around to each station and provides the station with authority to transmit. In CSMA/CD, multiple stations can try in an unsynchronized manner to access the medium. Collisions (attempts at simultaneous transmission by different stations) are detected and retransmission is accomplished by detection of the transmitting carrier and deference to a previously transmitting station. Both medium access control techniques will be covered in detail in following sections of this chapter. Specifically, if topology is considered, the 802 standard defines three medium access control methods, two incorporating token passing and one utilizing CSMA/CD. These methods are the token ring (IEEE Stand. 802.5 or ISO Stand. 8802/5) used in a ring topology, the token bus (IEEE Stand. 802.4 or ISO Std. 8802/4) used in a bus (broadcast) topology, and CSMA/CD (IEEE Stand. 802.3 or ISO Std. 8802/3).

The higher levels of the ISO OSI model will not be emphasized in this development because they are not part of the LAN model. In general, the higher level layers perform the operations necessary to have an intelligent operator utilize the network and implement applications. A brief summary of the individual level functions of the OSI model is given in Table 7.1.

Before leaving the higher level layers of the OSI model, it is interesting to relate the OSI model, the LAN-related layers, and the wide area network (*WAN*)-related layers. We define a WAN as one in which communications can occur among devices around the world through telephone and satellite links.

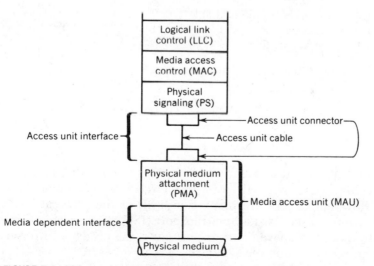

FIGURE 7.9 802 standard implementation reference model.

TABLE 7.1 Summary of ISO OSI Network Architecture Model Level Functions

ISO OSI Levels	General Characteristics
Application	User-written programs
	File transfers
	Resource sharing
	Network management
	Access to remote files
	Database management
	Planning network operation
	Control of network
	Operation of network
Presentation	Defines I/O procedures
	Controls network functions of application layer
	System-dependent process-to-process communication
	User application connections
Transport	Provides location-independent transport of packets
	Provides this end-to-end communications control, once data path has been established
Network	Distributed control policy that can allocate network management to different systems on the network
	Addresses messages
	Sets up paths between nodes
	Controls message flow between nodes
	Provides control and observation function for network planning and operation
Data link	Frames message packets
	Allocates channel capacity
	Determines station addresses to receive
	Error detection
	Establishes access to physical link
Physical	Provides electrical transmission of information, 1's and 0's, over the physical medium
	Encoding
	Decoding
	Physical connection
	Signaling

This relationship is illustrated in Figure 7.10. In Figure 7.10, the standards relating to individual layers are cited, along with levels that are common and levels that are different. There are many standards at the physical layer, reducing to one common standard at the transport (ISO 8073) layer. As one proceeds up into the layers above the transport layer, the needs of different applications again result in multiple standards at a given layer.

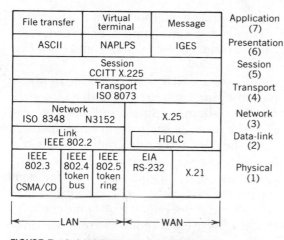

File transfer	Virtual terminal	Message	Application (7)
ASCII	NAPLPS	IGES	Presentation (6)
Session CCITT X.225			Session (5)
Transport ISO 8073			Transport (4)
Network ISO 8348 N3152	X.25		Network (3)
Link IEEE 802.2	HDLC		Data-link (2)
IEEE 802.3 CSMA/CD	IEEE 802.4 token bus	IEEE 802.5 token ring	EIA RS-232 X.21	Physical (1)

|←————LAN————→|←————WAN————→|

FIGURE 7.10 LAN/WAN layer relationships.

7.3.2 MEDIUM ACCESS OPTIONS

One of the fundamental considerations in the design or selection of a LAN is the means of accessing the network by multiple stations. There are two major techniques of access, namely *contention* and *polling* (noncontention). As the name implies, contention accessing involves multiple stations vying for the network in an unsynchronized manner. Polling, on the other hand, allots time for the individual stations to have temporary, but exclusive, use of the network for a period of time. Both versions of access have their respective champions. *Ethernet,*® a contention access method developed by *Xerox Corporation,* is widely used and a common version is being advanced in the marketplace by cooperation among *Xerox, Digital Equipment Corporation,* and *Intel Corporation.* Alternatively, *IBM Corporation* is promoting token passing, for its personal computer products.

Control of the polling can be either centralized or distributed. In centralized control, a master station polls each station, providing it with an opportunity to gain access to the network. Distributed polling can take on a number of forms. One form is a type of Time Division Multiplexing (TDM). In this technique, the stations are synchronized and each is allocated a specific and exclusive time slot in which to transmit. Another distributed polling access method is token passing. Since token passing is one of the principal access means used in LANs, it will be discussed in detail.

TOKEN PASSING

In the token-passing accessing scheme, a special message or token is passed around to each station, granting the station temporary, exclusive access to transmit on the network. Unless there is a malfunction in the network, there should be only one token circulating at any given time. The 802 committee has subdivided token passing into two categories: *token ring* and *token bus*.

Token Ring.   The token ring method is based on a physical ring in which a transmitted message travels around the ring from node to node. This message originates from a node that has the right to transmit based on its possession of the key (token) at the time. As the message circulates around the ring, each node examines the address on the message to determine if it is the intended recipient. If it is, it reads and processes the message. In any event, the node passes the message to the next node in the ring. The message circulates in this manner until it reaches the originating node. The message is then deleted by the originating node so that it does not circulate indefinitely. A node can also bypass itself so that it effectively is removed from the ring. A schematic representation of a token ring and its operation is given in Figure 7.11.

Specifically, the token code is contained in the *access control field,* one of a number of fields associated with a circulating message. A node will "claim" the token only if it has a message to send. If a station has a message to send, it monitors the access control field and source address field of the presently circulating message in the ring when it arrives. If the access control field contains the correct token code and the token has been sent from the previously assigned node as determined by a source address field check, the node alters the token field. The node thereby claims the token and inserts the data to be sent after the access control field. Thus, as the message circulates around the ring, other nodes examine the access control field and determine that the token is residing in another node. When the message completes the round trip and returns to the originating node, the proper token code is inserted into the access control field and the message, now carrying the token, is sent on to the next designated node. In summary, each node has a register for its own address (my address, MA) and a register for the address of the next node (next address, NA) in the ring to which the token is to be passed. In fact, the ring is formed by this sequence of addresses. The token circulates around the ring in a deterministic fashion based on the addresses in the MA and NA registers.

A token ring has the attributes of wiring simplicity (twisted pair, usually), performance, and low-cost implementation potential. Historically, token ring concepts have existed before those of CSMA/CD or the token bus. Some performance characteristics of the token ring based on simulations[1] are given in Figures 7.12 and 7.13. These figures plot the mean transfer delay time in milliseconds versus the information throughput rate in megabits per second. Mean transfer delay time is defined as the time from the generation of a

[1] W. Bux, "Performance Issues in Local-Area Networks," *IBM Systems Journal*, Vol. 23, No. 4, 1984, pp. 357–359.

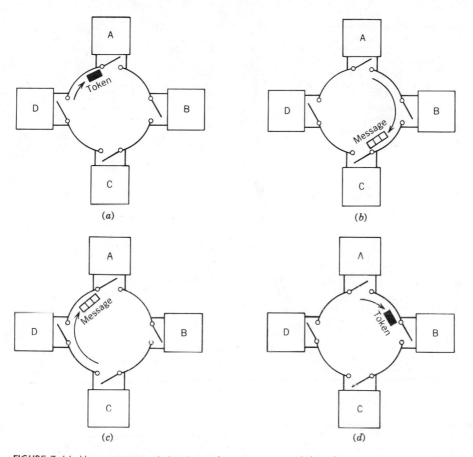

FIGURE 7.11 Message transmission in a token ring system: (*a*) node A receives token; (*b*) node A transmits message to node C. (*c*) Node C reads message and retransmits message; message eventually returns to A and is deleted from ring. (*d*) A sends token to B.

frame to be transmitted until its successful reception at the receiver. The information throughput rate is the number of bits in the information field of the transmitted frames per unit time. The parameters in Figures 7.12 and 7.13 are cable length and information field length.

Token Bus. A token bus still possesses a logical ring structure even though it physically is configured as a broadcast bus. Each node is tied to a common bus, but the token is passed around among the nodes and, thus, forms a virtual ring organization. Figure 7.14 illustrates a typical configuration with the ring developed by means of my address (MA), next address (NA) registers in each node. As in the token ring, access to the bus by a node is deterministic as opposed to a collision detection-type medium access method.

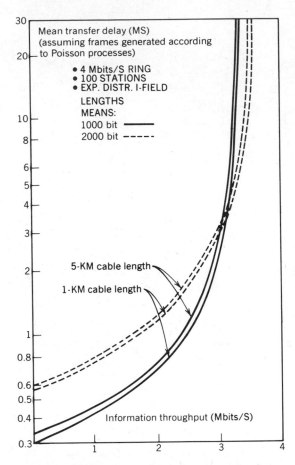

FIGURE 7.12 Token ring simulated performance characteristics (4 Mbits/s). (After Bux, "Performance Issues in Local Area Networks, *IBM Systems Journal*, Vol. 23, No. 14, 1984, p. 357.) Copyright 1984 International Business Machines Corporation. Reprinted with permission. *IBM Systems Journal, Vol. 23*, No. 4, 1984.

Western Digital Corporation has developed an NMOS *token access controller (TAC)* that implements a token bus configuration, but, at this writing, is not the IEEE standard. It was chosen for this example because its structure provides a good medium for illustrating token bus operation. In addition to the normal operation of the ring, this TAC handles the complex tasks of network initialization and recovery from failed nodes, the addition of nodes without shutting down the network, and recovery from the situation where two or more tokens are generated on the ring due to a malfunction. Two key elements in handling these exception situations are timers, referred to as TD and TA. TD is a relatively long time out unit and flags a long period of network

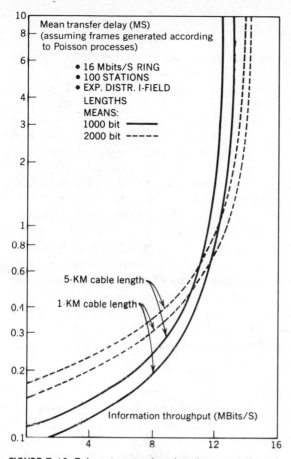

FIGURE 7.13 Token ring simulated performance characteristics (16 Mbits/s). (After Bux, "Performance Issues in Local Area Networks," *IBM Systems Journal.* Vol. 23, No. 14, 1984, p. 358.) Copyright 1984 International Business Machines Corporation. Reprinted with permission. *IBM Systems Journal,* Vol. 23, No. 4, 1984.

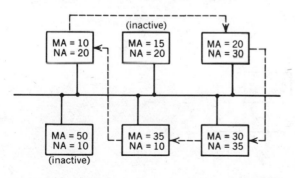

 Logical ring path.

FIGURE 7.14 Token bus configuration.

inactivity. TA times out after the maximum time for a station to acknowledge the receipt of a token transmitted by its predecessor in the ring.

A failed node can occur if the node fails to acknowledge a message from another node or if the token residing at a node is destroyed as the node malfunctions. The Western Digital TAC handles the first situation by giving the node that is awaiting the response the responsibility for recovery. If a response is not received by the waiting node before the time out of timer TA, this node seeks a successor to the failed node by polling. To accomplish this, the node sends a message to, and seeks a response from, nodes with successively higher addresses. For example, assume that node 15 has failed as shown in Figure 7.15. This failure was determined by its predecessor, node 12. After there is no response for a message to node 15 from node 12, the time out of TA in node 12 indicates a failure of node 15. Node 12 then transmits a message to node 16. If node 16 does not exist or is malfunctioning, TA in node 12 will time out before any response is received and the search will continue. Assume that node 20 is the next functioning node in the succession of addresses. Node 12 will receive an acknowledgment of the receipt of its message before TA times out and will, correspondingly, reset its next address register, NA, to 20. Thus the failed node 15 will be effectively removed from the network.

If a node fails while it is in possession of the token and the token is "lost," another type of recovery procedure is initiated. In this case, there will be no activity on the network and all stations have their long timers, TD, timing out. Stations that have a specific "INIT" bit set, will take recovery action when TD runs out. The TD timers in the stations can be preset to different values so that one station will time out prior to the others. The first station with a TD time out will either generate a token and pass it on to the next station in the sequence or claim the token and send a message. The network will then function in a normal fashion. If the station that failed and caused the loss of the token is still inoperative, it will be bypassed and taken out of the loop, as described earlier. Also, a similar procedure occurs when the network is initialized in a power-on situation.

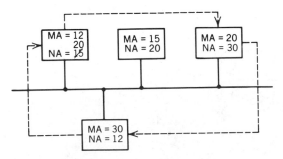

FIGURE 7.15 Recovery from a failed node.

There are a number of approaches to adding a node to the token bus network using the TAC. One method uses a master station to send a global message to all online nodes causing them to reset their next address (NA) registers. Then, on each node's token-passing activities, polling will take place to determine the next, on-line, active node and the NA register will be updated. In a more distributed alternative, each node could poll the address space between its address and the address in its NA register to see if a new node has been added. In a third, centralized approach, a master station could continuously poll the global address space in order to identify a new station requesting access to the network.

An idea of the complexity of the implementation of a token bus network to include initialization, fault recovery, and changing of node status can be obtained by noting that approximately 2350 lines of Ada code are required to represent the token bus logic.

The token-passing medium access method has the following distinguishing features:

1. A deterministic access delay for specific network configurations.
2. No collisions among nodes desiring to transmit on the same medium.
3. Management of medium access requires a very small percentage of the transmission capacity, even under overload conditions.
4. Nodes can be easily programmed to have different priorities of transmission.

MAP.   Token passing is particularly attractive for factory network applications. A notable example is the General Motors Corporation Manufacturing Automation Protocol (MAP). MAP is a token bus network that is compatible with the IEEE 802.4 token bus standard and allows dissimilar factory automation system elements to communicate. MAP provides a common syntax as well as a physical link for exchanging data among computers and intelligent manufacturing stations. Coaxial cable is GM's transmission medium for the MAP protocol as well as for video information that is transmitted throughout the plant environment. In addition, PBXs will be utilized to connect MAP networks at individual factories to a GM worldwide communications network.

MAP is based on the seven-layer ISO OSI communication model. A number of semiconductor manufacturers, particularly Intel and Motorola, have developed chip sets to implement levels 1 and 2 of the MAP levels. Companies such as Concord Data Systems and Industrial Networking Incorporated (INI) are supplying board level products to support MAP. For example, INI is providing a family of four board level products. These products support Multibus, VME Bus, IBM PC bus, and a general-purpose interface card. The INI board also implements levels 3 and 4 of the MAP system through the use of an 80186 microprocessor chip and 256K bytes of RAM.

History. In November 1980, a Manufacturing Automation Protocol (MAP) Task Force was chartered at General Motors by the Corporate Computers in

Manufacturing Steering Committee to investigate a common communications standard for plant-floor systems. The objectives of the Committee were

- To define a MAP message standard that supports applications to factory communications.
- To identify application functions to be supported by the message format standard.
- To recommend protocols that meet GM functional requirements.

Furthermore, the intent was to choose from existing documented and implemented procedures rather than implementing a new protocol.

The first MAP specification, published on October 23, 1982, discussed general network considerations along with implementation discussions. The second document, now known as Version 1.0, was released on April 6, 1984, and was an expanded version of the earlier document. The following document, Version 2.0, published on February 7, 1985, discussed GM efforts in the upper level OSI layers as well as new standards activities. The next release 2.1, dated March 31, 1985, was a major rewrite of Version 2.0 and added file transfer, network management, and MAP to PBX discussions. It also defines protocols for layers one through five of the OSI model. Version 3.0 of MAP is under development at this writing and is aimed at obtaining international agreement on OSI layers six and seven specifications. Communications regarding MAP can be addressed to:

MAP Chairman
General Motors Technical Center
Manufacturing Bldg., A/MD-39
30300 Mound Road
Warren, MI 48090-9040

MAP Layer Summary

The MAP protocol selections for the ISO/OSI layers are

Layer 1   IEEE 802.4 Broadband
Layer 2   IEEE 802.2 class I
Layer 3   Null
Layer 4   ISO Transport Class 4
Layer 5   ISO Session Kernel
Layer 6   Null
Layer 7   ISO CASE Kernel

The general plan for implementing a MAP network is given in Version 2.1 as follows:

- Installation of a broadband coaxial cable system as a backbone physical network.
- Implementation of IEEE 802.4 physical layer with IEEE 802.2 Class 1 data link layer. The features of 802.2 class 2 are not required because of the use of the National Bureau of Standards (NBS) Transport class 4 layer in the following step.
- Implementation of the NBS Transport Class 4 layer.
- Implementation of the additional higher layer MAP protocols as available.

Definition and References. Some important MAP-related definitions are needed at this point. They are

MAP End System. Implementation with protocol specifications at each of the seven layers of the OSI communications model. An End System is a peer device in that it can communicate with any other node on a MAP network.

Backbone Network. Collection of interconnected MAP End Systems.

Gateway. Dual architecture that supports the full MAP protocol as well as another non-MAP communications architecture. The interconnection between the MAP and non-MAP networks through the dual architecture is provided through a Gateway application.

MAP Network Relay (or Router). Interconnection between two physically distinct local area networks by implementing an OSI layer 3 protocol.

Bridge. Interconnection between two physically distinct local area networks (similar to Network Relay), without implementation of a layer 3 protocol. Thus both networks must utilize consistent frame sizes and address domains.

MAP Enhanced Performance Architecture System. Dual architecture system that supports the full OSI seven layer architecture as well as a "collapsed" architecture that bypasses the five upper OSI layers. The "collapsed" architecture can be used in time critical applications. A time critical segment of the MAP network is usually isolated from the Back bone by a MAP Network Relay.

References of interest relating to MAP include:

OSI Reference Model. "Information Processing Systems— Open Systems Interconnection—Basic Reference Model," Document #ISO 7498.

Layers 1 and 2. "Token Passing Bus Access Method and Physical Layer Specification," Document IEEE 802.4; "Logical Link Control," IEEE 802.2; "Network Management," IEEE 802.1 Section 5.

Layer 3. "Connectionless-Mode Internet," DIS/8475/N3154.

Layer 4. "Information Processing systems—Open Systems Interconnection—Transport Service Definition, ISO/IS8072, 1985; "Information Processing Systems—Open Systems Interconnection—Transport Protocol Specification," ISO/IS8073, 1985.

Layer 5. "Information Processing Systems—Open Systems Interconnection—Session Service Definition," ISO/IS8326, 1985; "Session Protocol," ISO/IS8327.

Layer 7. "Information Processing Systems—Open Systems Interconnection—Common Application Service Elements Service Definition—Part 2: Association Control, Context Control Information Transfer and Dialog Control (CASE)," ISO-TC9/SC21; "Information Processing Systems—Open Systems Interconnection—File Transfer, Access, and Management," DP8571; "Part I: General Description," SC16 N1669; "Part II: The Virtual Filestore," SC16 N1670; "Part III: The File Service Definition," SC16 N1671; "Part IV: The File Protocol Specification," SC16 N1672.

These references can be ordered from the following organizations:

American National Standards Institute, Inc., 1430 Broadway, New York, N.Y. 10018.

IEEE Standards Office, 345 East 47th Street, New York, N.Y. 10017.

Instrument Society of America, Standards and Practices Department, P.O. Box 12277, Research Triangle Park, N.C. 27709.

MAP Layer Descriptions. The Physical Layer provides a physical connection for data transmission between data link entities. The electrical and mechanical specifications for interfacing to the MAP network are given in the IEEE 802.4 token-passing document. Broadband coaxial cable is recommended as the medium for MAP networks and the transmission rate is 10M bits/s. The standard network will be composed of a backbone with gateways and routers providing connection to other MAP and non-MAP networks. An example of a MAP network with interconnections to other networks is given in Figure 7.16.

The Data Link layer manages and provides for the transmission of individual frames of data. The IEEE 802.2 specification is followed for this logical link control sublayer.

The Network Layer provides for message routing between end nodes. This function is performed if the nodes are on the same or different subnetworks.

The Transport Layer provides network independent transport of information to the session layer. The standard for this layer is the ISO compatible subset of the National Bureau of Standards (NBS) Class 4 Transport Protocol. Relative to the NBS Class 4 Transport, ISO does not support datagrams (transfer of data from transport user to a correspondent user without estab-

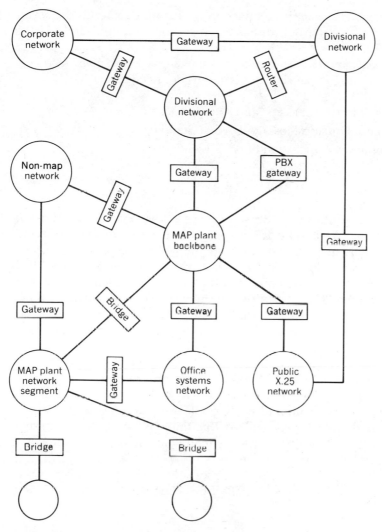

FIGURE 7.16 MAP, NON-MAP network Interconnection.

lishing a transport connection), "graceful-close" of a connection, and "status of connections."

The Session Layer enhances the transport service with mechanisms for managing and structuring reliable data transfer on a transport connection. This layer will follow the ISO Session Kernel.

The Presentation Layer will provide a negotiated standard data format for use by the application layer. This layer is null at this time.

The Application Layer supports access to the network by applications programs. For programs that desire to communicate with programmable devices, Common Application Service Elements (CASE) can be used.

MAP/Ethernet Comparison. A comparison of the approaches to the seven-layer OSI model with Ethernet (discussed in section to follow) and MAP is given in Table 7.2.

MAP Implementation Support. Prior to discussing specific manufacturer's products, the following expansion of acronyms are needed:

FTAM   File Transfer, Access, and Management

MMFS   Manufacturing Message Format Standard

CASE   Common Application Service Elements

MAC   Media Access Control

LLC   Logical Link Control

A number of manufacturers are producing chips, boards, systems, and software to support MAP. The RF modem needed to convert the digital signals to RF signals to be transmitted on the MAP coaxial cable poses a problem, however, since the RF devices and drivers are not easily amenable to integration into a silicon chip. Thus manufacturers are forced to use physically large RF modem "boxes" at this time to perform this function. Some innovative approaches are being utilized to address this problem. MAP product manufacturers include Industrial Networking Incorporated (INI), Concord Data Systems, Motorola, and Intel.

As stated earlier, INI provides four board level products, one each for use with the VME bus, Multibus, IBM PC bus, and one for use as a general-purpose interface board. Future plans include a Q-bus interface. INI utilizes an 80186 microprocessor chip and up to 256k bytes of RAM to implement levels 3 and 4 of the MAP protocol. At this time INI has the RF modem mounted as a "piggyback" card on the controller board so that only one slot of a backplane is needed.

The Concord Data Systems approach uses a data chip, protocol chip, and a DMA chip connected by a bus called the I bus. The DMA chip interfaces to

TABLE 7.2 MAP/Ethernet Comparison

ISO Model	ETHERNET	MAP
Application ⎫ Presentation ⎬ Session ⎭	RMX-NET(INTEL) XENIX-NET, MS-NET	ISO FTAM, ISO CASE, ISO SESSION (SUBSET), MMFS (EIA1393)
Transport	ISO 8073	ISO 8073
Network	ISO 8473	ISO 8473
Data/link ⎫ Physical ⎬	IEEE 802.3, 10M bits/s, baseband, CSMA/CD, probabilistic, long infrequent messages	IEEE 802.4, 10M bits/s, broadband, token bus, deterministic, short frequent messages

the host microprocessor; the protocol chip implements the IEEE 802.4 protocol with the data and DMA chips; the data chip interfaces to the RF modem, performing serial input decoding, output encoding, and various other functions required for interfacing to the modem. The Concord Data System board interfaces to both the Intel and Motorola buses.

Motorola MAP products presently include the MC68824 Token Bus Controller (TBC) chip, MC68184 Broadband Interface Controller (BIC), and the MC68194 Carrierband Modem (MC68194). The MC68020 32-bit microprocessor is utilized with the other MAP chips to implement VME bus-compatible board-level products. Some of these products are the MVME370SET MAP Interface controller, the MAP MVME 371 Broadband Modem, the MVME372 Advanced MAP Interface Controller, and the MVME370KIT-1/2 MAP Network Developers Kit.

The 68824 TBC implements the IEEE 802.4 Media Access Control sublayer of the ISO Data Link Layer. It supports communication data rates of 1, 5, and 10M bits/s and performs frame formatting and token management functions. The MC68824 incorporates four channel DMA with bus master capability, an 8- or 16-bit data bus, a 32-bit address range, and a 40-byte FIFO (First In, First Out) register array. Gateways to other local area networks are also supported by the TBC.

The Broadband Interface Controller (BIC), MC68184, interfaces to the TBC and implements most of the IEEE 802.4 Broadband Physical Layer specifications. It is involved with both data and control for the RF circuits. The BIC also performs error detection/correction, kicker insertion/deletion, data scrambling/descrambling, and internal/external loopback functions.

For the IEEE 802.4 Phase Coherent Physical Layer implementation, the MC68194 Carrier Band Modem (CBM) was developed. The CBM interfaces to the TBC, modulates the information from the serial interface, and transmits the information to the cable. It also receives signals from the cable, demodulates the signal, and transfers the information through the serial interface to the TBC.

In the board level area, the MVE370 Interface Controller supports the lower four layers of the MAP 2.1 Specification at a rate of 10M bits/s. A microprocessor board in association with the MVE370 is used to implement MAP layers 5, 6, and 7. The MVME371 Broadband Modem board interfaces with the MAP Interface Controller to the cable medium by implementing the required RF and logic circuits. A more comprehensive MAP interface implementation is provided by the MVME372 Advanced MAP Interface Controller. This board incorporates the MC68824 TBC and the MC68020 32-bit microprocessor. Portions of the MAP 2.1 Layer 1 and the Media Access Control Function of Layer 2 are accomplished by the TBC. Layers 3 through 7 and the Logical Link Control (LLC) of Layer 2 are implemented with the MC68020.

Another board level product available at this writing for MAP interfacing is the Intel iSXM 554 MAP Communications Engine. The board is Multibus

TABLE 7.3 iSXM 554 Modem Frequencies/Channel Pairs

| Part # | Modem Frequencies (MHz) | | Channel Pairs |
	Transmit	Receive	Transmit
SXM554-1	59.75 to 71.75	252 to 265	3 and 4
SXM554-2	71.75 to 83.75	264 to 276	4A and 5
SXM554-3	83.75 to 95.75	276 to 288	6 and FM1

compatible and incorporates a Token Bus Handler chip set, a Token Bus Modem, an 8-MHz 80186 microprocessor, ROM and RAM memory, and a MultiBus interface.

The iSXM 554 runs the Intel iNA 961 R2.0 Transport and Network software that implements the ISO 8073 Class 4 Transport Protocol and the ISO 8473 Internet Network Layer Protocol. In fact, the SXM 554 has the capability to provide layers 1 through 7 of MAP. Layers 5 through 7 are available through Intel for use on the 554 board. The CPU and associated memory (256k bytes of RAM and up to 160k bytes of EPROM) can accommodate the implementation of layers 3 through 7.

The Token Bus Handler (TBH) chip with the Token Bus Modem (TBM) form the interface to the cable medium. DMA transfers between the TBH and iSXM554 on-board memory are initiated by the TBH. The transfer of data to and from the token bus is accomplished through the TBM, which utilizes the IEEE 802.4 Token Bus Standard. Three versions of the 554 are offered with different modem frequencies/channel pairs. These are listed in Table 7.3. Cable connection from the iSXM554 is 75 Ω on a Type F female connector. A remodulator head end is required for the use of the iSXM554 on a token bus network.

The position of the iSXM554 board and associated software in the seven layer MAP implementation is as follows:

Application layer ⎫	ISO-FTAM	⎫
Presentation layer	ISO-CASE	Software
	MMFS (EIA 1393)	implementation
Session layer ⎭	ISO SESSION (SUBSET) ⎭	
Transport layer ⎫	INA 961	⎫ Firmware
	ISO 8073	implementation
Network layer ⎭	ISO 8473	⎭
Data Link	iSXM 554	⎫ Hardware
Physical	IEEE 802.4	⎭

TOP. At the 1985 AUTOFACT Convention held in Detroit, Michigan, from November 5 through November 7, Boeing Corporation introduced the

Technical and Office Protocol, TOP. TOP is a companion protocol to MAP and is aimed at technical, manufacturing, and office applications. As with MAP, TOP is a subset of the OSI protocols. At this time, TOP is an evolving specification that will address the following:

- File transfer.
- Electronic mail.
- Print, plot, file, and directory servers.
- Distributed database interfaces.
- Document, spreadsheet, and graphics exchange.

TOP utilizes the IEEE 802.3 CSMA/CD protocol for the Physical Layer implementation. A comparison of the MAP and TOP OSI layered implementations is given in Figure 7.17.

Additional information on TOP can be obtained from:

Boeing Computer Services
Network Services Group
P.O. Box 24346
M.S. 7C-16
Seattle, WA 98124-0346

CSMA/CD. An alternative to the deterministic, token-passing approach is the CSMA/CD (carrier-sense multiple access with collision detect) medium access method. CSMA/CD is a general name given to an Ethernet class of access methods originated by the Xerox Corporation. It has since been the object of a cooperative effort among Xerox, Digital Equipment, and Intel.[2] Since Ethernet is the principal CSMA/CD local computer network in use today, CSMA/CD will be discussed and explored using Ethernet.

Ethernet was developed at the Xerox Palo Alto Research Center in 1972 by Robert Metcalfe and David Boggs. This initial version is now referred to as the Experimental Ethernet System. The version of Ethernet that has been updated by Xerox, Digital Equipment, and Intel is called the Ethernet Specification. This specification was developed prior to the finalization of the IEEE 802.3 Standard but is fundamentally the same as the standard. In CSMA/CD, *carrier sense* refers to the ability of a node to listen for a carrier on the network before trying to transmit. If the medium is "quiet," the station will attempt a transmission. Because of propagation delay on the medium, another station may also have begun an undetected transmission. If this is the case, a collision will occur. This time interval in which a collision can occur after the start of a transmission is called the *collision interval*, or the *collision window* or *slot time* and

---

[2] The Ethernet, A Local Area Network: Data Link Layer and Physical Layer Specifications, Version 1.0, Digital Equipment Corporation, Intel, Xerox, September 30, 1980.

OSI LAYER	MAP	TOP
7. Application	Network Management Directory Service MMFS Subset FTAM ISO CASE Kernel	FTAM
6. Presentation	Null	Null
5. Session	ISO Session Kernel	ISO Session Kernel
4. Transport	ISO Transport Class 4	ISO Transport Class 4
3. Network	ISO CLNS	ISO CLNS
2. Datalink	IEEE 802.2, Link Level Control Class 1	IEEE 802.2, Link Level Control Class 1
1. Physical	IEEE 802.4 token access on Broadband Media	IEEE 802.3 CSMA/CD Baseband Media

FIGURE 7.17 TOP/MAP comparison.

is a slightly greater than the round-trip propagation delay of the network. If a collision does not occur during this period, a station is said to have acquired the communications path or "*Ether*."

Collision Detect in CSMA/CD is the capability of a transmitting node to detect a collision. Once a collision is detected by a transmitting node, the node will place a short burst of noise energy on the channel, to ensure that all other nodes attempting to transmit at this time detect the collision condition. This noise signal is called a *jam* signal and the jamming technique is a *collision consensus enforcement* procedure. After collision detection, a transmitting station will back off, wait for a random period of time, and then try again to transmit. Since the delay period before a retransmission attempt is random for each station, the probability of repeated collisions is reduced. If there is heavy activity on the Ethernet, the mean of the random retransmission delay for a node is increased as a function of the channel load. Channel load can be estimated by the number of collisions experienced by a node in trying to transmit a particular packet of data or information. Also, in CSMA/CD, the minimum time of a transmitted message must be greater than, or equal to, the slot time so that it can propagate throughout the network and all nodes, including the transmitting node, can detect a collision.

Carrier Sensing.   The sensing of a carrier on the network is important for a receiver to ensure that it can acquire a message that may be bearing its address and for a transmitter to determine if it is clear to transmit to other nodes on the network. One basic carrier sensing technique is to monitor the network for phase transitions. *Manchester* encoding (which will be discussed in detail later in this chapter) is used as the data representation means and by its nature ensures a 0 to 1 or a 1 to 0 transition during every bit time or *bit cell* time. The Ethernet Specification also defines another carrier sense technique, since a collision involving large numbers of nodes can saturate the medium and eliminate phase transitions.

Detection of Collision.   The efficiency of transmission of a network utilizing CSMA/CD is dependent on the time wasted on collisions. The sooner that a transmitter can recognize a collision and back off, the better. In general, collision detection is accomplished by comparing the transmitted signal with the received signal and noting differences. Collision detection may not always be certain, although most implementations have very high percentages of detection. A number of highly improbable, but possible, cases can be cited where collision detection will not occur. For example, in the Experimental Ethernet System, which is essentially a wired OR implementation, node locations and system delays can be such that a waveform transmitted by one node will exactly match that of a second node when the second node is comparing its transmitted and received signals. A third node that is the destination of a message may see the combined signals as an incorrect message. Also, another possibility that may occur when using coaxial cable or twisted pair is that a large number of nodes may try to transmit simultaneously and generate a collision situation where the medium is saturated and the impedance is such that additional current cannot be supplied by a transmitter. Thus the medium will sit at a dc level and no phase transitions will appear. With no apparent indication of line occupancy, additional nodes can attempt to initiate transmission, further deteriorating the situation. Each node, after attempting to transmit and comparing its transmitter signal with the dc level as received, would release the channel, back off, and try again later. The odds for such a situation are low, but the Ethernet Specification has incorporated an addition to the carrier sense technique to prevent such an occurrence.

As noted earlier, once a collision is detected by a node attempting to transmit, a collision consensus enforcement procedure is invoked that places a jam signal on the medium to ensure that other nodes attempting to transmit will recognize the collision regardless of propagation delays or the shortness of the original collision interval. In the Ethernet Specification, a jam is the transmission of 4–6 bytes of random data.

Retransmission.   When a node detects a collision on the medium, it enters a voluntary backoff procedure wherein it releases the channel, enters a delay routine, and then tries again to transmit. After 15 consecutive collisions, the transmitter reports an error condition to the station.

A *binary exponential back-off algorithm* is used to calculate the delay time before retransmission is attempted. Recalling that the collision interval or slot time is the time period, which, if free of collision, assures that the transmitter has acquired the channel, the delay calculation algorithm utilizes slot time as the minimum unit of delay. The slot time is usually set to be a little greater than the round-trip delay time of the network. The retransmission delay is then the product of a positive integer and the slot time. The appoach to retransmission is to try again quickly after the fist attempt, but then to back-off and permit the selection of increasingly longer delays after each unsuccessful attempt. Note that this approach **permits** the selection of an increasingly longer delay after each try but does not specify a longer delay. This is accomplished by specifying an interval of possible delays that increases after each unsuccessful attempt. Figure 7.18 shows the delay intervals as a function of retransmission attempt for the Ethernet Specification.

The intervals are defined to be 0 to 2 exp[min (retransmission attempt,n)] − 1, where 2^n is the maximum number of stations permitted on the network. For the Ethernet Specification, 1024 stations are permitted, therefore n = 10. The algorithm limits the doubling after n attempts as shown in Figure 7.18. For this reason, it is referred to as a *truncated binary exponential backoff algorithm*. The algorithm starts over at zero delay with the transmission of the next packet.

Ethernet Message Format.   A packet of information is transmitted on Ethernet in the format of Figure 7.19. As seen in Figure 7.19, a packet consists of a 64-bit preamble followed in sequence by a 48-bit destination address, a 48-bit source address, a 16-bit type field, an n-byte (8 bits) data field, and a 32-bit *Cyclic Redundancy Check* (CRC) field. The maximum size of a packet including the preamble is 1526 bytes and the minimum size is 72 bytes. The variant is the data field, which can range from 46 bytes to 1500 bytes. The sequence of transmission is preamble first, destination address second, and so on through the cyclic redundancy check field. Individual bytes are transmit-

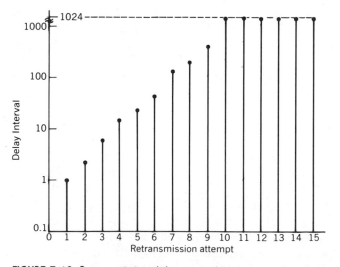

FIGURE 7.18 Retransmission delay interval versus retransmission attempt.

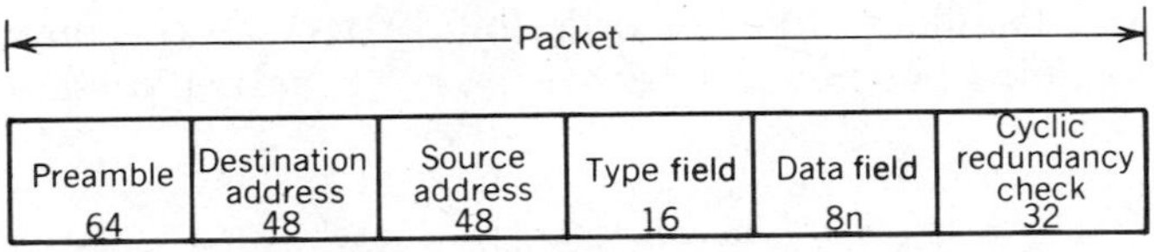

FIGURE 7.19 Ethernet packet format.

ted with the least significant bit transmitted first. The minimum spacing period between frames is specified as 9.6 μs to allow for message recovery and channel stabilization.

Preamble.   The 64-bit *preamble* is used for synchronization of the transmitter and receiver. The last two bits of the preamble are 1's and indicate the beginning of a frame. Version 1.0 of the Ethernet Specification defined the preamble as an alternating sequence of 1's and 0's beginning with a 1 and ending with 1's as the last two bits. In Hexadecimal format, the preamble is AAAAAAAAAAAAAAAB. Loss of carrier indicates the end of a frame.

Destination Address.   This 48-bit field specifies the address to which the frame is being sent. If the first bit transmitted is a 0, the remaining bits are the address of one destination. If the first bit transmitted is a 1, the remaining 47 bits are a *multicast* address representing a logical group of receiving stations. A special case of the multicast mode is all 1's in the 47-bit positions, specifying a *broadcast* mode to all stations. With 47 bits available in the destination address field, there are 2^{47} or over 1.407×10^{14} distinct addresses possible.

Source Address.   The source address is a 48-bit field representing the station sending the frame. Note that there are 2^{48} or 2.815×10^{14} combinations possible with this field.

Type Field.   The type field consists of 16 bits used to identify the higher-level protocol being used in the transmission. Examples of different protocols include IBM's SNA, DEC DECnet, and CCITT X.25.

Data Field.   The data field contains the data being transmitted. The minimum size of this field is 46 bytes (368 bits) to guarantee that a frame is distinguishable from a frame fragment truncated by a collision. The maximum size of the data field is 1500 bytes (12,000 bits).

Frame (Packet) Check Sequence.   The 32-bit FCS field contains a CRC (cyclic redundancy check) code that is computed from the contents of the destination address, source address, and type and data fields. The message polynomial is formed from these fields. The first transmitted bit of the destination address is the high-order bit of the message polynomial M(x) and the last transmitted bit of the data field is the low-order bit of M(x). M(x) is then divided by G(x), the generating polynomial of Ethernet. For Ethernet Version 1.0,

$$G(x) = x^{32} + x^{26} + x^{23} + x^{22} + x^{16} + x^{12} + x^{11} + x^{10} + x^8 + x^7 + x^5 + x^4 + x^2 + x + 1$$

The remainder, R(x), of M(x)/G(x), is inverted and transmitted as the CRC field. Thus the received message is M(x) with CRC appended. In order to check the validity of the received message, the polynomial formed by M(x) appended with CRC is divided by G(x) and must yield a polynomial of order 31 $(x^{31} + x^{30} + \cdots + x^0)$ with the corresponding coefficients:

$$1100011100000100110111010101111011$$

If this result is not obtained, the received message is in error and is not accepted.

Bit Encoding. Bits are represented on the coaxial cable medium by *Manchester encoding*. Manchester encoding is defined as having a 50% duty cycle with a transition in every *bit cell* (bit time). The complement of the bit value is present during the first half of the bit time, and the true value of the bit is present during the second half. A logic 1 is represented by 0 V and 0 mA in the cable while a logic 0 is −2.05 V with approximately −82 mA flowing in the cable. Positive current flows out of the center conductor of the cable. In a quiescent condition, the cable is in a logic 1 state. The average values of voltage and current on the cable should be −1.025 V and −41 mA, respectively, with a range of −0.9 volts to −1.2 V and −36 mA to −48 mA. At no time should the voltage on the cable go positive. The data transmission rate for the Ethernet Specification is 10M bits/s. Figure 7.20 summarizes the Ethernet Manchester encoding characteristics.

Figure 7.20 also illustrates the timing for carrier detect. A carrier is identified as being present if a transition is detected within a 0.75 to 1.25 bit time interval since the center of the last bit cell. If this transition is not detected, it is assumed that the packet has ended and/or the carrier has been lost.

Typical Implementation. Figure 7.21 relates the data link and physical layers of the ISO/OSI network architecture to physical components of the Ethernet local area network.

The five main physical categories in Figure 7.21 are the *station, Ethernet controller board, transceiver cable, transceiver,* and *coaxial cable transmission medium.*

Station. The station is the basic addressable device that is connected to the Ethernet. The station is typically a computer and provides a software and/or hardware interface between the Ethernet controller and the operating system environment.

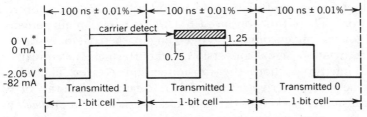

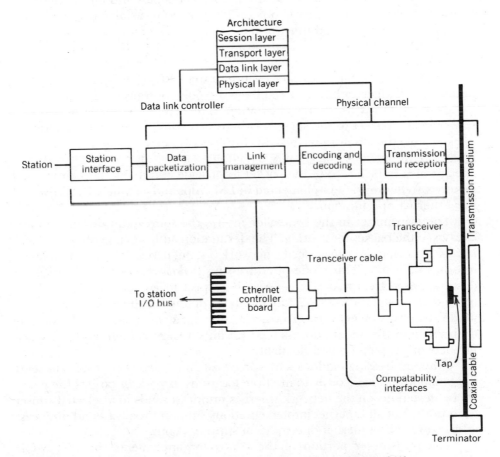

FIGURE 7.20 Ethernet manchester encoding characteristics.

FIGURE 7.21 Relationship to data link and physical components of Ethernet LAN.

247

Controller. The Ethernet controller performs functions needed to control access to the network. These functions are CSMA/CD channel management, serial-to-parallel conversion, encoding and decoding, packetization, error detection, address recognition, signal conversion, and buffering. The controller can be subdivided into two portions, the transmitter and receiver. The transmitter generates the Manchester encoded serial bit stream at a data rate of 10 MHz. Also the transmitter packetizes the data and inserts a 64-bit *preamble* at the beginning of the packet for synchronization by the receiving station. All packets must be multiples of 8 bits excluding the preamble, with the least significant bit in each byte transmitted first. The receiver portion of the controller performs the decoding of the packet by means of a *phase decoder*.

The decoder uses the preamble to establish synchronization by means of a *phase-locked loop* or equivalent timing circuits. Then the Manchester-encoded data are converted into a serial bit stream for the station.

Carrier sense is performed in the controller to determine when the medium is free for transmission and when a packet begins and ends. Similarly, collision detection is necessary to quickly terminate a transmission if a collision occurs and to keep wasted time on the medium to a minimum.

In order to detect errors in the transmitted packet, the transmitter portion of the Ethernet controller generates the *CRC* (*cyclic redundancy check*). The receiver portion of the controller examines the CRC of the received packet, determines if an error has occurred from the point of transmission, and sends the packet less the CRC code to the station. If an error is detected, the receiver can send the packet to the station with an error indication or eliminate the packet altogether. Naturally, the CRC cannot cover errors generated after the receiver CRC examination circuits. To implement broader coverage, software checksums can be added to the packet to provide end-to-end checking. CRC generator/checkers are implemented in LSI chips and a 32-bit CRC is used in the Ethernet Specification.

The transmitter in the controller inserts the source and destination addresses in the packet for routing. The destination address code can be *network relative*, that is, unique to a specific network but not necessarily unique relative to other networks. This mode of addressing is *network specific*. Conversely, a destination address that is unique with respect to all relevant networks is termed an *absolute* or *universal* address and is defined as the *unique* addressing mode. In terms of network growth and management, unique addressing is desirable and the 48-bit address field permits a large enough address space for network expansion and flexibility.

Other addressing modes such as *broadcast* and *multicast* are useful to send information simultaneously to multiple locations. Broadcast targets the packets to all stations on the network whereas multicast sends to a selected subset of stations. Not all Ethernet implementations support broadcast and multicast addressing. The Ethernet Specification supports both.

The transmitter portion of the controller implements the CSMA/CD channel management, that is, transmission, back-off, retransmission, and so on. Details of their functions are discussed in detail earlier under the heading "Retransmission". Similarly, the receiver in the controller determines if it is the addressed location, removes the preamble, and processes the bit stream.

Transceiver Cable. The transceiver cable runs from the controller on the station to the transceiver that taps into the coaxial cable transmission medium. The cable and associated connector specifications are given in Figure 7.22 and the overall relationship among transceiver components is shown in Figure 7.23.

Transceiver. The transceiver makes the connection from the controller to the transmission medium. It transmits data to and receives data from the transmission medium, furnishes power to the transmission system, and detects collisions for the controller. Electrically, the transceiver must appear as a high impedance to the medium in both the on and off states and must isolate the network from electronic failures in its associated circuitry.

Coaxial Cable Transmission Medium. The coaxial cable for the Ethernet Specification is a double-shielded, 50-Ω, solid center conductor, foam dielectric cable such as Belden 9880. Maximum end-to-end coaxial cable length is 2.5 km with a maximum segment length of 500 m. Maximum transmission rate is 10M bits/s. The shield of the coaxial cable should not be connected to alternating current ground or building ground.

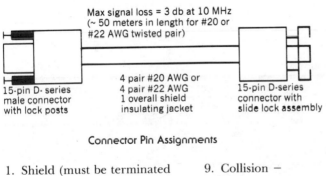

Max signal loss = 3 db at 10 MHz
(~ 50 meters in length for #20 or #22 AWG twisted pair)

15-pin D- series
male connector
with lock posts

4 pair #20 AWG or
4 pair #22 AWG
1 overall shield
insulating jacket

15-pin D-series
connector with
slide lock assembly

Connector Pin Assignments

1. Shield (must be terminated to connector shell)

2. Collision +

3. Transmit +

4. Reserved

5. Receive +

6. Power return

7. Reserved

8. Reserved

9. Collision −

10. Transmit −

11. Reserved

12. Receive −

13. Power positive

14. Reserved

15. Reserved

FIGURE 7.22 Transceiver cable and connector specifications. The power pair must have loop resistance $\leq$ 4 Ω.

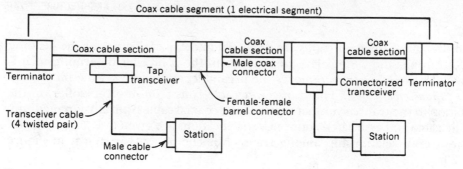

Note:

1. Connector shell and terminators must be insulated from building grounds. Usually a sleeve or boot on the connector will suffice.

2. Coax cable segments should be terminated with a female N series connector with an impedance of 50 Ω ± 1% with a power dissipation capability of 1 W

3. Coaxial cable sections must be terminated with male N series connectors. Sections can be joined with female-female adapters.

4. Transceivers should not be placed closer together than 2.5 m. No more than 100 transceivers should be placed on a coax cable segment.

5. Maximum segment length is 500 m.

6. Maximum end-to-end length is 2.5 km.

FIGURE 7.23 Overall transceiver/cable relationship.

TABLE 7.4 Typical Ethernet Products

Manufacturer	Controller	Encoder/Decoder	Transceiver
Intel	i82586(NMOS)	i82501(bipolar)	Development stage at this writing
Intel	iSBC 552 and iSXM 552 board level products interface with Multibus		
Advanced Micro Devices	AM7990(NMOS)	AM 7991(bipolar)	AM 7995(bipolar)
Fujitsu	MB 8795A(CMOS)	MB502(bipolar)	Development stage
National Semiconductor	DP 8390(CMOS)	DP 8391(bipolar)	DP 8392(bipolar)
Rockwell	R68802(NMOS)	—	—
Mostek	MK 68590(NMOS)	MK 3891(bipolar)	—
Seeq Technology	8001/8003(NMOS)	8002(CMOS)	Development stage at this writing

If a connection to one of these grounds is absolutely necessary for safety reasons, the connection should be made at only one point. For a cable that is to be used with a transceiver tap, the jacket O.D. should be 0.405 in.

Some Products.   Some typical IEEE 802.3 Ethernet products and manufacturers are summarized in Table 7.4.

FOR FURTHER READING

Bibbero R. J., and D. M. Stern, *Microprocessor Systems Interfacing and Applications,* John Wiley & Sons, Inc., New York, 1980.

Forbes, B., "Using the 8292 GPIB Controller," AP-66, Intel Corporation, 1980.

IEEE Standard 802.2, *Logical Link Control*; IEEE Standard 802.3, *CSMA/CD*; IEEE Standard 802.4, *Token Bus*; IEEE Standards Office, New York, 1984.

Parker, C. R. C. B., "The Influence of Local Networking Technology on Process Control Interfaces," *Interfaces in Computing*, Vol. 1, No. 1, May 1982.

8

SOFTWARE INTERFACING

Even though interfacing generally deals with hardware considerations and low-level (assembly language) programs, the designer of microcomputer systems must have at least a basic knowledge of high-level languages. In particular, high-level languages should be used, where possible. Where assembly language is required because of timing constraints, a higher level language should still be used to document the assembly language code and logic flow. The languages chosen for discussion in this section are BASIC, FORTRAN, Pascal, Ada, LISP, PL/M-86, and C. These languages cover a broad spectrum ranging from ease of use (BASIC) to processing lists (LISP), to real-time microprocessor applications (PL/M-86) and C. Since PL/M-86 is targeted for the Intel 8086 family of microprocessors and is one of the most widely used high-level languages for real-time and interfacing applications, it will be covered in much greater detail than the other languages. Similarly, because of its popularity and use with the Unix® operating system,[1] its portability, and broad usage, the language C will be developed along with real-time examples.

8.1

MODULAR PROGRAMMING[2]

Before discussing specific high-level languages, the concepts of modular programming will be presented. *Modular programming* is the art of breaking up a software package into a number of pieces where each one is

- Manageable in size.
- Separately solvable.
- Separately testable.
- Separately modifiable.
- Easily related to some portion of the problem.

[1] Unix® is a registered trademark of Western Electric Corporation.

[2] The material in Sections 8.1, 8.2, and 8.3 is taken from a course developed by Tom Maier and John Hudak of the Mellon Institute, Computer Engineering Center.

The objective of modular programming is to reduce the complexity of the problem. If achieved, the results are easier isolation of errors, increased programmer efficiency, enhanced testability, and ease of distribution of tasks among programmers. The importance of these results can be demonstrated by the graph of Figure 8.1.

Some rules for module design are

1. Develop and verify module specifications:
   - treat multiple entry points as separate modules.
   - define all passed parameters.
2. Select algorithms and set up data structures:
   - "simple and straightforward beats "compact and cute."
3. Define all temporary/local variables.
4. Write code and refine it in a "stepwise manner."
5. Verify the code.
6. Compile and test the module.

In specifying a module, the module *name* should be meaningful relative to the function it performs. The *function* of the module should be defined in terms of what it does, rather than how it does it. All passed parameters should be defined as well as in what order they are passed. Also, the name, format, size, units, valid range, and attributes of each input and output parameter should be given. The cause and effect relationships for all possible input conditions should be listed. Finally, any other pertinent information should be specified. Examples include other routines called by this module, routines

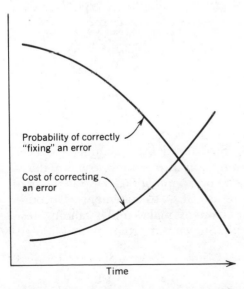

FIGURE 8.1 Importance of modular programming.

Module name-get_convrt_adc.

Function-The function of this procedure is to set up the
        desired analog mux channel, read the A/D converter,
        mask off the unused bits, multiply the result by
        a multiplier "mult" and return the converted result.
        For test purposes the "A/D Access" LED is turned on
        whenever this routine is called.

Parameter list-mux_chan, mult, result.

Inputs-mux_chan   byte   number of analog channel to be
                           read (range 0 to OFH)
        mult   byte   scaling multiplier
                           (range 23H to OEBH)

Output-Result   word   scaled analog input parameter
                           (range 0 to OFFFFH)

Cause and effect
1. If channel number is out-of-range and error message is
   printed.

2. If in range-normal operation.

Other phenomena
1. Calls select_mux to set up the correct analog input
   channel.

2. Calls get_adc to read and mask off A/D converter output.

3. Calls test_led to turn on "A/D Access."

4. Error message printed.

FIGURE 8.2 Sample module specification.

that call this routine, and treatment of error conditions. A typical module
specification that illustrates this approach is given in Figure 8.2.

8.2

STRUCTURED PROGRAMMING

Using modules in programming is part of an overall approach to programming known as *structured programming*. Structured programming has been
defined by Nicholas Wirth[3] as "the formulation of programs as hierarchical,
nested structures of statements and objects of computation." One attribute of
structured programming is to avoid the use of GOTO statements. Uncondi-
tional branching implemented by GOTO statements makes understanding, testing, and debugging of programs difficult. Structured programming also en-

[3] N. Wirth, "On the Composition of Well-Structured Programs," *Computing Surveys*, Vol. 6, No. 4,
December 1974, pp. 247–259.

courages *"top-down" design*, where larger concepts are broken down into successively, smaller, manageable modules. Structured programming enhances software development, documentation, maintenance, and trouble-shooting.

Some rules of thumb of structured programming are

1. Use standardized flow control statements:
   - DO loops.
   - IF . . . THEN . . . ELSE statements.
   - Sequential statements.
2. Avoid GOTOs, particularly "reverse" GOTOs.
3. Indent to accentuate breaks in execution sequence.
4. Use only one entry and exit point.
5. Keep modules small:
   - Preferably one page or less for readability.
   - Testable "by inspection."
6. Use straight-forward algorithms where possible.
7. Use "good programming style."

8.3

PROGRAMMING STYLE

Note item 7, referring to good programming style. Programming style is a combination of art and engineering principles. Programming style is easier illustrated than defined. A detailed list of manifestations of good programming style is as follows:

1. Use simple but meaningful variable names.
   Poor: X = 2.54 * Y
   OK: C = 2.54 * I
   Better: centimeters = 2.54 * inches
2. Avoid similar names.
   It is easy to confuse INPUT_BUFFER_POINTER
      with BUFFER_INPUT_POINTER
3. Avoid having too many temporary variables.
   Instead of

$$ITEM = A/(B*C)$$
$$TOTAL = Y + ITEM$$
$$VAR = X + ITEM$$

use

$$TOTAL = Y + A/(B*C)$$
$$VAR = X + A/(B*C)$$

4. Use parentheses to avoid ambiguity.
5. Avoid "hard coding" constants.
   Instead of output = input_word AND OFH
   use output = input_word AND mask1
6. Place only one statement per line.
   • For readability.
   • Simplifies block structure.
7. NEVER alter the value of the iteration variable inside a DO loop.
   • Later modification could result in "endless loops."
8. Avoid excessive use of statement labels.
   • When they are used, make sure that they are meaningful and dissimilar for maximum reader impact.
9. Learn a new language well.
   • Use the right constructs.
   • Use built-in library functions.
10. Avoid language tricks.
    • Just because you find a loophole, does not mean you have to use it.
11. Do not ignore warning messages.
    • Maybe the compiler is trying to tell you something.
12. Ignore efficiency suggestions until the program works.
    • Strive for functionality first, speed and compactness later.
13. Let the compiler optimize.
    • Do not fight it; let it work with you.
14. NEVER sacrifice readability for efficiency.
    • "Cute" algorithms may not be understandable or maintainable months after implementation.
15. Never optimize for the sake of optimization.
16. Tackle efficiency through "Macroefficiencies."
    • Optimize only where you need to.
    • NEVER optimize blindly; measure first to find the best places.
    • Likely candidates for "Macroefficiency" effort are
    a. I/O drivers.
    b. Sort and search algorithms.
17. Try to stick to the simplest algorithms.

18. Strike a good balance on the numbers and content of the comments.
    - Try to place yourself in the reader's shoes.
    - Try to answer potential questions.
    - Physically offset all comments.
    - Comment on all variables.
    - Do not make comments "echo the code."
19. Variables.
    - Explicitly declare all variables, attributes, and so on.
    - Avoid using multiple names for the same storage.
    - NEVER use a variable for more than one purpose.
    - NEVER use special values of variables for special meanings.
20. Consider using table-driven modules where appropriate.
    - Simplifies decision making for input data.
    - Makes future modifications easier.
    - Keeps the routine more general.
21. Make exhaustive decisions.
    - Do not ignore the "impossible cases."
    - Table-driven modules will help prevent this.
    - Use error messages, where appropriate.
22. Be cautious about using integers and floating point numbers.
    100.0 * .01 "hardly ever =" 1.
    2*I/2 "may or may not =" 1.
23. NEVER write self-modifying programs.
24. Avoid reinventing the wheel.
    - Establish and use in house libraries of modules.
    - Properly written and documented code can be used by many people.

8.4

BASIC

BASIC (Beginner's All Purpose Symbolic Instruction Code) was developed at Dartmouth College to provide a relatively easy way to program computers. BASIC is widely used in both home and office computer applications. The statements in BASIC are English-like and are written in free-format. Thus no special alignment of words is required. BASIC is usually interpreted and used interactively. Some examples of BASIC programs follow along with results of the program execution.

```
                              PRINT 7+5
(response)                    12

                              PRINT 1/2, 3*7
(response)                    .50   21

              10              INPUT Y
              20              PRINT 5*Y
                              RUN

(response)                    ?                request for input Y
                              50               enter Y
                              250              result

              10              INPUT Z
              20              IF Z > 0 THEN 50
              30              PRINT "INPUT LESS THAN OR EQUAL TO ZERO"
              40              GO TO 10
              50              PRINT "INPUT GREATER THAN ZERO"
              60              GO TO 10

(response)                    ?                        request for input
                              10                       enter Y
                              INPUT GREATER THAN ZERO   result

              10              NUMB = 2
              20              PRINT SQR(NUMB)
              30              NUMB = NUMB + 2
              40              IF NUMB = 50 THEN 60
              60              ----------
              70              -----------

(response)    1.41421         print out square root of
              2               all numbers in the sequence
              2.4494          2,4,6,8,10,12,14...50
              2.8284
              3.1622
              3.4641
              3.7416
              .
              .
              .
              7.0710
              --------
              ---------
```

A final example will be used to illustrate a more interesting program in BASIC. Also, an IF-THEN statement, a DATA statement, and the INT function will be used. The IF-THEN statement provides a conditional branch to another statement in the BASIC program and the DATA statement provides data to be read sequentially by a READ statement. The INT function, INT(X), usually determines the greatest integer not **greater** than X. The example program will determine the largest factor of a number N and also determine if N is prime. For this example, the values that N will assume are 11, 1009, and 1003. The result of running the program is given following the program statements.

```
   10    READ N
   20    FOR M = 2 TO SQR(N)
   30    IF N/M = INT(N/M) THEN 70
   40    NEXT M
   50    PRINT N; "IS A PRIME NUMBER."
   60    GOTO 10
   70    PRINT N/M;"IS THE LARGEST FACTOR OF"; N
   80    GOTO 10
   90    DATA 11, 1009, 1003
  100    END
```

```
(result)  11   IS A PRIME NUMBER.
          1009 IS A PRIME NUMBER.
          59   IS THE LARGEST FACTOR OF 1003
```

This program reduces the number of tries to determine the largest factor (if any) of N by trying only the values from 2 to the square root of N, since the largest factor must lie in this range.

BASIC language manuals are supplied by manufacturers of most minicomputer and microcomputer systems. There are usually differences in the BASIC language implementation on particular computers, but the examples given here serve to illustrate the nature of the language.

8.5

FORTRAN

FORTRAN (FORmula TRANslation) is a high-level language that was developed in 1956 by Backus and Ziller of IBM. It was developed primarily for scientific applications but is widely used in many other areas. Of the high-level languages, FORTRAN has one of the largest investments of person-hours in programming.

Before examples of FORTRAN can be presented, the following short list of some FORTRAN definitions of symbols is necessary.

LT.	less than
EQ.	equal to
GT.	greater than
NE.	not equal to
REAL	defines a variable as a real number
INTEGER	defines a variable as a real number
DIMENSION	declares and names an array
+	addition
—	subtraction
*	multiplication
/	division

Following is an example of a FORTRAN program that calculates the total amount in dollars of H half-dollars, Q quarters, D dimes, Y nickels, and P pennies.

```
10        FORMAT (5F4.0)
20        FORMAT (5(5X,F4.0),5X,F6.2)
30        READ(5,10)H,Q,D,Y,P
40        TOTAL = .5*H
          TOTAL = TOTAL + .25*Q
          TOTAL = TOTAL + .10*D
          TOTAL = TOTAL + .05*Y
          TOTAL = TOTAL + .01*P
100       WRITE(6,20)H,Q,D,Y,P,TOTAL
          STOP
          END
```

The format statements in the FORTRAN program specify the maximum number of quantities that can appear on a line, the maximum number of characters in each quantity, and whether each quantity is an integer, real without an exponent, or real with an exponent. In the example, the 5X in format statement 20 specifies that 5 spaces are to be left on the output form. In the same statement, F4.0 specifies that the number printed out will consist of 4 characters (including sign, decimal point, and blanks) with no digits to the right of the decimal point. Similarly, F6.2 will cause numbers to be printed out with 6 characters total, 2 of which will be to the right of the decimal point.

A second illustration of a FORTRAN program calculates the number of dollars, Y, that accumulate if D dollars are invested at an annual compound interest rate of R for N years. The standard formula for Y is

$$Y = D(1 + R)^N$$

The FORTRAN program that calculates Y is

```
10    FORMAT(2E12.5,I3)
20    FORMAT(2(3X,2E12.5),3X,I3,3X,E12.5)
      READ(5,10)D,R,N
      Y=D*(1.+R)**N
      WRITE(6,20)D,R,N,Y
      STOP
      END
```

Exponential and integer numbers are specified in the format statements of this example. The format E12.5 defines numbers with 12 characters, including sign, leading zero, decimal point, and 4-character exponent. The 5 in the format specification indicates 5 digits to the right of the decimal point in the mantissa; I3 indicates an integer of 3 digits.

8.6

PASCAL

Pascal is one of the relatively newer and popular programming languages in use today. It was developed by Professor Nicolas Wirth of the Eidgenhossiche Technische Hochshule in Zurich, Switzerland. The language was named after the seventeenth-century mathematician Blaise Pascal, who invented one of the earliest calculators. Professor Wirth used Pascal as a tool for teaching structured programming principles and techniques. One ramification of this goal is that Pascal requires that constants and variables be defined as to type, range of values, and so on, to enhance the probability of detecting errors during compile time rather than when the program is running.

Some examples of Pascal constant (CONST) and variable (VAR) declarations are

```
CONST
      MIN = 50;
      VOWEL ='A';
      LENGTH = 100;
      WIDTH = 20;
VAR
      AMOUNT:INTEGER;
      VALUE,MAX:INTEGER;
      LEVEL:BOOLEAN;
      LETTER:CHAR;
```

The following Pascal program calculates the circumference and area of a circle, given the radius.

```
PROGRAM   CALCCIRC(input, output);
      CONST
          PI= 3.14159;
      VAR
          RADIUS,CIRCUMFERENCE,AREA:REAL;
      BEGIN
          READ(RADIUS);
          CIRCUMFERENCE:=2*PI*RADIUS;
          AREA:=PI*SQR(RADIUS);
          WRITELN('CIRCLE DATA');
          WRITELN('RADIUS =', RADIUS);
          WRITELN('CIRCUMFERENCE =', CIRCUMFERENCE);
          WRITELN('AREA =', AREA);
      END.{CALCCIRC}
```

An example Pascal program that calculates the average value of N integers and illustrates iteration is shown as follows. Note that comments in Pascal are enclosed by (* *).

```
(*AVERAGE CALCULATION PROGRAM*)

PROGRAM   AVGCAL(INPUT,OUTPUT);
VAR  A,N:INTEGER;
     AVERAGE, NUMB, TOTAL:REAL;
BEGIN
     READ(N);
     TOTAL:=0;
     FOR A:=1 TO N DO
         BEGIN
              READ(NUMB);
              TOTAL:=TOTAL+NUMB
         END;
     AVERAGE:= TOTAL/N;
     WRITELN  ('AVERAGE IS', AVERAGE)
END.
```

The iteration portion of the Pascal program is initiated by the statement

```
FOR A:=1 TO N DO
```

and the statements enclosed between the BEGIN and END boundaries immediately following the FOR statement are repeated N times.

8.7

ADA

Ada is a high-level language developed under the sponsorship of the U.S. Department of Defense (DOD). Ada's features encourage structured programming and handle concurrent "tasks" defined by the programmer. Ada supports strong typing, separate compilation, tasking, and handling of exceptions. A key feature of Ada is the package, which allows the specification of logically related entities. In this framework, subprograms can be developed once and called from outside the package by users. Thus many common support algorithms can be accessed by a wide variety of software developers, saving them the time and effort to separately and independently develop these necessary programs on their own.

The use of Ada is required by DOD in defense projects, particularly in imbedded systems. Since approximately 55% of DOD applications involve imbedded systems, Ada is expected to have a significant impact on reducing DOD software development and maintenance costs.

Pascal was used as the starting point for Ada. Following a competition among CII Honeywell-Bull, Intermetrics, SRI International, and Softech, the CII Honeywell-Bull approach developed by Jean Ichbiah was chosen in 1979. The preliminary Ada Reference Manual was published in 1979.

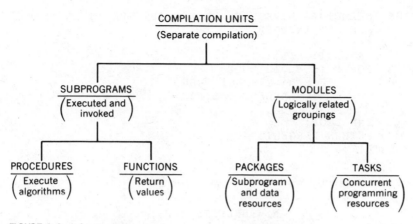

FIGURE 8.3 Ada structure.

The general structure of Ada is given in Figure 8.3. Some examples of Ada typing are

```
R:FLOAT;
I:INTEGER;
N:CONSTANT INTEGER:=0;
NUMBER,VAL:INTEGER;
BIT:BOOLEAN:=TRUE;
PALETTE:ARRAY(1..M)OF COLOR;
PI:CONSTANT:= 3.14159;
```

Ada also defines enumeration types such as:

```
TYPE DAY IS(MON,TUE,WED,THU,FRI,SAT,SUN);
TYPE COLOR IS (YELLOW,BLUE,RED,ORANGE);
TYPE COST IS (CAPITAL,EXPENSE,OVERHEAD);
```

An Ada procedure that multiplies two numbers is

```
PROCEDURE MULTIPLY IS
          I,M,N: INTEGER;
BEGIN
      GET(I);
      GET (M);
      N:= I*M;
      PUT (N);
END MULTIPLY;
```

Iteration can be illustrated in Ada by the following function example. When called, the function returns the largest number (LARGEST) of a list of 10 integers represented by the vector B(N).

```
FUNCTION    SELECTBIG(B:VECTOR) RETURN INTEGER IS
            LARGEST:INTEGER;
BEGIN

            LARGEST:=B(1);
            FOR N IN 2..10 LOOP
                IF B(N)>LARGEST THEN
                    LARGEST :=B(N)
                END IF;
            END LOOP;
            RETURN LARGEST;
END SELECTBIG;
```

The function is called by an expression of the form:

$$A:= SELECTBIG(W);$$

Tasks in Ada are entities that may operate in parallel. Tasks must be declared and developed and then they can be initiated by a procedure. An outline of a task initiation procedure is

```
PROCEDURE RUN_TASKS IS
     specification of tasks A,B,C, and D.
     specification of tasks 1,2,3, and 4.
     ------------------------------------
BEGIN
     INITIATE A,B,C,D;
     INITIATE 1,2,3,4;
     -----------------
END
```

In summary, concepts embodied in Ada are strong typing, packages, separate compilation, tasking, exceptions, and standard library I/O.

8.8

LISP

LISP (LISt Processing) was developed by John McCarthy of MIT circa 1960 and was designed for manipulation of symbols. Symbols could be characters, words, or sentences. LISP is a functional language in that it accomplishes symbol manipulation through functions.

LISP is an interactive language, providing rapid response and a good editing and debugging environment. It is widely used in the following applications:

- Learning—concepts "learned" by computer.
- Expert solving of problems—medical diagnosis, calculus, computer configuration.

- Speech and vision understanding.
- Natural language understanding—programs interacting with user in English, and so on.
- Education—provides information to people.

There are a number of LISP dialects. Some of the dialects along with their developers and approximate dates of development are

```
         LISP  McCarthy, 1960, MIT
      MACLISP  Moon, 1974, MIT
    INTERLISP  Teitelman, 1975, XEROX PARC/BBN
   Franz LISP  Foderaro, 1980, UC Berkeley
Standard LISP  Marti et.al, 1979
  Common LISP  Fahlman, 1983, CMU/DEC
```

8.8.1 DEFINITIONS

The fundamental entity in LISP is an *atom*. For example, in the following LISP function, there are three atoms: PLUS, 4.78, and 3.19.

```
(PLUS 4.78 3.19)
```

LISP functions are in Cambridge Polish form as seen in the example. This function returns the sum of 4.78 and 3.19. A *list* in LISP is a group of atoms. Examples of lists are

```
( PLUS 4.12 5.21 )
( THIS IS A LITTLE LIST )
( W X Y Z )
( (CD) (EF) )
```

An *S-expression* consists of atoms and lists taken collectively.
Some *S-expression* examples are as follows:

```
        MIN          atom,  S-expression
 (A LITTLE LIST)     list,  S-expression
        5.21         atom,  S-expression
 ( QUOTIENT 15 5)    list,  S-expression
```

Thus S-expressions are the data objects manipulated in LISP.

8.8.2 EXAMPLES

The following examples are LISP functions and the values returned when the functions are called. A LISP program is a series of such expressions, some more complex than others.

```
(TIMES 6 3 )                                18
(TIMES(MIN 4 5 6) (QUOTIENT 18 6))          12
(DIFFERENCE 6 4)                             2
(DIFFERENCE (PLUS 9 8) (TIMES 4 3))          5
```

Two interesting functions that serve to illustrate the power of LISP in manipulating symbols are the CAR and the CDR (coo'der). CAR returns the first element in a list while CDR returns a list containing the list without the first element. In using the CAR and CDR functions as well as other functions in LISP, some means are necessary to indicate whether the contents of a list should be evaluated (e.g., interpreted as a function) or treated strictly as an element not to be interpreted. In some LISP dialects, a quote mark, ', is used for this purpose. The quote mark prevents a list from being evaluated. For example, the function

(CAR ' (I   J   K   L))

returns I, the first element of the list and does not attempt to interpret the elements of the list. Similarly,

(CDR ' (I   J   K   L))

returns (J   K   L), all but the first element of the list. If functions are nested, however, as in

(CAR (CDR ' (I   J   K   L)))

the CDR function is performed first, returning (JKL). Then, CAR(JKL) is evaluated, returning J. If the quote mark were placed in a different position, a different result would be returned. For example, in the function

(CAR' (CDR(W   X   Y   Z)))

the quote mark indicates that all elements to its right are not to be evaluated. Thus the element CDR is seen no differently than elements W, X, Y, and Z. The function therefore returns the element CDR, which is the first element of the list.

Another related function in LISP is CONS. CONS(CONSTRUCTOR) inserts a new first element in a list. If the element W is to be inserted in the list (X   Y   Z), the function

(CONS 'W' (X   Y   Z))

would be used. It returns (W   X   Y   Z). For further illustration,

(CDR(CONS 'I' (J   K   L)))

returns (J K L). Conversely,

(CAR (CONS 'I' (J K L)))

returns the element I.

Functions can be developed by the user in LISP by using the function definition:

(DEFUN ⟨ function name⟩ (⟨formal parameter list⟩) ⟨ expression⟩)

where the brackets ⟨ ⟩ indicate the description that is to be placed between them. To illustrate, a function called AVERAGE to calculate the average of four numbers will be defined according to the function definition.

(DEFUN AVERAGE (A B C D)
     (QUOTIENT (A B C D) (4.0)))

The function AVERAGE, to calculate the average of the four numbers 10.0, 20.0, 30.0, and 40.0, is invoked by

(AVERAGE 10.0 20.0 30.0 40.0)

and the value 25.0 will be returned.

A function, DISTANCE, to calculate the distance covered by an automobile traveling at a speed, V, for a period of time, T, can be defined as

(DEFUN DISTANCE (V T)
    (TIMES V T))

Function DISTANCE is invoked to calculate distance traveled by an automobile moving at a rate of 55 miles per hour for a period of 3.5 hours by writing

(DISTANCE 55.0 3.5)

The function DISTANCE will return 192.5 miles.

8.8.3 CONTROL

The two main methods of program control in LISP are recursion and iteration. In recursion, a function uses itself in its own definition. The use of recursion in LISP is one of its most powerful features. A general example of recursion is illustrated with the function example.

(DEFUN EXAMPLE (W X)
    (. . . . . .(EXAMPLE W ( ))))

In iteration, the repetition of function execution occurs without recursion, such as in a FORTRAN DO loop.

To summarize, a program in LISP is a series of expressions. LISP is a functional language that manipulates symbols in an iterative manner.

8.9

PL/M-86[4]

The high-level language, PL/M-86, developed by Intel Corporation, will be discussed in more depth than the other high-level languages, since it is aimed at the popular Intel family of microprocessors and is widely used in interfacing applications.

8.9.1   BACKGROUND

PL/M-86 is a block-structured language and encourages structured programming techniques. The language is strongly typed, that is, it will flag multiple-type expressions at compile time to prevent potential run-time problems. Since PL/M was developed for microprocessor applications, it provides access to the features of the target microprocessor. It also provides facilities for data structures and pointer-based dynamic variables.

8.9.2   DESCRIPTION

The character set of PL/M-86 includes all upper- and lower-case alphabetic characters, all numerals, all special characters, horizontal tab, space, carriage-return, and line-feed. Upper and lower case letters are interpreted identically, except when used in strings.

PL/M-86 special characters are given in Table 8.1. Comments are delimited by /* */ and may contain any printable ASCII characters including space, carriage-return, tab, and line-feed. As an example, the following line is a comment.

/*This is a comment*/

8.9.3   IDENTIFIERS

Identifiers are used to name procedures, variables, constants, and statement labels. Identifiers can have a length of up to 31 characters with the first

---

[4] This material is taken from a course developed by Tom Maier and John Hudak of the Mellon Institute, Computer Engineering Center.

TABLE 8.1 PL/M-86 Special Characters

Symbol	Name	Use
$	Dollar sign	Number and identifier spacer
=	Equal sign	Two distinct uses:
		1. assignment operator
		2. relational test operator
:=	Assign	Embedded assignment operator
@	At-sign	Location reference operator
.	Dot	Three distinct uses:
		1. decimal point
		2. structure member qualification
		3. address operator
/	Slash	Division operator
/*		Beginning-of-comment delimiter
*/		End-of-comment delimiter
(	Left paren	Left delimiter of lists, subscripts, and expressions
)	Right paren	Right delimiter of lists, subscripts, and expressions
+	Plus	Addition operator
−	Minus	Subtraction or unary minus operator
'	Apostrophe	String delimiter
*	Asterisk	Multiplication operator
<	Less than	Relational test operator
>	Greater than	Relational test operator
<=	Less or equal	Relational test operator
>=	Greater or equal	Relational test operator
<>	Not equal	Relational test operator
:	Colon	Label delimiter
;	Semicolon	Statement delimiter
,	Comma	List element delimiter

character being alphabetic. The remaining characters must be either alphabetic, numeric, an underscore, or a dollar sign. Any words can be used as identifiers except for PL/M-86 reserved words as shown in Table 8.2. Some valid identifiers are

```
NEW
CALCULATE
check
CALC_Routine
NEW$Routine
scoop_up
```

8.9.4   DATA TYPES

PL/M-86 supports operations in both unsigned and signed arithmetic. In unsigned arithmetic, BYTES, WORDS, and DWORDS are defined. A byte is an

TABLE 8.2 PL/M-86 Reserved Words

These are the reserved words of PL/M-86. They may not be used as identifiers.

ADDRESS
AND
AT
BASED
BY
BYTE
CALL
CASE
DATA
DECLARE
DISABLE
DO
ELSE
ENABLE
END
EOF
EXTERNAL
GO
GOTO
HALT
IF
INITIAL
INTEGER
INTERRUPT
LABEL
LITERALLY
MINUS
MOD
NOT
OR
PLUS
POINTER
PROCEDURE
PUBLIC
REAL
REENTRANT
RETURN
STRUCTURE
THEN
TO
WHILE
WORD
XOR

TABLE 8.3 Valid Constants in PL/M

Real	Hexadecimal	Decimal	Octal	Binary
1.68	4AF6H	123D	4570	11101111B
2046	BF68H	4596D	1240	11000011B
.18E+3				
.15E−5				

8-bit unsigned binary number (0-255), a word is a 16-bit unsigned binary number (0-65,535), and a DWORD is a 32-bit unsigned binary number (0-4,294,967,295). For signed arithmetic, a 16-bit signed binary number, or INTEGER, representing values from −32,768 to 32,767 is used.

Floating point (REAL) numbers are made up of a 1-bit sign, 8-bit exponent (offset by 128), and a 23-bit significand with implied significant bit and binary point, totaling 32 bits. The form of a floating point number is

|←————————————— 32 bits ————————————→|

1	8	23
sign	actual exponent +127	binary number with implied significant bit and exponent

The range of numbers that can be represented by this format is

$$-1.17 * 10^{-38}, 0, 1.17 * 10^{-38}$$
$$-3.37 * 10^{38} - 3.37 * 10^{38}$$

Real, hexadecimal, decimal, octal, and binary constants are valid in PL/M-86. Examples of each type of constant are shown in Table 8.3.

Another classification is a *pointer*. A pointer is a 32-bit address of a memory storage location of the form,

```
selector:offset
```

where the selector is a base address and the offset is a displacement from the base address.

8.9.5 OPERATORS

The arithmetic operators +, −, *, and / operators are supported in PL/M-86. The rules for using these operators are

1. If both operands are of similar type, the result is the same type, except in the case of BYTEs where the * and / operations produce WORD results.

2. If the operands are dissimilar, BYTE and WORD are converted to WORD or DWORD by adding leading zeroes as required, the result being the largest type size encountered.

3. If one of the operands is an unsigned whole number and the other is a constant, the type of the constant is determined by the value of the constant (i.e., a BYTE if the value is ≤ 255, a WORD if the value is ≤ 65,535, etc.).

4. If one of the operands is an INTEGER and the other is a constant, the operation is treated as if it were between INTEGER values. If the whole-number constant exceeds 32,767, the operation is invalid.

5. If one of the operands is a whole-number constant and the other is of type REAL, POINTER, or SELECTOR, the operation is invalid.

6. Arithmetic operations on POINTER- or SELECTOR-type variables are invalid.

The MOD operator is permitted only with BYTE, WORD, DWORD, and INTE-GER operands. The result of a MOD operator on an operand is a remainder. For example, in the PLM-86 statements

```
DECLARE (W, X, Y) INTEGER DATA (18,4);
Y= W MOD X;
```

the value of Y is 2.

Defined relational operators in PL/M-86 are

< less than	> greater than
< = less than or equal to	> = greater than or equal to
<> not equal to	= equal to

These operators are binary operators and are used to compare operands of the same type. Real numbers can be compared with reals, pointers with pointers, and so on. Mixed whole-number comparisons, such as BYTE/WORD, BYTE/DWORD, and so on, are permitted, following the same rules as for arithmetic operators. The result of a comparison using the relational operators is a BYTE equal to OFFH if the result is true and equal to OOH if the relationship is false.

Logical operators in PL/M-86 are NOT, AND, OR, and XOR (Exclusive OR). The logical operators can be used only with BYTE, WORD, and DWORD operands. Mixed operations are permitted following the same rules as for arithmetic operators.

8.9.6   DECLARATIONS

Variables, procedures, macros, and constants are declared in PL/M-86. A variable can either be a scalar, an array, or a structure. A *scalar* is a single object whose value may or may not be known at compile time and may change during execution. An *array* is a list of scalars with the same identifier that are identified by a subscript. A *structure* is a list of scalars and/or arrays that use the same identifier and are differentiated by MEMBER-IDENTIFIERS separated from the identifier by a period.

Examples of typical declarations in PL/M-86 are

Scalar:
```
DECLARE VALUE REAL;
DECLARE SUM WORD, CHECK BYTE;
DECLARE FEE BYTE;
DECLARE (FEE, FI) BYTE;
```
Array:
```
DECLARE BOOK(12) REAL;
DECLARE GAMMA(10) BYTE, RAY(150) DWORD;
DECLARE (FIRST,SECOND,THIRD)(10) BYTE;
```
Structure:
```
DECLARE RECORD STRUCTURE (ONE BYTE, TWO WORD);
(this structure has two members, RECORD.ONE and RECORD.TWO)
DECLARE  GROUP  STRUCTURE (ELEMENT NO(130)BYTE,
                TOTAL REAL)
```
Procedure:
```
Procedure_name:    PROCEDURE[(parameters)][type][attributes];
        [declaration statements;]
        [executable statements;]
        [return[expression];]
        END;
```
Substitution:

In text substitution, a declared word is replaced in-line with another string. The declaration for text substitution is
```
DECLARE identifier LITERALLY 'string';
```
For example,
```
DECLARE STATE LITERALLY 'GOOD';
```
causes the string GOOD to replace the identifier STATE, anywhere it is used, during compilation.

8.9.7   INITIALIZATION

Constants in PL/M-86 must be initialized before they are encountered during execution. Two methods of initialization of constants are

1. `DECLARE VALUE REAL;`
   `VALUE = 5.17362;`
2. `DECLARE VALUE REAL DATA (5.17362);`

Method 2 is a compile-time initialization.
    Similarly, variables are initialized as:

1. `DECLARE NEW REAL;`
   `NEW = 15.7;`
2. Using the INITIAL attribute for compile-time initialization,
   `DECLARE SUM REAL INITIAL (5.72);`
   `DECLARE TOTAL(4) BYTE INITIAL (1,46,8,10);`
   `DECLARE ALPHA(4) BYTE INITIAL ('FULL');`

An array can also be initialized by the implicit dimension specifier (∗) to have the same number of elements as the value list. The implicit declaration should not be used to specify an array whose elements are structures, an array that is a member of a structure, or in a factored declaration. An example of an implicit declaration is

`DECLARE TTL_CHIP(*) BYTE INITIAL ('7400');`

In this example, array TTL_CHIP will have four elements.

8.9.8   UTILIZATION OF DATA AND INITIAL ATTRIBUTES

The following rules apply to the use of DATA and INITIAL attributes in PL/M-86:

1. DATA attribute can ONLY be used with constants, not variables.
2. DATA initializations can occur at ANY block level, INITIAL can only be used at the module level.
3. If the DATA attribute is used in a PUBLIC declaration when compiling with the ROM option, DATA (without any value list) must be used in all EXTERNAL declarations that reference the constant.
4. DATA and INITIAL CANNOT be used in the same declaration.
5. NO initializations are permitted with based variables or with the EX-TERNAL attribute.

6. Either initialization can be used with the AT attribute but if this causes multiple initializations, the results CANNOT be predicted.

7. Identifiers with DATA attributes must NOT appear in the left side of an assignment statement. This will be flagged as an error.

8.9.9   PROCEDURES

A PROCEDURE is a subroutine with a carefully designed interface through which parameters are passed. A result may or may not be returned. Procedures may be typed or untyped. A *typed procedure* has a type such as WORD, BYTE, REAL, INTEGER or POINTER in its PROCEDURE statement and always returns a value of this type to the calling routine. The body must always contain a RETURN statement with an expression. An *untyped procedure* has no type specified, does not return a value, and is invoked by a CALL statement.

Termination of a procedure is accomplished by execution of a RETURN statement, reaching an END statement, or by executing a GOTO to a statement outside the procedure body. The RETURN statement has the form

```
RETURN;
```

in an untyped procedure and the form

```
RETURN expression;
```

in a typed procedure. The value of the expression is the value returned by the procedure.

A general outline of a procedure is

```
name:PROCEDURE[(parameter list)][type][attributes];
statement 1;
statement 2;
statement 3;
       .
       .
       .
statement n;
END[name];
```

The procedure name is not a label and procedures should not be labeled. Procedure names and other program labels should not be the same.

An example of a short procedure, CHECK, is as follows:

```
CHECK:PROCEDURE (amount) BYTE PUBLIC;
     IF AMOUNT<'0' CALL SPECIAL;
     IF AMOUNT>'9' CALL ERROR;
     END CHECK;
```

Parameters passed to procedures must be declared as nonbased scalar variables in a DECLARE statement preceding the first executable statement in the procedure. When a procedure has more than one parameter, PL/M-86 does not guarantee the order in which actual parameters are evaluated when the procedure is activated. The CALL statement or function reference that activates a procedure with formal parameters contains a list of actual parameters. Each actual parameter is an expression whose value is assigned to the corresponding formal parameter in the procedure, before the procedure begins to execute. An example that illustrates a procedure and its calling sequence is

```
RANGE_CHECK:PROCEDURE (PTR, N, LOWER, UPPER);
            DECLARE PTR POINTER;
            DECLARE (N, LOWER, UPPER, I) BYTE;
            DECLARE ITEM BASED PTR (1) BYTE;

            DO I=0 to N-1;
            IF (ITEM (I)<LOWER) OR (ITEM(I)>UPPER)
               THEN CALL ERRORSET;
            RETURN;       /*optional RETURN*/
            END RANGE_CHECK
```

```
/*errorset is a procedure that is defined elsewhere*/
```

The procedure is called as follows:

```
CALL RANGE_CHECK(@QUANTS, 25, LOW, HIGH);
```

The @ preceding QUANTS in the CALL expression forms a location reference of type POINTER. Thus @QUANTS is the address of the scalar QUANTS.

The following three examples illustrate a typed procedure, untyped procedure, and a typed procedure utilizing arrays.

Typed Procedure:

```
TORQUE: PROCEDURE (HP,S) REAL;
        DECLARE (HP,S) REAL;
        RETURN (HP*5252)/S;
        END TORQUE,
```

This procedure is called by:

```
TORQUE_EXT = TORQUE (HP_RATED, S_FL);
```

```
/* external torque is returned in lb-ft*/
```

Untyped Procedure:

```
ESTOP:   PROCEDURE (SPEED);
         DECLARE SPEED INTEGER;
         IF SPEED >1750 THEN MAIN=ZERO;
         RETURN;
         END STOP;
```

The procedure CALL is

```
CALL ESTOP(TACHVAL);
```

Typed Procedure Utilizing Arrays:

```
BATTERY_VOLTAGE: PROCEDURE (INDEX, BATMAX) INTEGER;
                 DECLARE INDEX POINTER,
                 ARRAY BASED INDEX (1) INTEGER,
                 (BATMAX, TOTAL, I) INTEGER;

                 TOTAL=0;
                       DO I=0 TO BATMAX;
                       TOTAL=TOTAL+ARRAY(I);
                       END;
                 RETURN TOTAL;
                 END BATTERY_VOLTAGE;
```

This procedure is called by:

```
TOTAL_VOLTS=BATTERY_VOLTAGE(@TEST1,10);
```

The purpose of this procedure is to obtain the total voltage of the set of batteries pointed to by TEST1. The array contains the voltage readings of the batteries (from the 0th to the max).

8.9.10 ATTRIBUTES

As a background to discussing attributes in PL/M, the following definitions relative to block structuring in a language are given:

1. "Outer level" or "Exclusive extent" is defined as including the labels and statements contained in a block but NOT contained in any nested blocks.
2. "Inner level" or "Inclusive extent" is defined as including all of the outer level and all of the nested blocks.

3. "Scope" of an object is defined as those parts of a program where its name, type, and attributes are recognized.

4. An "object" is defined as a variable, label, procedure, or symbolic constant.

The attributes PUBLIC and EXTERNAL in PL/M-86 permit the user to extend the scope of an object, except modules, so that its name is usable in other modules. A name of a module cannot be declared with an attribute. This capability makes the name available for use in modules other than the one in which in the name was defined. Specifically, the statement

```
DECLARE BUS BYTE PUBLIC;
```

allocates a byte known as BUS whose address is known to any other module in which the declaration

```
DECLARE BUS BYTE EXTERNAL
```

is made. In the following example, a procedure NEW_PROC in one module is invoked by another module. The defining procedure is

```
NEW_PROC: PROCEDURE (ALPHA, BETA) WORD PUBLIC;

    DECLARE (ALPHA, BETA) WORD;

    /* other declarations and statements */

        .

        .

        .


END NEW_PROC;
```

The other module invokes NEW_PROC by means of a "dummy" declaration before activating the procedure.

```
NEW_PROC: PROCEDURE (GAMMA, ZETA) WORD EXTERNAL;

    DECLARE (GAMMA, ZETA) WORD;

END NEW_PROC;
```

Within each program, each procedure with extended scope must have exactly one defining declaration. In other words, the procedure statement must contain the PUBLIC attribute. The PUBLIC attribute may only be used at the outer level of a module. For EXTERNAL attributes, the following rules should be used:

1. The EXTERNAL attribute may only be used at the outer level of the module, that is, the main program.
2. The EXTERNAL attribute may only be used if the procedure is declared PUBLIC in another module of the same program.
3. The EXTERNAL attribute may not be used in the same procedure statement as PUBLIC, INTERRUPT, or REENTRANT attributes. Note, however, that the defining declaration of a procedure may have the INTERRUPT and REENTRANT attributes.
4. A usage declaration of a procedure should have the same number of parameters as the defining declaration.
5. The procedure body of a usage declaration may not contain anything except the declarations of the formal parameters. The formal parameters must be declared with the same types as in the defining declaration.
6. The END statement of a usage declaration may not be labeled.

Similarly, the following guidelines should be applied relating to PUBLIC and EXTERNAL attributes:

1. Must be declared in the outermost level of a module, that is, NEVER in a nested block.
2. Only one may appear in any declaration.

       ILLEGAL Usage DECLARE BOZO WORD PUBLIC EXTERNAL;

3. A name can be declared PUBLIC only ONCE.
4. A name may be declared EXTERNAL if and only if it is declared PUBLIC somewhere else.
5. Where a name is declared EXTERNAL, it MUST be given the same type as was used where it was declared PUBLIC.
6. A name declared EXTERNAL must NOT be given a location, for example, by using the AT attribute.
7. A name declared EXTERNAL CANNOT be initialized using the DATA or INITIAL attribute.
8. Neither PUBLIC or EXTERNAL attributes are permitted with a name that is based.

Of particular interest in interfacing is the handling of interrupts. The INTERRUPT attribute provides for developing a procedure to handle interrupts. The procedure is activated when a corresponding interrupt is received by the microprocessor. An interrupt can also be initiated by the PL/M-86 statement:

```
CAUSE INTERRUPT(constant);
```

where constant is in the range 0 to 255.

An `INTERRUPT` attribute is of the form

```
INTERRUPT n
```

where n is any whole number from 0 to 255. Each number can only be used once in a program. This `INTERRUPT` attribute is used in conjunction with the `ENABLE` and `DISABLE` statements. Since the 8086 microprocessor always starts in the disabled state, the `ENABLE` statement must be used to enable interrupts. A `HALT` statement in PL/M-86 also will enable interrupts. Conversely, the `DISABLE` statement is used to disable interrupts.

The following sequence of events occurs when an external device interrupts the microprocessor (assuming that the interrupts are enabled);

1. The machine instruction currently being executed will be completed.
2. Interrupts are disabled.
3. The interrupt procedure whose number corresponds to the number sent by the interrupting device is activated. If no procedure by that number exists, the results are `UNDEFINED`.
4. When the interrupt procedure terminates, interrupts are automatically reenabled, and control returns to the point in the program where the interrupt occurred.

   Note: It is possible for the procedure to terminate by executing a `GOTO` with a target coutside the procedure in the outer level of the main program module. In this case, control will never be returned to the point where the program was interrupted and interrupts will not automatically be enabled.

When using the `INTERRUPT` attribute, the following rules apply:

1. The `INTERRUPT` attribute may not be used in combination with the `EXTERNAL` attribute. It may be used with the `PUBLIC` attribute.
2. It may only be used in a `PROCEDURE` statement in the outermost level of a program module.
3. The range of n is 0 to 255 inclusive.
4. The procedure must be untyped and may not have any parameters.

Note: A procedure with the `INTERRUPT` attribute may also be activated by means of a `CALL` statement. Bear in mind that interrupts are not automatically disabled upon activation of the procedure. Two things must be taken into consideration if this is done. First, the procedure called could have a DISABLE attribute as its first executable statement. If not, it will run with interrupts enabled. If this is desired, this module must be reentrant. This is accomplished by means of the REENTRANT attribute. If not made reentrant, the results and failure modes will be indeterminant.

If interrupts are disabled in the called routine, they are not automatically re-enabled by means of the RETURN or END statement.

An example of an interrupt routine that is activated when an external interrupt numbered 50 occurs is

```
EXT1:       PROCEDURE INTERRUPT 50;

      /* procedure statements */

      .

      .

      .

END EXT1;
```

The REENTRANT attribute allows a procedure to temporarily suspend execution, begin again with new parameters, and then complete the suspended segment. A procedure with this capability is said to be *reentrant*. If a procedure calls itself, this is known as *direct recursion*. A procedure that calls another procedure that, in turn, calls the first procedure before it runs to completion is an example of *indirect recursion*.

If a procedure will be used in a reentrant fashion, the REENTRANT attribute should be used with the following rules:

1. Any procedure that may be interrupted and is also activated from within an interrupt procedure should have the REENTRANT attribute.
2. Any procedure that calls itself (directly or indirectly) should use the REENTRANT attribute.
3. Any procedure that is activated by a reentrant procedure should also have the REENTRANT attribute.

8.9.11   PL/M-86 FLOW CONTROL

Flow control statements are used to alter the sequence of execution of program statements. These statements fall into the general categories:

1. DO statements.
2. IF... THEN...ELSE statements.
3. GOTO statements.

4. HALT statement.
5. CAUSES INTERRUPT statement.

As discussed earlier, the use of GOTO statements should be avoided, in accordance with structured programming practice.

There are four categories of DO statements in PL/M-86. These are

1. Simple DO.
2. DO WHILE.
3. Iterative DO.
4. DO CASE.

The simple DO statement is of the form:

```
BLOCK_NAME:DO;
        statement 0; /* execute each statement in order */

        statement 1;

            .

            .

            .

        statement n;

    END [BLOCK_NAME];
```

DO statements of this type can also be nested as shown in the following example:

```
LABL1:    DO;
              statement 0;
              statement 1;
LABL2:            DO;
                      statement a;
                      statement b;
                  END LABL2;
              statement 2;
              statement 3;
              statement 4;
              statement 5;
          END LABL1;
```

The DO CASE statement follows the logic outlined in Figure 8.4. The general DO CASE statement is

```
BLOCK_NAME: DO CASE select_expression
               statement 0; /* execute if select_expression =0 */
               statement 1; /* execute if select_expression +1 */
               .
               .
               .
               statement n; /* execute if select_expression =n */
           END [BLOCK_NAME];
```

In this DO CASE statement, the select_expression must yield a BYTE, WORD, or INTEGER value. Also, if the select_expression evaluates to a value that is greater than n or is negative, the result of the DO CASE statement is undefined.

The DO WHILE statement flow chart is given in Figure 8.5. The general form of the DO WHILE statement is

```
BLOCK_NAME: DO WHILE expression_true;
               statement 0;   /* execute each statement in order */
                              /* as long as the expression is true *
                              /* when it first tests false, */
                              /* go to END */
               statement 1;
               .
               .
               .
               statement n;
           END [BLOCK_NAME];
```

The iterative DO statement logic is shown in Figure 8.6. The general iterative DO statement form is as follows:

```
BLOCK_NAME: DO counter = start_exp TO limit_exp [BY step_exp];
               statement 0; /* execute each statement in order */
               statement 1;
               .
               .
               .
               statement n;
           END [BLOCK_NAME];
```

An example of an iterative DO block is

```
DO I = 1 TO 5;
        CALL HORN;
END;
```

In this example, if HORN is a procedure that blows a horn, the horn will be activated five times. The index variable in the iterative DO block can also be used within the block. The following example uses the index variable to calculate the product of the numbers from 1 to 20 and put the result in PRODUCT.

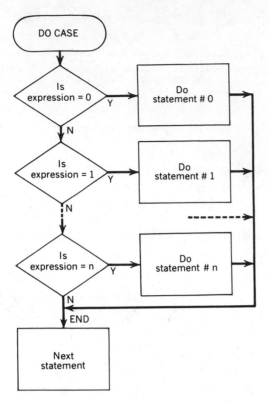

FIGURE 8.4 DO CASE statement logic.

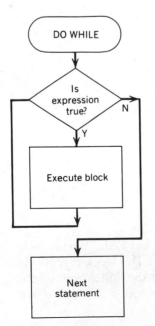

FIGURE 8.5 DO WHILE statement logic.

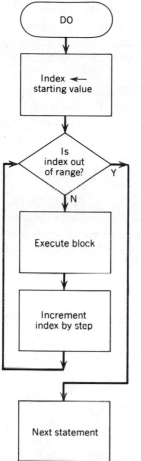

FIGURE 8.6 Iterative DO statement logic.

```
PRODUCT = 1;
DO I = 2 TO 20;
        PRODUCT = PRODUCT * I;
END;
```

The following rules are observed for the iterative DO statement:

1. The counter MUST be a BYTE, WORD, or INTEGER value.
2. The start_exp, the limit_exp and the step_exp may be any valid PL/M-86 expression types, BYTE, WORD, or INTEGER. If the type of the counter or any of the other expressions is INTEGER, then they must all be INTEGER.
3. For BYTE or WORD counters, the loop is exited immediately if:
   a. The counter exceeds the limit_exp.
   b. The counter "rolls over" due to modulo arithmetic.

4. For INTEGER counters, the loop is exited immediately in 3 cases:
   a. If the counter "exceeds" the limit_exp in the desired direction.
   b. If the step_exp goes from + to − and the counter is < limit_exp.
   c. If the step_exp goes from − to + and the counter is > limit_exp.

Another class of statement, the IF... THEN... ELSE statement can also change the sequence of execution of PL/M-86 statements. The logic of this type of statement is given in Figure 8.7. The form of the IF... THEN... ELSE statement is

```
IF expression THEN statement #1 [ELSE statement #2];
```

The expression is evaluated as if it were a BYTE and is evaluated as TRUE if the least significant bit (lsb) is a 1 and FALSE if the lsb is a 0. The statements may be any valid PL/M-86 statements or blocks of statements. The ELSE statement #2 is optional. A general example of the IF... THEN... ELSE statement incorporating DO statements is as follows:

```
IF  X=Y  THEN
          DO;
                statement 0;
                statement 1;
                    .
                    .
                    .
                statement n;
          END;
ELSE
          DO;
                statement a;
                statement b;
                    .
                    .
                    .
                statement m;
          END;
```

Another interesting example uses DO statements in "sequential" IF statements. This type of construction is useful whenever a series of tests is to be made, but the tests are to be terminated when one of the tests in the sequence succeeds.

```
IF  LETTER = 'E' THEN statement_1;

ELSE DO;

      IF LETTER = 'G' THEN statement_2;

      ELSE DO;
```

```
        IF LETTER = 'H' THEN statement_3;

        ELSE statement_4;

    END;
```

The GOTO statement in PL/M-86 will be discussed in this section, but its indiscriminate use is discouraged. Improper use of GOTO statements will result in a program that is difficult to debug, modify, and comprehend. Instead of GOTO statements, DO, DO WHILE, DO CASE, IF, or procedure activation statements should be used.

The GOTO statement transfers control of the program directly to the specified statement. It is of the form:

```
GOTO label;
```

Examples are

```
GOTO NEW:
GOTO CHECK;
GOTO BEGIN; /* embedded blank is allowed */
            /* between GO and TO */
```

The use of the GOTO statement is restricted to:

1. From a GOTO in the outer level of a given block to a labeled statement in the outer level of the same block.
2. From a GOTO in an inner block to a labeled statement in an enclosing block. If the inner block is a PROCEDURE, then the transfer is confined to the main program module.

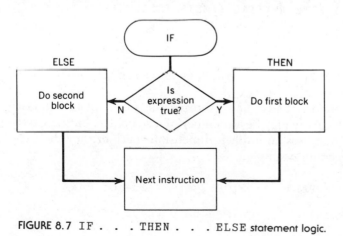

FIGURE 8.7 IF . . . THEN . . . ELSE statement logic.

3. From a GOTO in a program module to a labeled statement in the outer level of the main program module, provided that the label is declared PUBLIC in the main module and EXTERNAL in the module containing the GOTO.

The HALT statement is of the form

```
HALT;
```

and causes the microprocessor to come to a halt with interrupts enabled.

The CAUSEINTERRUPT statement is of the form CAUSEINTERRUPT (constant) and causes a software interrupt to be generated. Constant is a whole number in the range 0 through 255 and indicates a particular interrupt type to the microprocessor. Program control is then transferred to the appropriate interrupt vector.

8.9.12   PL/M-86 BUILT-IN PROCEDURES

PL/M-86 provides built-in procedures and functions that are available to the user and act as if they were declared in a global block. The three built-in procedures LENGTH, LAST, and SIZE take variable names as parameters and return information based on the variable declarations.

Length is a WORD function that returns the number of elements in an array. The array may be a member of a structure but may not be an EXTERNAL array that uses an implicit dimension specifier. Some examples of the use of the LENGTH procedure are

```
DECLARE BOOK STRUCTURE (CHAPTER WORD, PARA(3) WORD);

DECLARE TEXT1(*) BYTE INITIAL ('THIS IS TEXT');


VAL1 = LENGTH (BOOK.PARA) /* VAL1 gets value of 3 */

VAL2 = LENGTH (TEXT1)     /* VAL2 gets value of 12*/
```

Another WORD function is LAST. LAST returns the subscript of the last element of an array. The array may be a member of a structure, but may not be an EXTERNAL array that uses an implicit dimension specifier. For the two declarations in the preceding example,

```
STOR1 = LAST(BOOK.PARA) /* STOR1 gets value of 2 */

STOR2 = LAST(TEXT1)     /* STOR2 gets value of 11 */
```

SIZE is a third word function and returns the number of bytes occupied by an array or structure. Using the same declarations as in the previous two examples:

```
MEM1 = SIZE(BOOK) /* MEM1 gets value of 8 */

MEM2 = SIZE(TEXT1) /* MEM2 gets value of 12 */
```

Ten functions that provide conversion from one type to another and from signed values to or from absolute magnitudes are given in Table 8.4. These functions are invoked by writing

```
function_name(expression)
```

where the function has a type and the expression a value.

Examples of the 10 conversion functions are as follows:

1. ```
   DECLARE ALPHA BYTE, BETA WORD,
   GAMMA DWORD;
   RSLT1 = LOW(ALPHA);          /* RSLT1 gets ALPHA */
   RSLT2 = HIGH(ALPHA);         /* RSLT2 gets 0 */
   RSLT3 = LOW(BETA);           /* RSLT3 gets lsbyte of BETA */
   RSLT4 = HIGH(BETA);          /* RSLT4 gets msbyte of BETA */
   RSLT5 = LOW(GAMMA);          /* RSLT5 gets lsword of GAMMA */
   RSLT6 = HIGH(GAMMA);         /* RSLT6 gets msword of GAMMA */
   ```

2. ```
   DECLARE ALPHA BYTE, BETA WORD;
   RSLT1 = DOUBLE(ALPHA);       /* RSLT1 gets 00:ALPHA */
                                /* 8 high order 0 bits are */
                                /* appended to ALPHA value */
   RSLT2 = DOUBLE(BETA);        /* RSLT2 gets 0000:BETA */
                                /* 16 high order bits are */
                                /* appended to BETA value */
   ```

3. ```
   RSLT1 = FIX(3.6);            /* RSLT1 gets 4 */
   RSLT2 = FIX (5.3);           /* RSLT2 gets 5 */
   RSLT3 = FLOAT (5);           /* RSLT3 gets 5.0 */
   ```

4. ```
   RSLT1 = INT (00001010);      /* RSLT1 gets 10 */
   RSLT2 = SIGNED (00001000);   /* RSLT2 gets 8 */
   RSLT3 = UNSIGN (8);          /* RSLT3 gets 0000000000001000 */
   ```

5. ```
   RSLT1 = ABS(-5.36);          /* RSLT1 gets 5.36 */
   RSLT2 = IABS(-4);            /* RSLT2 gets 4 */
   ```

6. ```
   ROL(10011100B,2) returns 01110010B
   ROR(10011100B,1) returns 01001110B
   ```

TABLE 8.4 Explicit Type and Value Conversions

Procedure Name	Type	Parameter Expression Type	Function	Result Returned
LOW	BYTE WORD	WORD DWORD	Converts WORD value to BYTE value Converts DWORD value to WORD value	BYTE value unchanged Low-order byte of WORD Low-order word of DWORD
HIGH	BYTE WORD	WORD DWORD	Converts WORD value to BYTE value Converts DWORD value to WORD value	0 (zero) High-order byte of WORD High-order word of DWORD
DOUBLE	WORD DWORD	BYTE WORD DWORD	Converts BYTE value to WORD value Converts WORD value to DWORD value	WORD value, by appending 8 high-order zero bits DWORD value, by appending 16 high-order zero bits DWORD value unchanged
FLOAT	REAL	INTEGER	Converts INTEGER value to REAL value	Same value of type REAL
FIX	INTEGER	REAL	Converts REAL value to INTEGER value (rounds toward zero)	INTEGER value modulo 32768, i.e., within range ± 32767
INT	INTEGER	BYTE WORD	Converts BYTE or WORD to INTEGER; interprets parameter as positive	Corresponding INTEGER value within range 0 to 32767
SIGNED	INTEGER	BYTE WORD	Converts a WORD value to an INTEGER value	BYTE value is extended with 8 high-order zeros; WORD value unchanged
UNSIGN	WORD	INTEGER	Converts an INTEGER value to a WORD value	The bit pattern is unchanged but can now be used in WORD expressions
ABS	REAL	REAL	Converts negative value to positive	Absolute value of expression supplied. If positive, returned unchanged; if negative,—(expression) is returned
IABS	INTEGER	INTEGER	Converts negative value to positive	Absolute value of expression supplied. If positive, returned unchanged; if negative,—(expression) is returned

There are additional built-in procedures in PL/M-86 that will not be covered here. The built-in procedures discussed to this point serve to illustrate the capabilities and operation of typical PL/M-86 procedures.

8.9.13 EXAMPLES

Two examples will be presented to consolidate the PL/M-86 high level language material. These examples are taken directly from the Intel PL/M-86 User's Guide.

The following sample program implements a straight insertion sort algorithm based on Knuth's "Algorithm S" in *The Art of Computer Programming,* Vol. 3, page 81. Readers who look up Knuth's algorithm should note the following differences:

- The algorithm has been adapted to PL/M-86 usage by using an array of structures to represent the records to be sorted. The sort key for each record is a member of the structure for that record.
- The algorithm has been modified by using a DO WHILE block to achieve the same logical effect as the GOTOs implied in steps S3 and S4 of Knuth's algorithm.
- The index I is used in a slightly different manner (it is initialized to J instead of J–1).

The effect of the algorithm is to arrange 128 records in order according to the values of their keys, with the smallest key at the beginning (lowest location, and the largest key at the end (highest location).

The sorting method is as follows. Assume that the records are all in memory, stored as an array of structures. The key for each record is a member of the structure.

Now we go through the array from the second record (record number 1) upwards. When we reach any given record (the "current" record), we will already have sorted the preceding records. (The first time through, when we look at record number 1, record number 0 is the only preceding record.)

We take the current record, store it temporarily in a buffer, and look backwards through the preceding records until we find one whose key is not greater than that of the current record. Then we put the current record just after this record.

The sample program and a detailed explanation follow. Please study the program and the explanation until you understand how the program works (especially the DO WHILE block, which is controlled by a more complex condition expression than we have seen up to this point).

```
M: DO;                          /* Beginning of module */
   DECLARE RECORD (128) STRUCTURE (KEY BYTE, INFO WORD);
   DECLARE CURRENT STRUCTURE (KEY BYTE, INFO WORD);
   DECLARE (J,I) WORD;
   /* Data are read in to initialize the records. */
   SORT:   DO J = 1 TO 127;
      CURRENT.KEY = RECORD(J).KEY;
      CURRENT.INFO = RECORD(J).INFO;
      I = J;
      FIND:   DO WHILE I > 0 AND RECORD(I-1).KEY > CURRENT.KEY;
                 RECORD(I).KEY = RECORD(I-1).KEY;
                 RECORD(I).INFO = RECORD(I-1).INFO;
                 I = I-1;
      END FIND;
      RECORD(I).KEY = CURRENT.KEY;
      RECORD(I).INFO = CURRENT.INFO;
   END SORT;
                 /*Data are written out from the records. */
   END M;        /*End of module*/
```

Let us now consider the text of this program. First we declare the following variables:

- RECORD, an array of 128 structures to hold the 128 records. Each structure has a BYTE member that is the sort key, and a WORD member that could contain anything (in a working program, this would be the data content of the record).
- CURRENT, a structure used as a buffer to hold the current record while we look back through the records already sorted. Its members are like those of one structure element of RECORD.
- J, which will be used as an index variable in an iterative DO statement. J is always the subscript of the current record. When J becomes greater than 127, the sort is done.
- I, which will be used like an index variable in controlling a DO WHILE block. I−1 is always the subscript of a previously sorted record.

A working program would include code at this point to read data into the array RECORD. At the end of the program, there would be code to write out the data from RECORD. In this example, we omit this code because it would make the example too lengthy and because the method used for I/O would depend on the particular system used to execute the program. Comments have been inserted in place of this code.

The executable part of the program is organized as two DO blocks, one nested within the other. The outer block (labeled SORT) is an iterative DO block that goes through the records one at a time. The record selected by the index variable J each time through this block is the "current record." (Notice that J is never 0—because of the way the algorithm is defined, we must have a preceding element to look back at, and so we start with the second element of the array and look back at the first.)

The first two assignment statements in the block transfer the current record into CURRENT. The next statement sets the initial value for I, which will be used to control the inner block.

The inner block (labeled FIND) is the one that looks back through previously sorted records to find the right place to put the current record. The way this block is controlled is worth examining. The variable I is used like an index variable in an iterative DO, but it is changed explicitly inside the block, instead of automatically as in an iterative DO statement. The DO WHILE construction is used instead of an iterative DO because it allows two or more tests to be combined—in this case, by means of an AND operator.

I is set to J before the first time through the DO WHILE block, and decremented each time through. As long as I remains greater than 0, the first half of the DO WHILE condition is satisfied.

The value I−1 is the subscript of the record being "looked back at." The second half of the DO WHILE condition is that the key of this record must be greater than the key of the current record.

We are looking for a previously sorted record whose key is not greater than the key of the current record. Thus the condition in the DO WHILE statement will cause the DO WHILE block to be repeatedly executed until such a record is found, or until I reaches 0 (meaning that all previously sorted records have been examined).

Each time the DO WHILE block is executed, it moves the I−1st record "up" into the Ith position, and then decrements I.

When the condition in the DO WHILE statement is *not* met, one of the following is true:

- I = 0, because we have looked through all the previously sorted records without finding one whose key is not greater than that of the current record. All of the previously sorted records have been moved "up" by one.
- I−1 is the subscript of a record whose key is not greater than the key of the current record. All of the previously sorted records whose keys *are* greater than that of the current record have been moved "up" by one.

In either case, the failure of the DO WHILE condition means that the current record (being held in CURRENT) belongs in the Ith position. It is transferred into this position by the two assignment statements that form the remainder of the outer DO block.

Now the outer DO block repeats with an incremented value of J, to consider the next unsorted record.

Notice that the entire program is contained within a simple DO block labeled M.

The second example is a rework of the sort program of the first example. In the following sample program we first declare a procedure called SORTPROC. This procedure makes only the following assumptions about the records it is to sort:

- Each record occupies a contiguous set of storage locations. Therefore, by using based variables each record can be handled as a sequence of bytes, even though the parts of a record are not necessarily BYTE scalars.
- The records themselves are also stored contiguously, so the entire set of records can be regarded as a single sequence of bytes. The location of the first byte of the first record is specified by the POINTER parameter PTR.
- All records are the same size; that is, each occupies the same number of bytes. This size is specified by the WORD parameter RECSIZE and may not exceed 128.
- In each record, the value of 1 byte is used as the sort key. Within each record, this byte is always in the same relative position, that is, the first byte in the record, or the third, and so on. This relative position (or offset) is specified by the WORD parameter KEYINDEX, which resembles an array subscript; that is, it is 0 if the key is the first byte in the record, 1 if the key is the second byte, and so on.
- The number of records is specified by the WORD parameter COUNT.

The program is followed by a detailed explanation.

```
SORT$MODULE: DO; /* Beginning of module. */

SORTPROC: PROCEDURE (PTR, COUNT, RECSIZE, KEYINDEX);
    DECLARE PTR POINTER, (COUNT, RECSIZE, KEYINDEX) WORD;

/*Parameters:
    PTR is pointer to first record.
    COUNT is number of records to be sorted.
    RECSIZE is number of bytes in each record, max is 128.
    KEYINDEX is byte position within each record of a BYTE
        scalar to be used as sort key.*/

    DECLARE RECORD BASED PTR (1) BYTE,
            CURRENT (128) BYTE,
            (I, J) WORD;

    SORT:
        DO J = 1 TO COUNT-1;
            CALL MOVB (@RECORD(J*RECSIZE), @CURRENT, RECSIZE);
            I=J;
```

```
        DO WHILE I > 0
                AND (RECORD((I-1)*RECSIZE+KEYINDEX) >
                                CURRENT(KEYINDEX));
            CALL MOVB (@RECORD((I-1)*RECSIZE),
                        @RECORD(I*RECSIZE), RECSIZE);
            I=I-1;
        END;
        CALL MOVB (@CURRENT, @RECORD(I*RECSIZE), RECSIZE);
    END SORT;
END SORTPROC;

/* Program to sort two sets of records, using SORTPROC. */

DECLARE SET1 (50) STRUCTURE (
    ALPHA WORD,
    BETA (12) BYTE,
    GAMMA INTEGER,
    DELTA REAL,
    EPSILON BYTE);

DECLARE SET2 (500) STRUCTURE (
    ITEMS(21) INTEGER,
    VOLTS REAL,
    KEY BYTE);

    /* Data are read in to initialize the records.  */

    CALL SORTPROC (@SET1, LENGTH(SET1), SIZE(SET1(0)),
            SIZE(SET1(0).ALPHA));
        CALL SORTPROC (@SET2, LENGTH(SET2), SIZE(SET2(0)),
            .SET2(0).KEY - .SET2(0));

    /* Data are written out from the records.  */

END SORT$MODULE; /* End of module. */
```

After the PROCEDURE statement and the declaration of the parameters, we declare a based BYTE array called RECORD. This array is based on the parameter PTR, which points to the beginning of the first record to be sorted. Therefore, a reference to a scalar element of RECORD will be a reference to some byte within the set of records to be sorted, as long as the subscript used with RECORD is less than the total number of bytes in all the records (i.e., the subscript must be less than COUNT * RECSIZE).

Note that a dimension specifier of 1 is used in declaring RECORD. We need to use a nonzero dimension specifier here, in order to use subscripts later in the procedure. However, the value of the dimension specifier is unimportant because RECORD is a based array and does not have any actual storage allocated to it. The value 1 is chosen arbitrarily.

Next we declare CURRENT, an array of 128 BYTE elements. Like the structure CURRENT in sample program 1, the array CURRENT will be used to store the "current" record. Note that the dimension (size) of the array CURRENT is what establishes the maximum size of the records that this procedure can handle. We have chosen 128 here, but in principle any dimension could be specified.

As in sample program 1, the WORD variables I and J are used to control the DO WHILE and iterative DO loops. They have the same meaning as before. However, here they are also used to calculate subscripts for the based array RECORD.

In the statement following the iterative DO, we used the built-in procedure, MOVB to copy a sequence of byte values from one storage location to another.

In the first activation of MOVB, the parameter @RECORD(J*RECSIZE) is the location of the beginning of the Jth record, and @CURRENT is the location of the beginning of the array CURRENT. Thus the effect of this CALL statement is to copy the Jth record into the array CURRENT.

To understand the DO WHILE statement, consider that RECORD ((I-1)*RECSIZE) would be the first byte of the (I-1)st record, so RECORD ((I-1)*RECSIZE + KEYINDEX) is the byte that is to be used as the sort key of the (I-1)st record. Similarly, CURRENT(KEYINDEX) is the sort key of the "current" record. Therefore, this DO WHILE is logically equivalent to the corresponding DO WHILE in sample program 1.

The second CALL statement activates MOVB to copy the (I-1)st record into the position of the Ith record, and the third CALL on MOVB copies the "current" record into the position of the Ith record.

Thus the sorting method of this procedure is identical to that of sample program 1. To illustrate the way this procedure can be used, it is set in a program that declares two sets of records, SET1 and SET2, and sorts them. As in the previous sample program, comments are inserted in place of the code that would be used in a working program to read data into the records and write it out after they are sorted.

SET1 is a set of 50 structures, each of which represents one record. Each structure contains a WORD scalar, an array of 12 BYTE scalars, an INTEGER scalar, a REAL scalar, and another BYTE scalar. We want to sort the records using the first element of the 12-byte array as the sort key. Since the key is the second element within the structure, its offset is just the number of bytes occupied by the first element of the structure. Therefore, we will use the built-in function SIZE to calculate the offset of the key and use that as the value of the parameter KEYINDEX.

SET2 is a set of 500 structures, each containing an array of 21 INTEGER scalars, a REAL scalar, and a BYTE scalar that is to be used as the key. This time, the key is deep within the structure. We can calculate the offset of the key manually, but this is both error prone and inflexible. If the elements within the structure are changed, then we must remember to recalculate the offset of the key. A better method is to let the compiler calculate the offset for us. This can be accomplished by subtracting the location of the key element from the location of the start of the record. The result of this subtraction is the offset of the key, which becomes the value of the parameter KEYINDEX.

In the two CALL statements used to activate SORTPROC, we used the built-in function LENGTH to determine the number of records that are to be sorted, and we used the built-in function SIZE to calculate the number of bytes occupied by each record. These two values are used for the COUNT and RECSIZE parameters for an activation of SORTPROC.

8.9.14 CONCLUSION

This introduction to PL/M-86 is intended to provide an overview of the language that is widely used with the iAPX86 family of processors. It supports modular, structured programming and provides access to interrupts, bits, and assembly language routines.[5]

8.10

THE C PROGRAMMING LANGUAGE

C is a programming language that was developed at Bell Laboratories by Dennis Ritchie to assist with the development of the Unix[6] operating system. It is a language with a relatively small number of operators and is aimed at general purpose applications. It is not as "high-level" a language as the others discussed in this chapter but rather can be considered intermediate between high-level languages and assembly languages. The fact that C is "closer" to the hardware can provide the programmer with a greater degree of control of the hardware.

Some of the characteristics of C are

1. Single- and double-precison integers, single- and double-precison floating point numbers, and pointers to these data types are supported.
2. Both arithmetic and logical operations are provided.
3. Operations on sets, lists, arrays, and character strings are not provided.
4. There are no READ, WRITE, file-access, file-control or I/O statements inherent in C. These operations must be implemented by calling specific functions, which are usually provided by the operating system.

---

[5] For more information, refer to the PL/M-86 User's Guide, Intel Corporation, 3065 Bowers Avenue, Santa Clara, Calif. 95051.

[6] Unix is a trademark of Bell Laboratories.

5. Single-thread control flow constructs such as loops, subprograms (function calls), and relational tests are provided. Parallel operations and multiprogramming are not supported.

6. Functions can be utilized when repetitive operations are to be performed. These functions can be called recursively. Program modules residing in separate files can be compiled separately.

7. C is independent of any particular machine architecture and has the potential for a high degree of portability.

8. C is not a strongly typed language, as are Ada and Pascal. A separate program, called "lint," can be used to perform a more rigorous type-checking.

8.10.1   GENERAL DESCRIPTION

A C program begins executing at the function main( ). The empty parentheses indicate that the function "main" has no arguments. Main( ) usually invokes other functions and must always exist in a C program. C programs that are invoked from an operating system may receive parameters, such as an argument count and the address of an array of pointers-to-character strings comprising the arguments. These would be written in a program as: main(argc,argv).

8.10.2   VARIABLES

All variables in a C program must be declared before they can be used. Declared variables may also be initialized. The fundamental data types in C are as follows:

int	integer
short	short integer
long	long integer
float	single-precision floating point
double	double-precision floating point
char	character

In addition to single variables of a particular type, arrays of variables can also be declared. The size of a particular data type is dependent on the particular machine on which the C compiler is implemented. For example, on a DEC PDP-11, both "int" and "short" are 16 bits; a "long" and a "float" are 32 bits; a "double" is 64 bits; and a "char" is 8 bits.

The actual allocation or use of a declared variable can be controlled by the use of modifiers. These modifiers include:

Register   allocate the variable in a CPU register, if possible.

Extern   the variable is actually declared externally to this module.

Auto   the variable will exist and is accessible only in the context of the currently active block.

Static   the variable will exist for the duration of the program but will be accessible only in the context of the function or module in which it was declared.

Unsigned   the integer will be treated as an unsigned value during arithmetic operations.

Some examples of variable declaration and initialization are

int   alpha = 10,   /* alpha declared initialized to 10 */
        beta = 0XAF,   /* beta declared, initialized to hexadecimal AF (175_{10}) */
        gamma = 0376,   /*gamma declared, initialized to octal 376 (254_{10}) */
        delta;   /*delta declared, but not initialized */

float   value = 5.9,
        total;

int   group[5],   /* an array of 5 integers */
        member[4] = {5,3,17,312}; /* an initialized array of 4 integers */

char   ch,
        nch = 'a',
        str[6] = {'a', 'b', 'C', 'd', 'Y', 0X40}
        nstr[ ] = "a string";

long   time = 147L;

The above examples include several important ideas worth mentioning. Ints, longs, and chars can be initialized by constants represented in any of three radices: decimal, octal, or hexadecimal. Octal constants are prefixed by a 0 (zero), while hexadecimal constants are preceded by 0X (zero, X). Long integer constants, in any radix, are indicated by a postfixed L. Character constants are generally of the form 'a.' Certain common nonprintable (control) characters can be represented in a "visible shorthand" using a (backslash) as an escape character:

\b   backspace
\n   newline (ASCII linefeed)
\r   carriage return
\t   horizontal tab
\f   formfeed

In addition, this escape mechanism allows one to represent character constants as arbitrary bit patterns in octal format. It is also used to specify the backslash and single quote as character constants:

\\   backslash
\'   single quote
\0   ASCII Null—often used as a string terminator

Character arrays can also be initialized with character string constants, as in the above example (nstr[ ] = "a string"). The important characteristic of the character string constant is that the C compiler will always add an invisible NULL character to terminate the string. As a result, the string requires one more byte for storage than the number of characters appearing within the double quotes.

One other type of variable exists in the C language, the pointer. The pointer is a variable that contains the **address of a variable of the specified type.** For example,

```
    int *ptr;
char *chrptr;
```

declare, respectively, a pointer to an integer, and a pointer to a character. Pointers provide a very powerful method of manipulating variables, and especially arrays of variables. In addition, it is often more convenient to pass pointers as function parameters.

8.10.3   PREPROCESSOR

The C compiler's preprocessor provides two important capabilities. The first is the ability to specify other text files to be included in the compilation of a particular module, via the #include "filename" compiler control line. The second, is a macro substitution facility that includes the ability to define symbolic names or constants. The #define construct allows a programmer to equate an arbitrary string of characters to another arbitrary character string (macro definition) or to a constant (symbolic constant definition). As an example,

1. #define skipwhite( ) while (isspace(*cptr)) cptr++ defines a macro to ignore "white space" in a character stream being processed by the character pointer variable cptr, while
2. #define LINELENGTH 132 /*printer line length */ defines a symbolic constant equal to the maximum length of, for example, a line printer line. Comments in C source code are delimited by /*...*/, and comments may stretch over more than one line.

8.10.4 OPERATORS

The binary arithmetic operations available in the C language are: + (addition), − (subtraction), * (multiplication), / (division), and % (modulo). The binary logical operations available include: | (bitwise inclusive OR), ^ (bitwise exclusive OR), & (bitwise AND), ≫ (shift right), and ≪ (shift left). The supported unary operators are: − (negation), and ~ (bitwise complement). Relational (logical) operations include ! (logical not), && (logical AND), ‖ (logical OR), == (equal), != (not equal), > (greater than), >= (greater than or equal), < (less than), and <= (less than or equal).

Expressions can be constructed from the binary operations in the normal fashion, for example, y = a + b. There is however, a special shorthand available for certain expression formats. When one variable appears on both sides of the assignment statement,

```
b = b & MASK,
```

the statement can be shortened to the equivalent:

```
b &= MASK.
```

Two interesting operators in C are "++" and "−−" that, respectively, increment and decrement the operand. If the operand is a pointer, it is increased or decreased by the appropriate amount to point to the next element of that type. These operators may be used both as either prefix or postfix operators, with the following results

```
++new     increments "new" before using its value as the operand
new++     uses the value of "new" as the operand and then increments
          "new"
```

To illustrate:

```
int m,  new=2; /* new is initialized with 2 */
    m= new++; /* after execution, m will be 2, new will be 3 */
    m= ++new; /* after execution, m will be 3, new will be 3 */
```

8.10.5 PROGRAM STRUCTURE AND CONTROL

Braces, { }, in a C program perform the function of delimiting blocks (sequences of statements). They are thus similar to the BEGIN−END statements in PASCAL. A block could be the statements comprising the body of a for or while loop, or it could be all of the statements that comprise a function.

Control statements in C include: if-else, for, while, while-do, do-while, switch, break, and goto. The uses and effects, but not necessarily the syntax, of most of these statements are similar to those described for the other languages discussed in this chapter. The exceptions are switch and break, which are used as follows. The switch statement, in concert with the case statement, is used to implement a multiway branch by matching the value of variable with one of a number of constant values. Break causes the innermost enclosing loop (or switch) to be exited immediately. When combined with an if statement, the break statement can provide an alternate or early exit from a do, while, or for loop.

8.10.6   LIBRARY FUNCTIONS

There are usually several standard libraries of functions available on a system supporting C, including routines for mathematical functions (such as evaluating transcendental functions), and routines for manipulating characters and character strings. Among the latter category the more commonly used functions include:

getchar()   returns the next available character from the character stream defined as "standard input" (usually the terminal keyboard). Returns EOF (end-of-file) at the end of the input stream.

putchar(c)   sends the character, c, to the character stream defined as "standard output" (usually the terminal screen or paper).

gets(buf)   returns an entire string from "standard input," placing the characters in a character array pointed to buy buf.

printf()   formats and prints its arguments according to information provided by embedded control strings. As an example:

printf ("character count = %d\n",ccnt);
would cause the following to be printed on the "standard output":

character count = 137

assuming that the value of ccnt was 137 when the printf() was executed. Two items to note: First, the \n would cause the terminal to be advanced to the left margin on the line following the message; second, the %d causes the variable ccnt to be printed in signed decimal format.

| scanf() | accepts formatted input from the "standard input," parsing it according to embedded control string instructions and storing the input in the variables named by the function's arguments. |

In addition to these, there is an entire complement of routines for dealing with null-terminated character strings, including functions such as string copying, string concatenation, and string comparison.

8.10.7   REFERENCE

Since this discussion of the C language is intended as an overview, there are naturally many aspects of C that have not been covered. An excellent reference that provides the detailed information required by the serious programmer is *The C Programming Language* by B. Kernighan and D. Ritchie, Prentice-Hall, Englewood Cliffs, N.J., 1978

8.10.8   INTERFACING EXAMPLE

To illustrate the usefulness of C in real-time interfacing applications, a relevant example is presented.[7] This example demonstrates the use of C to implement a user interface, via a keypad and a liquid crystal display (LCD), to a microprocessor based, real-time control system.

The functional requirements for this interface call for a 16-key keypad containing the symbols 0 through 9, A through D, *, and #, and a $4\frac{1}{2}$-digit, 7-segment LCD with the following additional, individually controllable symbols: LO BATTERY, − (minus sign), and CONTINUITY. As the user presses keys on the keypad, the appropriate characters must be scrolled into the display from the right. In addition, the correct ASCII value of the key's character must be inserted into an input buffer for parsing by a command interpreter. (This requirement exists so that a common command interpreter and input buffer can be used for both the keypad input and input from a remote ASCII terminal.) The * key is used to indicate an operator entry error, which will cause the LCD to be cleared and the input buffer to be (effectively) cleared. The # key is used to indicate the end of an entry, which causes a 0 to be placed in the command buffer to act as a delimiter.

Depending on the state of specified bits in certain alarm and flag bytes, the LO BATTERY and/or CONTINUITY segments of the LCD may be illuminated, or an audible alarm may be sounded. At certain times, the LCD may be

---

[7] The software for this example was developed by Tron McConnell of the Mellon Institute, Computer Engineering Center, Pittsburgh, Pa. 15213.

used to display information that must be continuously updated, such as a voltage or temperature. If this mode of operation, known as "multimeter mode" is in effect, a certain flag bit will have been set. If the user begins to enter a command when the display is in multimeter mode, the display must exit that mode, the display must be blanked, and the pressed key values must be scrolled in. The LCD will also be used to signal specific error conditions to the user by displaying ErXX, where XX represents a hexadecimal number between 00 and FF, with the display ErFF being used as a general purpose error indication.

The hardware to perform these tasks is shown in the accompanying schematic diagram (Figure 8.8). Briefly, it consists of an 8-bit 6502 microprocessor, an R65C21 CMOS PIA, the COP472 Liquid Crystal Display controller chip, the 16-switch keypad, and the audio alarm. The keypad software must scan the keypad switches, detect and debounce switch closures and openings, and interpret the switch closures. The display software must convert an array of hexadecimal numbers to an array of 7-segment bit patterns, build a 40-bit control vector, and serially transmit the vector to the display controller.

The high-level language software to perform these functions consists of three routines for keypad control, six routines for display control, and one "include" file containing definitions of constants and global data space declarations. The C code was hand-compiled to the 6502 assembly code shown in the listing. This hand compilation was performed to minimize the size of the object code and to maximize the execution speed. (The actual size of the assembled object code is 487 bytes of executable code with 113 bytes of data space required.) In addition, comments were added to the assembly language to make the task of maintaining the code much easier for programmers other than the original author.

Hardware

1. The 6502 microprocessor is an 8-bit device with an accumulator register (A), two index registers (X and Y), and a stack pointer register (SP.) These registers are 8-bits in width. The accumulator is the source and destination of data operated on by the 6502; the X and Y registers are provided for intermediate storage of data and memory location indexing. The SP register is used for "pushing" and "popping" data to and from a specific series of RAM locations operating as a stack memory.

2. The keypad consists of a 4 × 4 matrix of columns and rows of switches. It is scanned by selectively bringing "low" one column line at a time and then examining the row lines to see if any of them are low. If they are low, a key has been pressed, and it must be debounced and interpreted. The use of this technique is made possible, in part, by the internal pullup resistors on the Port A inputs of the PIA.

3. Port B of the PIA in this application also controls interrupts coming from an external UART. Because of this, care must be taken when

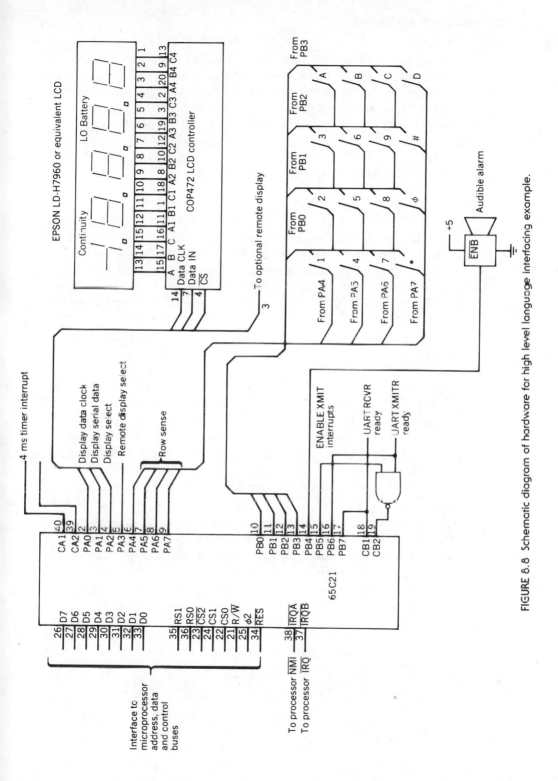

FIGURE 8.8 Schematic diagram of hardware for high level language interfacing example.

reading and writing this port not to inadvertantly clear, enable, or disable interrupts unintentionally. Thus a copy of the data most recently written to Port B is maintained in a Zero-Page location for use by these and other routines.

4. The COP472 Liquid Crystal Display Controller is controlled by a select (enable) line, a serial data line, and a data clock line. These lines are all connected to PIA Port A to facilitate transmitting the 40-bit control vector to the display controller.

Software

1. The 5-byte array, dpy_word, contains the hexadecimal representation of the numbers to be displayed on the LCD, with the first byte in the array corresponding to the rightmost digit. The highest byte of the array is specially encoded as follows: Bit 7 = turn on "–" segment, Bit 6 = turn on the high-order 1 segment, Bits 5 : 3 = unused, and Bits 2 : 0 = decimal point position.

2. If the high-order bit (bit 7) is set in any of the four lowest elements of the dpy_word array, the corresponding digit is displayed as a blank.

3. The routines dpy_hndlr and display_digits communicate via a 1-bit flag. Dpy_hndlr sets the "dsply_pend_flag" bit when it completes, thereby signalling the background task manager to call display_digits. The display_digits routine will clear the "dsply_pend_flag" bit when it completes processing the serial bit stream.

4. In the assembly language code, where possible, the X Index Register in the 6502 microprocessor is used as an iteration counter (e.g., for loop control) and the Y Index Register is used as an array index.

5. In the C code, the calls to save_regs and unsave_regs are actually implemented in the assembly language code by an in-line sequence of pushes or pops, not an actual subroutine call.

6. The complete, real-time, control system, of which these routines are but a part, is assumed to be controlled by periodic interrupts from a 4-ms clock. All routines in this control system are called, either directly (as in the case of kybd_scan), or indirectly (as in the case of dpy_hndlr and display_digits) at varying intervals, as a result of these clock interrupts. Thus scheduling of various events can be adjusted by having the appropriate routines called at user-selected time intervals.

ASSEMBLY LANGUAGE VERSION OF EXAMPLE PROGRAM

```
Include File "defs.h"
***********************************************************************
Variable and Address Declarations
***********************************************************************
* = $0000

i_o_flags:       .byte    $00      ; holds all of the I/O flags
                                   ; bit 0 - multimeter_flag
                                   ; bit 4 - kybd_cmd_flag
                                   ; bit 7 - dsply_pend_flag
i_o_flags2:      .byte $00         ; Continuation of i_o_flags
                                   ; bit 0 - key has been parsed
                                   ; bit 7 - string_fault flag
alarm_flags:     .byte    $00      ; Warning LED indicators, and bell
                                   ; enabling flag:
                                   ; bits 0 to 5 - Unused
                                   ; bit 6 - Enable Lo Bat LCD sign
                                   ; bit 7 - Enable Bell Flag
column_number:   .byte    $08      ; copy of keybrd scan outputs
row_number:      .byte    $00      ; copy of keybrd sense inputs
bounce_count:    .byte    $00      ; key bounce timeout indicator
write_portb:     .byte    $00      ; holds the current contents of pia
                                   ; data reg B so we don't have to read
                                   ; it again and risk clearing the
                                   ; interrupt
cmd_in_ptr:      .byte    $00      ; command buffer input pointer
dpy_word:        .byte    $00,00,00,$00,$00
                                   ; 5 BYTE ARRAY, holds the unpacked BCD
                                   ; value to be placed into the front
                                   ; panel display
digits:          .byte    $00,00,00,00,00
                                   ; 5 BYTE ARRAY, holds the bit patterns
                                   ; to be displayed on the front panel
***********************************************************************
Buffer Base Address and I/O Device Address Definitions
***********************************************************************
command_buffer  = $0300           ; address of the command interpreter
                                  ; buffer - 32 bytes long
dsply_port      = pia_porta       ; alternate name for PIA port A
kybd_rd_port    = pia_porta       ; alternate name for PIA port A
kybd_wr_port    = pia_portb       ; alternate name for PIA port B
pia_ddra        = $188C           ; PIA Port A Data Direction Register
pia_porta       = $188C           ; PIA Port A Data Port
pia_csra        = $188D           ; PIA Port A Control Register
pia_ddrb        = $188E           ; PIA Port B Data Direction Register
pia_portb       = $188E           ; PIA Port B Data Port
pia_csrb        = $188F           ; PIA Port B Control Register
***********************************************************************
Constant Declarations
***********************************************************************
multimeter_flag = $01             ; BIT in i_o_flags, set whenever
                                  ; multimeter_function is to repeatedly
                                  ; output the contents of a particular
                                  ; data channel
kybd_cmd_flag   = $10             ; BIT in i_o_flags, "1" indicates that
                                  ; a command has been received from the
                                  ; front panel keypad
```

```
dsply_pend_flag = $80           ; BIT in i_o_flags, "1" indicates that
                                ; there is data in the array digits[]
                                ; waiting to be sent out to the COP472
lcd_dsply_dun   = $7F           ; mask complement of dsply_pend_flag
isolate_lo_nyb  = $0F           ; MASK for isolating lower nybble
isolate_hi_nyb  = $F0           ; MASK for isolating upper nybble
enab_xmit_ints  = $20           ; to enable the xmit interrupt gate
                                ; bit 5 of the PIA
dis_xmit_ints   = ~enab_xmit_ints
neg_sign        = $01           ; enable bit for display's "-" sign
hi_order_one    = $01           ; enable bit for display's MS digit
lo_bat_ind      = $01           ; enable bit for display's "Low Battery"
string_fail_ind = $01           ; enable bit for display's "Continuity"
dspy_sync_ctrl  = $30           ; COP472 mode control bits
r_bit_pattern   = $0A           ; 7-segment bit pattern for "r"
dsply_clk_on    = $01           ; COP472 Data Clock High
dsply_clk_off   = $FE           ; COP472 Data Clock Low
dsply_sel_dsb   = $0C           ; COP472 Display Controller Disable
dsply_sel_enb   = $F3           ; COP472 Display Controller Enable
send_1_bit      = $02           ; Data to COP472 - High
send_0_bit      = $FD           ; Data to COP472 - Low
bell_on         = $10           ; Turn on audible alert buzzer
enable_bell     = $80           ; Flag used to determine whether or
                                ; not to assert "bell_on" in
                                ; kybd_wr_port
enable_lobat    = $40           ; Flag in alarm_flags used to
                                ; determine whether to turn on LCD's
                                ; Low Battery indication
disable_lobat   = $BF           ; mask to turn off enable_lobat flag
set_neg_sign    = $80           ; bit to be set in display_word[4] to
                                ; turn on negative sign in LCD display
set_hi_one      = $40           ; bit to be set in display_word[4] to
                                ; turn on high order one in LCD display
CMDBUFSIZ       = $20           ; Size of command buffer
CTRLU           = $15           ; ASCII NAK
key_parsed_flag = $01           ; we parsed the pressed key
*****************************************************************************
SYSTEM PARAMETERS
*****************************************************************************
max_bounce_count: .byte $20     ; keyboard debounce constant
include "defs.h"
```

The following routine is called from the 4 msec interrupt handler
to control keyboard scanning. If no key is pressed, reset the
debounce counter and skip to the next keyboard column. If a key
is pressed, decrement the debounce counter, and if we've reached
zero, interpret the key. After the key has been interpreted,
parse it: if it's a NAK (Control-U) clear the cmd_in_ptr; if it's
a NULL (EOF), null terminate the command buffer and set the
kybd_cmd_flag bit; otherwise just shove the character in the
command buffer, update the cmd_in_ptr (adjust it for overflow if
necessary) and continue scanning.

```
scan_kybd:
        pha                     ; Save Accumulator
        txa
        pha                     ; Save X-Register
        tya
        pha                     ; And, save Y-Register
        lda     kybd_rd_port    ; get keybrd sw array row sense lines
```

```
        eor     #$FF              ; complement row sense data
        and     #isolate_hi_nyb   ; mask for PIA port keybrd sense lines
        sta     row_number        ; save copy for (possible) call to key_intrp
        beq     _no_key_pressed   ; go to update scan output
        dec     bounce_count      ; update bounce-time counter
        bne     _still_bouncing   ; What do you think?!
        lda     #key_parsed_flag
        bit     i_o_flags2        ; see if we've already parsed this key
        bne     _still_bouncing   ; yes, wait for first release of key
        ora     i_o_flags2        ; no, but we're about to; set flag saying we did
        sta     i_o_flags2
        lda     row_number        ; prepare to call key_intrp
        jsr     key_intrp         ; create ASCII value from row-col numbers
        beq     _kybd_cmd_dun     ; key_intrp returns ASCII value in Accum
        bmi     _kybd_err_fix     ; if negative value, sumpin's worng!
        cmp     #CTRLU            ; see if user NAK'ed previously entered keys
        bne     _insert_key       ; No? well then let's shove it in command buf
_cir_cmd_in_ptr:
        ldy     #$00              ; NAK calls for just clearing cmd buf ptr
_update_cmd_ptr:
        sty     cmd_in_ptr
        jmp     _still_bouncing   ; continue scanning
_kybd_cmd_dun:
        ldy     cmd_in_ptr        ; get cmd buf index for inserting EOF ($00)
        sta     command_buffer,y
        sta     cmd_in_ptr        ; we also clr cmd_in_ptr (for later parsing?)
        lda     i_o_flags         ; we also have to set flag saying that a
        ora     #kybd_cmd_flag    ; keybrd command has been entered
        sta     i_o_flags
        bne     _still_bouncing   ; always keep scanning (keeps Romulans away)
_insert_key:
        ldy     cmd_in_ptr        ; get cmd buf index for inserting new char
        sta     command_buffer,y
        iny                       ; bump cmd buf index
        cpy     #CMDBUFSIZ        ; have we overrun the command buffer
        bne     _update_cmd_ptr   ; no, continue processing (branch always taken)
        jsr     dpy_huh           ; tell user he blew it
        jmp     cir_cmd_in_ptr    ; yes, undo the damage (flush input)
_kybd_err_fix:
        and     #$7F              ; see if key_intrp called with no key pressed
        beq     _no_key_pressed   ; if so just continue
        cmp     #$01              ; see if scan control error was signalled
        bne     _no_key_pressed   ; ignore error (this shouldn't happen!)
        sta     column_number     ; set column_number to overflow on next shift
_no_key_pressed:
        lda     #~key_parsed_flag & $FF
        and     i_o_flags2
        sta     i_o_flags2        ; clear key parsed flag bit
        lda     max_bounce_count
        sta     bounce_count      ; reset key bounce timeout counter
        lda     column_number     ; prepare to look at next kybd column
        lsr     a                 ; scanning from left to right (bitwise)
        bcc     _no_col_ovflw     ; don't have to reset column scan pattern
        lda     #$08              ; start again with high column bit
_no_col_ovflw:
        sta     column_number     ; update zpg copy
        lda     write_portb       ; get the xmit_int bit...plus some others
        and     #enab_xmit_ints   ; ...and get rid of the others...
        ora     column_number     ; ...back to where we were...except we got
                                  ; another bit...maybe...
```

```
        bit       alarm_flags      ; we have to see if bell is supposed to be on
        bpl       _no_bell_on      ; no, it isn't so just output kybd col scan
        ora       #bell_on         ; enable bell also
_no_bell_on:
        eor       #$0F             ; complement column drive bits (active low)
        sta       kybd_wr_port     ; assert keybd column scan and (possibly) bell
        sta       write_portb      ; save a copy of this
_still_bouncing:
        pla                        ; prepare to go home
        tay                        ; Recover Y Register
        pla
        tax                        ; Recover X Register
        pla                        ; Also get back Accumulator
        rts                        ; Return to 4 msec intrp handler
```

The following routine is called with a (hopefully) non-zero value
in the accumulator which represents the sensed keyboard row mask
(in bits <7-4>). This row number and the corresponding column
number are mapped into an index into an array of ASCII values.
The appropriate ASCII value for the key pressed is returned in
the accumulator. If a zero row mask was provided to this routine,
$80 is returned. If the column number was out of its allowable
range $81 is returned. If more than one key is pressed (row mask
has more than one bit set), the "first key" (first bit set) is
the one that is mapped to an ASCII value. The "#" key is mapped
to the ASCII value $00 so that it can easily be recognized as the
EOF character.

If the mapped ASCII value is a number (0-9) then the front panel
display is updated by scrolling the new digit in from the right
(least significant digit). If the "*" key was pressed indicating
an operator entry error, the display is cleared.

```
key_intrp:
        pha                        ; Save Accumulator
        txa
        pha                        ; Save X-Register
        tya
        pha                        ; And, save Y-Register
        tsx                        ; Copy SP for fancy stack indexing
        lda       $0103,x          ; Recover copy of saved accumulator
        lsr       a                ; Need to move row mask down to low nybble of
        lsr       a                ; accumulator so we can use it next.
        lsr       a
        lsr       a
        tay                        ; We're going to use key row mask as an index
        lda       row_map,y        ; into an array: it's faster and easier to find
                                   ; first set bit this way.
        bmi       _scan_error      ; if y=0 (no key pressed) array has neg value
        ldy       column_number    ; get col mask to use as an index, as with row
        clc                        ; prepare for addition.
        adc       col_map,y        ; add col scan value to row sense value to
        tay                        ; create a combined row-col number to use as an
        lda       keybrd_map,y     ; index to get key's equivalent ASCII value.
        bmi       _scan_error      ; if y has more than 1 bit asserted then array
                                   ; has a negative value to signal error
        sta       $0103,x          ; set up return value in acc. stacked position
        beq       _was_not_alpha   ; return at once if EOF (#) is pressed
        cmp       #CTRLU           ; if it's a NAK we'll clear the display (cute)
        bne       _cvt_to_hex      ; it wasn't so continue with normal processing
```

```
        ldy     #$03                ; set up array index/loop counter
        lda     #$80                ; Set up display blanking character
_clr_display_loop:
        sta     dpy_word,y
        dey                         ; bump pointer
        bpl     _clr_display_loop
        lda     #$00
        sta     dpy_word+4          ; blank DP "-", "1"
        lda     #CTRLU              ; reset the accumulator to $15
        bpl     _display_it         ; this branch always taken
_cvt_to_hex:
        lda     cmd_in_ptr          ; if this is start of new command we must blank
        bne     key_intrp1          ; the display from the continuous update fnctn.
        lda     #$80                ; blanking character for use in dpy_word[]
        sta     dpy_word+2          ; don't have to clear dpy_word[3] because
        sta     dpy_word+1          ; shift below will blank it for us.
        sta     dpy_word
        lda     #~multimeter_flag       ; clear the flag
        and     i_o_flags
        sta     i_o_flags
key_intrp1:
        lda     keybrd_hex,y        ; get hex value of key pressed, not ASCII
        ldy     dpy_word+2          ; we will now perform a left shift on the
        sty     dpy_word+3          ; four unpacked BCD digits in dpy_word[0-3]
        ldy     dpy_word+1
        sty     dpy_word+2
        ldy     dpy_word
        sty     dpy_word+1
        sta     dpy_word            ; entering the newest digit in the LCD Display
        lda     #$00
        sta     dpy_word+4          ; setting A = 0, we can use it to clear out
                                    ; byte controlling sign, DP and hi order 1
                                    ; of the display.
        beq     _display_it         ; This branch always taken.
_scan_error:
        sta     $0103,x             ; put it on stack in place of accumulator
        bmi     _was_not_alpha      ; Head for home
_display_it:
        jsr     dpy_hndlr           ; display newly entered digit
                                    ; returned from call to dsply_hndlr intact
                                    ; (and flags are properly set by PLA instr.)
_was_not_alpha:
        pla
        tay                         ; Recover Y-Register
        pla
        tax                         ; Recover X-Register
        pla                         ; Recover Accumulator (has ASCII equiv of key)
        rts                         ; return to scan_kybd
```

KEY_INTRP TABLES

The following array provides the mapping for converting
the sensed keyboard row number into the next level array
index. This array also performs the function of finding
the "first set bit" (in case multiple keys are pressed)
and returns an error value ($80) for no key pressed at
all.

```
row_map:
        .byte $80,$03,$02,$02
        .byte $01,$01,$01,$01
        .byte $00,$00,$00,$00
        .byte $00,$00,$00,$00
```

The following array provides the mapping for converting
the "column scanned" mask into an array index that can be
added to the index from the row_map in order to find out
which one key has been pressed. For index values (i.e.,
column numbers) into this array that have none or more
than one bit set, a condition which should not occur
under normal circumstances, a "next-level" index is
returned that will map into an invalid ASCII key value
(i.e., the key will be $81).

```
col_map:
        .byte $10,$00,$08,$10
        .byte $04,$10,$10,$10
        .byte $00,$10,$10,$10
        .byte $10,$10,$10,$10
```

This array is the one that actually produces an ASCII
value corresponding to the key pressed. It also produces
error indications for malformed column scan values.

```
keybrd_map:
        .byte $44,$43,$42,$41    ; "D","C","B","A"
        .byte $00,$39,$36,$33    ; "#","9","6","3"
        .byte $30,$38,$35,$32    ; "0","8","5","2"
        .byte $15,$37,$34,$31    ; "*","7","4","1"
        .byte $81,$81,$81,$81    ; Column scan error return
```

This array is analogous to the one above except that it
contains hexadecimal values of the keys pressed to speed
up the display update function. Saves having to convert
ASCII back to HEX.

```
keybrd_hex:
        .byte $0D,$0C,$0B,$0A
        .byte $00,$09,$06,$03
        .byte $00,$08,$05,$02
        .byte $00,$07,$04,$01
```

The following routine takes the BCD or hexadecimal number
located in the five byte array dpy_word[5] and converts
it to the appropriate 7-segment bit patterns required to
control and enable the LCD display. A 40-bit (5-byte)
output array is filled with the bit patterns, and the
routine display_digits is called to actually transmit the
contents of the output array to the display controller.

```
dpy_hndlr:
        pha                             ; Save accumulator
        txa
        pha                             ; Save X-Register
        tya
        pha                             ; Save Y-Register
        ldx     #$03                    ; Set up X as index into digits to be displayed
_dspy_hndlr1:
        lda     dpy_word,x              ; Get the digit and
        bpl     _dspy_hndlr2            ; check for blanking (hi-bit set)
        lda     #$00                    ; if digit is blanked set no segment enable bits
        beq     _dspy_hndlr3            ; this branch always taken
_dspy_hndlr2:
        and     #isolate_lo_nyb         ; make sure we only look at low nybble.
        tay                             ; Use Y as index into segment map array.
        lda     segment_map,y           ; Get 7-segment bit pattern and
_dspy_hndlr3:
        sta     digits,x                ; put it into output array.
        dex                             ; bump array index
        bpl     _dspy_hndlr1            ; if not continue with normal mapping routine
        bit     dpy_word+4              ; get copy of bits 6 & 7 in V and N bits
        bpl     _pos_number             ; negative sign not displayed
        lda     digits+2                ; get byte with negative sign control bit
        ora     #neg_sign               ; set the negative sign control bit and
        sta     digits+2                ; save the updated copy of the bit pattern
_pos_number:
        bvc     _no_msd                 ; see if the MS ONE is to be turned on.
        lda     digits+3                ; get byte with MS ONE control bit
        ora     #hi_order_one           ; turn it on, and...
        sta     digits+3                ; restore it.
_no_msd:
        bit     alarm_flags             ; check for low battery indication
        bvc     _no_lo_bat
        lda     digits+1
        ora     #lo_bat_ind
        sta     digits+1                ; enable "low battery" symbol on LCD
_no_lo_bat:
        bit     i_o_flags2              ; check for string failure indication
        bpl     _no_lo_string
        lda     digits
        ora     #string_fail_ind
        sta     digits
_no_lo_string:
        lda     dpy_word+4              ; prepare to set DP
        and     #$07                    ; mask off all but DP position bits
        beq     _set_sync               ; if no DP just set SYNC and MODE bits
        tay                             ; otherwise use DP position as an index
        lda     dp_map,y                ; into a decimal point position array
_set_sync:
        ora     #dspy_sync_ctrl         ; Set SYNC and MODE control bits for COP472
        sta     digits+4                ; and save it away
        lda     i_o_flags               ; set bit for background routine display_digits
        ora     #dsply_pend_flag
        sta     i_o_flags               ; telling it that display data is waiting
        pla
        tay                             ; Unsave Y-register
        pla
        tax                             ; Unsave X-register
        pla                             ; Unsave Accumulator
        rts                             ; Return to sender (address unknown).
```

7 SEGMENT BIT PATTERN TABLE

The 7-segment bit patterns for the various hexadecimal
characters that are stored in the array segment_map are
organized as follows: Each byte contains a bit for each
segment of a standard 7-segment display digit, with
segment-to-bit mapping as follows: <7> Seg. A, <6> Seg.
B, <5> Seg. C, <4> Seg. D, <3> Seg E, <2> Seg F, <1> Seg
G, <0> Special Character or Display Element (digit
dependent).

```
segment_map:
        .byte   $FC,$60,$DA,$F2  ; characters "0", "1", "2", "3"
        .byte   $66,$B6,$BE,$E0  ; characters "4", "5", "6", "7"
        .byte   $FE,$F6,$EE,$3E  ; characters "8", "9", "A", "b"
        .byte   $9C,$7A,$9E,$8E  ; characters "C", "d", "E", "F"
dp_map:
        .byte   $00,$01,$02,$04
        .byte   $08,$00,$00,$00
```

The following routine will display an error message on
the LCD display in the format "ErXX" where XX is the
contents of the accumulator upon entry into this routine.

```
dpy_error:
        pha                             ; save the Accumulator
        txa                             ; and also save the index registers
        pha
        tya
        pha
        tsx                             ; Now, cleverly point at the saved accumulator
        lda     $0103,x                 ; to get it back. What a hack!
        pha                             ; save copy of error number
        and     #isolate_lo_nyb         ; look at only low order nybble
        tay                             ; use Y as index into 7-seg bit pattern array
        lda     segment_map,y           ; get pattern and store it in
        sta     digits                  ; lo-order display location
        pla                             ; get back saved copy of error number
        lsr     a                       ; move hi nybble into lo nybble
        lsr     a
        lsr     a
        lsr     a
        tay
        lda     segment_map,y           ; as above
        sta     digits+1                ; except for next display location
        lda     #r_bit_pattern          ; Shove out bit pattern for "r"
        sta     digits+2
        lda     segment_map+$0E         ; and bit pattern for "E"
        sta     digits+3
        lda     #$00                    ; Need a zeroed accumulator for jump to
        jmp     _set_sync               ; code to work correctly.
```

The following routine will display "ErFF" on the LCD
display to indicate the system's confusion with the
operator's action(s) and clear out the dpy_word array to
blanks.

```
dpy_huh:
        pha                             ; Save Accumulator
        txa
        pha                             ; And X-Register also
        ldx         #$03                ; Set up array index
        lda         #$80                ; source of 0
_clr_disp:
        sta         dpy_word,x          ; clear array
        dex
        bpl         _clr_disp           ; do while not done
        asl         a                   ; fast way to clear accumulator
        sta         dpy_word+4          ; make sure we blank "-","1", and DPs.
        txa
        jsr         dpy_error
        pla                             ; Clean up stack
        tax                             ; Restoring X
        pla                             ; and Accumulator
        rts                             ; back to caller
```

This routine will set up the array dpy_word with
"blanking" information in all digits, and then call
dpy_hndlr() in order to clear the LCD display.

```
dpy_clr:
        pha                             ; Save Accumulator
        lda         #$00                ; Clear out DP, "-", & "1" indicators
        sta         dpy_word+4
        lda         #$80                ; Set blanking bit in other digits
        sta         dpy_word+3
        sta         dpy_word+2
        sta         dpy_word+1
        sta         dpy_word+0
        jsr         dpy_hndlr           ; call dpy_hndlr to blank the display
        pla                             ; Recover Accumulator
        rts
```

The following routine, called indirectly from
dply_hndlr, dpy_huh, and dpy_error is the routine that
ships out 40 bits serially to the COP472, asserting and
reasserting the clock, enabling and disabling the
COP472's chip select, and keeping things organized.

```
display_digits:
        pha                             ; Save Accumulator
        txa                             ; and other index registers
        pha
        tya
        pha
        lda         #dsply_clk_off      ; dsply_sel_dsb
                                        ; make sure display clk off and not selected
        sta         dsply_port          ; and send command to display port
        and         #dsply_sel_enb      ; now enable display
        sta         dsply_port          ; Display now enabled to receive new data
        ldx         #$00                ; set up output array index
    dply_digits1:
        clc                             ; make sure restoration of output array is OK
        ldy         #$08                ; set up shift counter
```

```
_out_shift_loop:
        ror     digits,x          ; get bits, starting with low-order
        bcc     _send_a_0         ; if carry clear, then send a 0, else send a 1
        ora     #send_1_bit       ; set data bit high in control port
        bne     _send_it          ; this branch always taken
_send_a_0:
        and     #send_0_bit       ; set data bit low in control port
_send_it:
        sta     dsply_port        ; data bit now sent to COP472
        inc     dsply_port        ; data now strobed into COP472
        dec     dsply_port        ; data clock to COP472 now reset
        dey                       ; update shift counter
        bne     _out_shift_loop   ; continue until counter is zero
        ror     digits,x          ; complete the data restoration

        inx                       ; bump output array index/counter
        cpx     #$05              ; and see if we're done
        bne     dsply_digits1     ; if not, then continue
        ora     #dsply_sel_dsb    ; can deselect the display chip
        sta     dsply_port        ; COP472 now disabled, and we're finished
        lda     i_o_flags         ; have to turn off display data waiting flag
        and     #lcd_dsply_dun
        sta     i_o_flags
        pla                       ; Restore Y, X and A registers
        tay
        pla
        tax
        pla
        rts                       ; Return to caller.
```

'C' VERSION OF EXAMPLE PROGRAM

```
/*
 * Include file "param.h"
 */
#define ISOLATE_LO_NYB   0x0F
#define ISOLATE_HI_NYB   0xF0
#define KEY_PARSED_FLAG  0x01
#define KYBD_CMD_FLAG    0x10
#define CMDBUFSIZ        0x20
#define BELL_ON          0x10
#define MULTIMETER_FLAG  0x01
#define CTRLU            0x15
#define NEG_SIGN         0x01
#define HI_ORDER_ONE     0x01
#define LO_BAT_IND       0x01
#define STRING_FAIL_IND  0x01
#define DSPY_SYNC_CTRL   0x30
#define DSPLY_PEND_FLAG  0x80
#define R_BIT_PATTERN    0x0A
#define DSPLY_CLK_OFF    0xFE
#define DSPLY_CLK_ON     0x01
#define DSPLY_SEL_ENB    0xF3
#define DSPLY_SEL_DSB    0x0C
#define LCD_DSPLY_DUN    0x7F
#define SEND_1_BIT       0x02
#define SEND_0_BIT       0xFD
```

```
int     row_map[] = {0x80, 0x03, 0x02, 0x02, 0x01, 0x01, 0x01, 0x01,
                     0x00, 0x00, 0x00, 0x00, 0x00, 0x00, 0x00, 0x00},
        col_map[] = {0x10, 0x0C, 0x08, 0x10, 0x04, 0x10, 0x10, 0x10,
                     0x00, 0x10, 0x10, 0x10, 0x10, 0x10, 0x10, 0x10},
        keybrd_hex[] = {0x0D, 0x0C, 0x0B, 0x0A, 0x00, 0x09, 0x06, 0x03,
                        0x00, 0x08, 0x05, 0x02, 0x00, 0x07, 0x04, 0x01},
        row_number,
        column_number,
        bounce_count,
        max_bounce_count,
        i_o_flags,
        i_o_flags2,
        cmd_in_ptr,
        alarm_flags,
        write_portb,
        *kybd_rd_port = { 0x188C },
        *kybd_wr_port = { 0x188E },
        dpy_word[5],
        digits[5],
        segment_map[] = {0xFC, 0x60, 0xDA, 0xF2, 0x66, 0xB6, 0xBE, 0xE0,
                         0xFE, 0xF6, 0xEE, 0x3E, 0x9C, 0x7A, 0x9E, 0x8E},
        dp_map[] = {0x00, 0x01, 0x02, 0x04, 0x08, 0x00, 0x00, 0x00},
        *dsply_port = { 0x188C };

char    command_buffer[CMDBUFSIZ],
        keybrd_map[] = { 'D', 'C', 'B', 'A', 0x00, '9', '6', '3',
                         '0', '8', '5', '2', 0x15, '7', '4', '1',
                         0x81, 0x81, 0x81, 0x81 };
/*
 * Source Code
 */
#include "param.h"
```

This routine is called from the 4-ms interrupt handler to
control keyboard scanning. If no key is pressed, reset
the debounce counter and skip to the next keyboard
column. If a key is pressed, decrement the debounce
counter, and if we have reached zero, interpret the key.
After the key has been interpreted, parse it; if it is a
BAK (Control-U) clear the cmd_in_ptr; if it is a NULL
(EOF), null terminate the command buffer and set the
kybd_cmd_flag bit; otherwise just shove the character in
the command buffer, update the cmd_in_ptr (adjust it for
overflow if necessary) and continue scanning.

```
extern  int     save_regs(),
                unsave_regs();
scan_kybd()

register         int     a;
```

```c
        save_regs();
        row_number = (~*kybd_rd_port) & ISOLATE_HI_NYB;
        if (row_number == 0)
                no_key_pressed();
        bounce_count--;
        if (bounce_count == 0)
                {
                if ((i_o_flags2 & KEY_PARSED_FLAG) == 0)
                        {
                        i_o_flags2 = i_o_flags2 | KEY_PARSED_FLAG;
                        a = key_intrp(row_number);
                        if (a < 0)
                                {
                                if (a == 0x81)
                                        column_number = 0x01;
                                no_key_pressed();
                                }
                        else if (a == 0)
                                {
                                command_buffer[cmd_in_ptr] = a;
                                cmd_in_ptr = a;
                                i_o_flags = i_o_flags | KYBD_CMD_FLAG;
                                }
                        else if (a == CTRLU)
                                cmd_in_ptr = 0;
                        else
                                {
                                command_buffer[cmd_in_ptr] = a;
                                if (cmd_in_ptr++ == CMDBUFSIZ)
                                        {
                                        dpy_huh();
                                        cmd_in_ptr = 0;
                                        }
                                }
                        }
                }
        unsave_regs();
        return;
}
no_key_pressed()
{
        i_o_flags2 = i_o_flags2 & ~KEY_PARSED_FLAG;
        bounce_count = max_bounce_count;
        if ((column_number >> 1) == 0)
                column_number = 0x08;
        if (alarm_flags < 0)
                {
                write_portb = (~column_number | (0x20 & write_portb) | BELL_
                *kybd_wr_port = write_portb;
                }
        else
                {
                write_portb = (~column_number | (0x20 & write_portb));
                *kybd_wr_port = write_portb;
                }
        unsave_regs();
        return;
}
```

The following routine is called with a nonzero value in
the accumulator, which represents the sensed keyboard row
mask (in bits <7-4>). This row number and the
corresponding column number are mapped into an index into
an array of ASCII values. The appropriate ASCII value for
the key pressed is returned in the accumulator. If a zero
row mask was provided to this routine, $80 is returned.
If the column number was out of its allowable range, $81
is returned. If more than one key is pressed (row mask
has more than one bit set), the "first key" (first bit
set) is the one that is mapped to an ASCII value. The "#"
key is mapped to the ASCII value $00 so that it can
easily be recognized as the EOF character.

If the mapped ASCII value is a number (0-9) or a letter
(A-D), then the front panel display is updated by
scrolling the new digit in from the right (least
significant digit). If the "*" key was pressed indicating
an operator entry error, the display is cleared.

```
key_intrp(row_value)
int     row_value;
{
register        int     a,
                y;
int     i;

        save_regs();
        if ((a = row_map[(a=row_value) >> 4])< 0)       /* a = 0x80 */
                return(a);
        a = a + col_map[column_number];
        if ((a = keybrd_map[(y=a)]) <= 0)       /* a = 0x81 or a = 0 */
                return(a);
        row_value=a;
        if (a == CTRLU)
                {
                for(y = 3;y >= 0; y--)
                        dpy_word[y] = 0x80;
                dpy_word[4] = 0;
                dpy_hndlr();
                }
        else
                {
                if (cmd_in_ptr == 0)
                        {
                        for(i = 2;i >= 0; i--)
                                dpy_word[i] = 0x80;
                        i_o_flags = i_o_flags & ~MULTIMETER_FLAG;
                        }
                a = keybrd_hex[y];
                for(i = 3;i >= 0; i--)
                        dpy_word[i] = dpy_word[i-1];
                dpy_word[0] = a;
                dpy_word[4] = 0;
                dpy_hndlr();
                }
        unsave_regs();
        return(row_value);
```

The following routine takes the BCD or hexadecimal number located in the 5-byte array dpy_word[5] and converts it to the appropriate 7-segment bit patterns required to control and enable the LCD display. A 40-bit (5-byte) output array is filled with the bit patterns, and the routine display_digits is called to actually transmit the contents of the output array to the display controller.

```
dpy_hndlr()
{
register        int        a, x;

        save_regs();
        for (x=3; x >= 0; x--)
                {
                if (dpy_word[x] >= 0)
                        digits[x] = segment_map[dpy_word[x] & ISOLATE_LO_NYB
                else
                        digits[x] = 0;
                }
        if (dpy_word[4] & 0x80)
                digits[2] |= NEG_SIGN;
        if (dpy_word[4] & 0x40)
                digits[3] |= HI_ORDER_ONE;
        if (alarm_flags & 0x40)
                digits[1] |= LO_BAT_IND;
        if (i_o_flags2 & 0x80)
                digits[0] |= STRING_FAIL_IND;
        if (a = (dpy_word[4] & 0x07))
                set_sync(dp_map[a]);
        else
                set_sync(a);
}


set_sync(dp_msk)
int     dp_msk;
{
        digits[4] = dp_msk | DSPY_SYNC_CTRL;
        i_o_flags |= DSPLY_PEND_FLAG;
        unsave_regs();
        return;
}
```

The following routine will display an error message on the LCD display in the format "ErXX," where XX is the contents of the accumulator upon entry into this routine.

```
dpy_error(errno)
int     errno;
{
        save_regs();
        digits[0] = segment_map[errno & ISOLATE_LO_NYB];
        digits[1] = segment_map[errno >> 4];
        digits[2] = R_BIT_PATTERN;
        digits[3] = segment_map[0x0E];
        set_sync(0);
}
```

The following routine takes the BCD or hexadecimal number
display to indicate the system's confusion with the
operator's action(s) and clear out the dpy_word array to
blanks.

```
dpy_huh()
{
register        int     x;

        save_regs();
        for(x = 3;x >= 0; x--)
                dpy_word[x] = 0x80;
        dpy_word[4] = 0x00;
        dpy_error(0xFF);
        unsave_regs();
}
```

This routine will set up the array dpy_word with
"blanking" information in all digits, and then call
dpy_hndlr() in order to clear the LCD display.

```
extern  int     save_accum(),
                unsave_accum();
dpy_clr()
{
        save_accum();
        dpy_word[4] = 0;
        dpy_word[3] = 0x80;
        dpy_word[2] = 0x80;
        dpy_word[1] = 0x80;
        dpy_word[0] = 0x80;
        dpy_hndlr();
        unsave_accum();
```

The following routine, called indirectly from
dsply_hndlr, dpy_huh, and dpy_error is the routine that
ships out 40 bits serially to the COP472, asserting and
deasserting the clock, enabling and disabling the
COP472's chip select, and keeping things organized.

```
display_digits()

register        int     x,
                y;
int     temp;

        save_regs();
        *dsply_port = DSPLY_CLK_OFF;
        *dsply_port = *dsply_port & DSPLY_SEL_ENB;
        for(x = 0; x < 5; x++)
                {
                temp = digits[x];
                for(y = 0;y < 8; y++)
                        {
```

```
                    if (digits[x] & 0x01)
                            *dsply_port |= SEND_1_BIT;
                    else
                            *dsply_port |= SEND_0_BIT;
                    *dsply_port |= DSPLY_CLK_ON;
                    *dsply_port &= DSPLY_CLK_OFF;
                    digits[x] >>= 1;
                    }
            digits[x] = temp;
            }
    *dsply_port |= DSPLY_SEL_DSB;
    i_o_flags &= LCD_DSPLY_DUN;
    unsave_regs();
    return;
```

8.10.9   CONCLUSION

The language C is widely used for real-time interfacing applications. C compilers are available on a variety of computers, including the IBM PC and Personal System/2 family as well as the Digital Equipment Corporation MicroVax II minicomputer. The portability of C along with its support of port and register manipulation make it a popular choice for implementing microcomputer interfaces.

FOR FURTHER READING

Ada Reference Manual, U.S. Department of Defense, July 1980.

Applesoft II Basic Programming Reference Manual, Apple Computer Inc., 1981.

Dahl, O. J., E. W. Dijkstra, and C. A. R. Hoare, *Structured Programming*, Academic Press, New York, 1972.

Gottfried, B. S., *Programming with FORTRAN IV*, Quantum Publishers, 1972.

Jensen K., and N. Wirth, *PASCAL User Manual and Report*, 2nd ed., Springer Verlag, New York, 1976.

Kemeny J. G., and T. E. Kurtz, *BASIC Programming*, John Wiley & Sons, Inc., New York, 1967.

PL/M-86 Programming Course Manual, 2nd ed., Mellon-Institute, Computer Engineering Center, 1983.

PL/M-86 User's Guide, Intel Corporation, 1980, 1981.

The C Programming Language, Prentice-Hall, Inc., Englewood Cliffs, N.J., 1978.

Winston P., and B. K. Horn, *LISP*, Addison-Wesley, Mass., 1981.

A

ASCII CHARACTER SET

Graphic or Control	ASCII (Hexa-decimal)	Graphic or Control	ASCII (Hexa-decimal)	Graphic or Control	ASCII (Hexa-decimal)
Tape Aux. Off	14	<	3C	L	4C
Error	15	=	3D	M	4D
Sync	16	>	3E	N	4E
LEM	17	?	3F	O	4F
S0	18	[	5B	P	50
S1	19	/	5C	Q	51
S2	1A	]	5D	R	52
S3	1B	↑	5E	S	53
S4	1C	←	5F	T	54
S5	1D	@	40	U	55
S6	1E	blank	20	V	56
S7	1F	0	30	W	57
				X	58
				Y	59
				Z	5A
NULL	00	ACK	7C	1	31
SOM	01	Alt. Mode	7D	2	32
EOA	02	Rubout	7F	3	33
EOM	03	|	21	4	34
EOT	04	"	22	5	35
WRU	05	#	23	6	36
RU	06	$	24	7	37
BELL	07	%	25	8	38
FE	08	&	26	9	39
H. Tab	09	'	27	A	41
Line Feed	0A	(	28	B	42
V. Tab	0B	)	29	C	43
Form	0C	*	2A	D	44
Return	0D	+	2B	E	45
SO	0E	,	2C	F	46
SI	0F	-	2D	G	47
DCO	10	.	2E	H	48
X-On	11	/	2F	I	49
Tape Aux. On	12	:	3A	J	4A
X-Off	13	;	3B	K	4B

B

INTERFACE-RELATED READINGS

REAL-TIME OPERATING SYSTEM NEEDS, *Interfaces in Computing*, May 1982, Vol. 1, No. 1.

DIM: A NETWORK COMPATIBLE INTERMEDIATE INTERFACE STANDARD, *Interfaces in Computing*, May 1983, Vol. 1, No. 2.

VME BUS INTERFACING: A CASE STUDY, *Interfaces in Computing*, August 1983, Vol. 1, No. 3.

REAL-TIME OPERATING SYSTEM NEEDS

P. N. CLOUT

Los Alamos National Laboratory, P.O. Box 1663, Los Alamos, NM 87545 (U.S.A.)
(Received November 1981)

SUMMARY

When attempting to standardize real-time features for computer languages, one quickly runs up against the large differences between the operating systems under which the processes will run. One is then left with the choice of doing only part of the job or attempting to cope with the differences at the compiler implementation level. The latter approach leaves something to be desired in efficiency and denies access to some desirable features that are only available within the operating system.

In this paper the needs of real-time systems are discussed and attempts are made to establish the need for standards for real-time operating systems.

Reprinted with permission of P. N. Clout, "Real-Time Operating System Needs," in *Interfaces in Computing*, Vol. 1, No. 1, May 1982, and North-Holland Publishing Company, Amsterdam, 1982.

1   INTRODUCTION

There are many definitions of "real-time" with respect to computing. Often, speed is the emphasis of such definitions, and indeed speed is important in many systems, but it is not a unique test for a real-time system. Before we attempt a better explanation of real-time, perhaps we should first consider some aspects of non-real-time or conventional computing. In this case the code will always have predictable results regardless of when it is run or the lapsed time from start to finish. This lapsed time can vary considerably in shared systems, depending on how much the system is being used. Examples of such codes are scientific calculations and commercial data processing. For these, the central processing unit (CPU) clock always shows approximately the same CPU time for reruns of the programs, even though the lapsed times between start and finish can differ markedly. The program has no knowledge of the passing of clock time, only of CPU time.

Real-time computing, in contrast, is associated with responding to events occurring outside the computer system and controlling devices, external to the system. The clock that governs the behavior of the external devices and events is not the CPU clock but the real-time clock. Hence the computer system has to respond and to act in a fixed lapsed time; this fact is the essential aspect of real-time systems.

The external device can be anything from a computer user sitting at a time-sharing terminal, in which case the response time is important but not critical, to an expensive high speed device being set up, monitored, and controlled by the computer system. Here the response time is critical. Most connections between such a device and the computer system are simple analog or binary connections. In the former the data being transferred are represented by a voltage, say 0–10 V, or an electrical current. Binary connections are simple two-state on–off connections, usually monitoring a value, (e.g., open or shut) or controlling a device, (e.g., on or off).

This paper is not concerned with the needs of time-sharing systems but addresses the needs of computer systems involved with data acquisition and control. First, we establish these needs, and then we consider how they can be met by computer operating systems. However, the key to this paper is not that these facilities are not now fully available but rather that operating systems implement them in different ways, so that it is difficult, or impossible, to use the facilities from *computer-independent* high level languages.

2   THE REQUIREMENTS

2.1   ACCESS TO REAL-TIME

Clearly, a program running as part of a real-time system needs to be able to know the time, with a resolution at least comparable with the system's reaction time. Typically, this means a 1–10 ms resolution.

2.2 ACCESS TO PROCESS-SPECIFIC INPUT–OUTPUT DEVICES

By definition, every real-time system has attached to it the device that it monitors or controls. This connection is almost always such that the device (or devices) looks to the computer to be quite different from the conventional input–output (I/O) devices associated with computer systems. In addition, the interface operation by the controlling program cannot be at the same level as blocks of data from files or lines of text for a line printer. The intimate control of I/O devices of real-time systems from high level languages is a requirement. Often this I/O is at a fairly simple level, reading or setting binary or analog points, but there also arise more complex requirements, such as the driving of motors and the reading of digitized waveforms.

To achieve this access in a standardized way, one clearly needs to standardize the I/O interface; here CAMAC, IEEE 488, and so on, are successful. However, the control program needs to express the I/O requirements in CAMAC- and IEEE 488-like terms.

The other problem is to determine by which route the process should access the I/O devices. There are two choices for this: Either the operating system can allow the program direct access to the I/O device, or the program accesses the I/O device through the operating system. The former solution lowers all the protection barriers provided by operating systems but is the most time efficient. However, this approach will necessitate extra design, debugging and maintenance; thus it will cost more in programming effort. The second method involves adding a driver to the operating system to bring the I/O interfacing scheme into the operating system. This maintains all the protection and services of the operating system and, indeed, can extend the protection into the I/O system, but it does involve the price of slower I/O because of all the checks and queueing code associated with these services.

2.3 THE CONTROL OF CONCURRENT PROCESSES

Multiprocessing operating systems are designed to schedule the execution of processes by the CPU. Of all the processes ready and waiting for execution, the operating system can use a number of criteria, either alone or in combination, to select the next process to execute. Examples of these criteria are by priority and by round-robin scheduling. The priority is usually determined by the system manager and can be adjusted by the operating system. Clearly, this is adequate for conventional computing systems, but for real-time systems some further dynamic control is needed over these priorities.

These considerations lead to further needs; one must be able to start and stop other processes, to synchronize with them, to communicate data between processes and to wait for a given time or event.

These capabilities allow the real-time system designer to write separate processes for each process of the system being monitored or controlled. Indeed, an analysis of the problem might well indicate that these processes should be split up further. The important thing to remember is that a process should only be waiting for one or a very few events at a time. If it needs to wait for more than one event, it needs to test which event has occurred and then to do a multiple branch to the appropriate code. That code, in servicing that event, might need to wait for another event. Does it branch back to the original wait, so that all possible events are waited for, or does it have its own wait, in which case the process will not respond to other events for the duration of this wait? If, in the main wait, more than one event has occurred, how does the code choose which to process first? One can quickly see how we soon need a mini operating system under the vendor-supplied operating system. Clearly, this is not desirable.

2.4   ACCESS TO FILES

The processes running in the real-time computer system and cooperating in monitoring and controlling will often need common access to files associated with the process. If all access is read-only access, there is no problem; however, if one or more processes need to write to the file, there are some classical conflicts that need to be resolved. Apart from the ability to open, to read from and to write to existing files, real-time systems need the ability to create and delete files, as well as to modify the access mode.

2.5   ACCESS TO ERROR TRAPS

A real-time process needs to handle all the likely errors that might occur; clearly, it simply cannot be canceled by the operating system. These features are implemented in most systems already, and the differences between computer systems usually can be accommodated in the compiler code generation phase.

3   THE PROBLEM

Having established the operating system's requirements for real-time systems, one now needs to ask what the problem is, because most operating systems now available do provide these facilities. What is being sought is an improvement in two areas. These are standardization and extension.

3.1   THE STANDARDIZATION NEED

To be able to express real-time needs in a high-level language, the operating system interface should be standardized. This is because the high-level language is standardized and one wishes to avoid the insertion of unnecessary or impossible conversions by the compiler. In addition, the trend in modern languages is toward strong type checking: Calls to procedures or subroutines have to have the correct number of arguments, and each argument in the call has to be of the expected data type. This directly precludes the calling of operating system routines where, for example, the value of the first parameter infers the number and data types of the remaining parameters.

So that real-time features can be brought into the language directly, the semantics need to agree not only across all real-time operating systems, but across many aspects of the syntax as well. Clearly, the actual call to the operating system (the software interrupt) will be different between various computer architectures, but the compiler can deal with these differences. In the following sections, we examine in some detail what aspects of the operating system interface need to be standardized to achieve the aim of computer-independent real-time programs. If this standardization can be achieved, then the programmer will need only the language manuals and the manuals for the I/O plug-ins that connect the computer devices under control.

4   A SOLUTION

4.1   ACCESS TO REAL-TIME

All that is necessary for access is to agree on the order and data type of the parameters from year to milliseconds. This agreement implies that the resolution of the system clock is equal to or better than a millisecond and that real-time programs do not need to know the time with any greater resolution. Processes also need to be able to specify a wait for a given period, expressed in relative or absolute time. Once again, the order and data type need agreement.

4.2   PROCESS-SPECIFIC INPUT–OUTPUT

4.2.1 Direct Input–Output Access.   Clearly, the detailed I/O access, the actual I/O instructions, cannot be a part of any computer-independent standard at present, however, desirable this might be. However, for high speed I/O access that need not be intimately part of the operating system, the code can be allowed direct access, either to the I/O address space or to the I/O reserved instructions. An important consideration, however, is that the code should be able to access only the I/O devices that it needs, thus isolating the error and making correction much simpler. This protection requirement is only possible if the architecture of the computer system allows it, and for most current systems this represents an extension of the system. One I/O service normally provided by the operating system is the queueing of I/O requests. In this case, the real-time processes will need to queue their access, using synchronization facilities.

4.2.2 Indirect Input–Output Access.   With indirect I/O access, code is added to the operating system that provides the required I/O access as part of the normal I/O system. Usually, this code has full privileges and thus can crash the system if it contains an error. The I/O access from the user code is slower than with the previous example, but that access is far more flexible. The price paid, however, is once per transfer; therefore, block transfers do not have such a high overhead per word. Unfortunately, much real-time I/O is quite unlike standard computing I/O in that one or a very few words are transferred at a time, making the overhead a real consideration. This approach does provide good protection between users and should be considered when many different users need I/O access and protection from each other. However, in single-user systems, such protection will have the highly desirable feature of being able to isolate faults.

4.2.3 Input–Output Access in High-Level Languages.   Access can be provided for simple data types such as analog or binary I/O. Such access does not make any assumptions about the details of the hardware connection. Attempts have been made in the past to standardize I/O calls, and the *ISA Standard S61.1* [1] was the first to do this. This work has now been incorporated and developed by Purdue Europe, now renamed the European Workshop on Industrial Computer Systems, in the standard proposal Industrial Real-time FORTRAN (IRTF) [2], with calls

```
CALL   AISQW   ( )
CALL   AIRDW   ( )
CALL   AOW     ( )
CALL   DIW     ( )
CALL   DOMW    ( )
```

that cover analog and digital input and output to sequential addresses or devices except for an analog random input, the second call. All digital I/O is word oriented and the output is a pulse output where the corresponding bit is a 1.

Anything more complicated must be either controlled with new high-level calls or controlled directly with machine-specific calls.

This picture changes completely if a computer-independent interface standard is used for the I/O connection. This standard, such as CAMAC or IEEE 488, will have the detailed I/O functions defined independently of the computer to which it is connected. Thus these functions can be brought into the high-level language as has been done with the Subroutines for CAMAC [3]. The details of the CAMAC connection, in that case, are then handled by the compiler and/or the operating system CAMAC driver.

4.3   CONTROL OF PROCESSES

Most real-time operating systems recognize the need for this control and do indeed provide facilities for such control. However, once again, there is no standard model, let alone a standard implementation. Different computer architectures and operating systems do not agree on the number of priority levels, how they are broadly assigned and what access individual processes have to them. IRTF does address this and has a scheduling model but does not address the various ways of communicating data between processes. It does include a model for synchronizing processes but assumes that data are exchanged between processes through a common memory area. The IRTF model could form a start for a standard real-time operating-system model but it does need to be extended. With such a model, control of processes and data communication between processes can become part of a high-level language.

For data communication between processes, one must consider the rapidly growing use of computer communications so that, if possible, the model does not make the assumption that the communicating processes execute in the same computer or share common memory. This indicates a message-oriented scheme of communication.

4.4   FILE ACCESS

A file is a collection of data stored sequentially, and possibly with an imposed structure and indexes to aid information retrieval. The process is not concerned with where this information is actually stored, as long as performance criteria are achieved. Although in conventional computing systems usually only one process accesses a file at a time, in real-time systems many processes need simultaneous access. For write access, these processes need to serialize their access, at least to the file if not at the level of the individual record of the file, to avoid conflicts and to ensure that correct information is written.

Thus, as before, a standard model for a file system and access facilities through the operating system is needed. This extends even to file-naming and device-naming conventions. If these facilities are available, including memory-resident files, then much effort currently expended for each application to provide safe access will be saved, and file access can become a fully defined part of high-level languages.

4.5   ERRORS

Although differences in error handling can be dealt with by the compiler, there is a need to standardize the set of errors that can be detected. Also, for efficiency, as many errors as possible should be detected by the hardware to avoid the need for the compiler to generate code that can make all the necessary checks.

5   CONCLUSION

Before the full benefits of standard programming languages for real-time systems can be enjoyed, the real-time facilities provided by the computer system need to be standardized. In some instances, this might mean only a common data format, for time and date, for example. In other areas, a standard model must be developed; then standard access to the facilities of that model must be determined.

An identified area for real-time computer system extension is that of providing efficient interprocess protection in the I/O system.

The advantages of establishing such a scheme are that the implementation of an operating system will be much simplified and that the user will find the system easier to learn and more efficient to use.

Many say that it is a disadvantage that new concepts in software and hardware will be slow to be included in the standard. However, those who must can extend the system before a new revision of the standard, whereas others, who do not need the improvement, continue their work within the standard.

ACKNOWLEDGMENTS

While the views expressed in this paper are the personal views of the author, they have been formed during many discussions on standardization with colleagues on the European Workshop on Industrial Computer Systems, Technical Committee 1, and the Subroutines for CAMAC Working Group of the European Standards on Nuclear Electronics Committee.

This work was done in part under the auspices of the U.S. Department of Energy.

REFERENCES

1 Industrial computer system FORTRAN procedures for executive functions and process control input–output, *ISA Stand. S61.1*, 1972 (Instrument Society of America).

2 IRTF—industrial real-time FORTRAN, *Int. Purdue Workshop European Workshop on Industrial Computer Systems, Rep. III/437/81 EN, IPW–EWICS TC1, 2.2/80*, October 1980.

3 Subroutines for CAMAC, ESONE/SR/01, *IEEE Stand. 758, IEC Doc. 45(CO) 142*, September 1978 (International Electrotechnical Commission).

DIM: A NETWORK COMPATIBLE INTERMEDIATE INTERFACE STANDARD

R. C. S. MORLING

Division of Engineering, Polytechnic of Central London, 115 New Cavendish Street, London W1M 8JS (Gt. Britain)
(Received July 5, 1982)

SUMMARY

The salient features of DIM, the standard interface used by the MININET instrumentation network, are described. In its most basic form this interface supports all the normal interactions between a computer and a peripheral or another computer, either directly or via a transparent communication link or network. It consists of 16 parallel bidirectional data lines, a data class flag to distinguish between end-user data and interface control information, and two pairs of handshake lines to control transfers in either direction. The basic interface may be extended with up to six address lines and separate parity lines for address and data. The computer–peripheral convention defines the initialization, flow control, and exception procedures and assigns the format of the control class transfers to allow the design of peripheral-independent computer interfaces.

1   INTRODUCTION

The advantages of an intermediate interface standard for the interconnection of computers and peripherals are well known. It avoids the necessity of designing a large number of special-purpose computer and peripheral-specific interfaces. Normally, only one peripheral-independent computer interface needs to be designed for each computer type and only one computer-independent peripheral interface needs to be designed for each peripheral type. Furthermore, an intermediate interface becomes almost mandatory when computers and peripherals (or, for that matter, computers and computers) are connected via a communications network.

The DIM interface has been specifically designed to facilitate computer–peripheral and computer–computer transfers either directly or via the MININET transparent communication network [1, 2]. (For those who believe that behind every acronym is a pertinent meaning, we can offer not one, but two, expansions: devilishly inappropriate mnemonic or Dik's interfacing madness.)

Reprinted with permission of R. C. S. Morling, "DIM: A Network Compatible Intermediate Interface Standard," *Interfaces in Computing*, Vol. 1, No. 2, May 1983, and North-Holland Publishing Company, Amsterdam, 1983.

2 INTERFACE REQUIREMENTS

Very early in the specification of MININET it was necessary to adopt a flexible and economical standard interfacing technique that was compatible both with the requirements of its application areas (e.g., laboratory instrumentation, process control, medical applications, etc.) and with the transparency, cost and speed of goals of MININET. These requirements can be summarized as follows.

(i) A single interface must be able to support data transfers in both directions and, in addition, the transfer of control information in either direction. That is, in terms of conventional computer terminology a single interface should support read data, write data, read status, and write command operations.

(ii) Especially in process control and medical applications, it may be necessary to isolate a maverick device that is threatening to disrupt the entire system by some form of unsociable behavior. This rules out the use of a bus system (e.g., CAMAC [3] or the International Electrotechnical Commission IEC-625 bus [4]) because a device has only to short a transfer control line to halt all operations on the bus or to short a data line to corrupt the data transfers. An interface between single devices, or a group of closely associated devices, allows each device to be isolated. Furthermore, it allows a communications network to apply flow control to each device individually, thus avoiding the danger that one user floods the network with data.

(iii) The transfer rate should be asynchronously controlled by handshake signals in the interface to enable data throughput that is as fast as can be comfortably accommodated by both the sender and the receiver. This freedom, however, may well be qualified by some relatively long time-out period to detect whether the other party is powered down, faulty, or otherwise not at home.

(iv) The interface should be able to transfer in a single operation the "basic unit of data" handled by the majority of devices. For example, the basic unit for a teletypewriter is one character, for an analogue-to-digital converter it is one sample and so on. This requirement simplifies the interface complexity in that data units do not have to be split up for transmission or reassembled at the other end of the interface. This requirement implies that the width of the interface should be greater than 8 bits (which is perhaps a first choice as a suitable value) as analog-to-digital and digital-to-analog converters frequently have a sample size of 10 or 12 bits. A natural and convenient choice for the maximum basic data unit is therefore 16 bits as it is large enough to accept the output of most data acquisition equipment and is the word size of most minicomputers. A larger data unit would not be convenient because few peripherals would fully use it and most minicomputers would not know what to do with it if they did.

(v) Protection against both internal and external electrical interference should be provided to avoid false transfers, lost transfers, or corruption of the information during transfer. Handshake-driven interfaces are notoriously prone to false transfers and the early abortion of transfer cycles because of the appearance of impulsive noise on the handshake control lines. Thus special attention should be given to the protection of these lines.

Whether or not the data itself should be encoded for error protection across the interface is less certain, as most instrumentation systems and computer–output systems have no such protection. Computer–computer block transfers would usually include end-to-end error protection of the entire data block to cover all parts of the data's journey. Thus the extra complexity of error encoding in the interface does not seem to be justified. Nevertheless, there may be some very sensitive applications where at least a parity check should be made. Consequently, the interface should provide the *option* of including a parity bit with the data.

(vi) The maximum transfer rate should be greater than the maximum user throughput requirements. At the same time, it should not be so fast as to require any exotic logic technology in the interfaces. This requirement implies a maximum throughput in the area of 10–20 Mbits/s. This speed is more than sufficient for most instrumentation applications. Almost the only well-known application area where this may be too slow is block transfers between number crunchers and their massive file storage systems. With the huge capital costs of such systems they can well afford the cost of interface systems specially designed for that application, e.g., HYPER channel [5] or IDANET [6]. Of course, in an unsympathetic noise environment the maximum transfer rate may be reduced in order to allow longer validation intervals.

(vii) Under normal conditions the interface should be able to operate over distances of at least 10 m at maximum transfer rates and up to 30 m at reduced transfer rates. These distances are much greater than the normal distances expected within one application area. It is anticipated that greater separation would be serviced by a communications system such as MININET.

(viii) The interface should operate in such a way that a transparent communications network can be inserted between two devices that were communicating directly without any change to the interface or higher level protocols. The most important consequence of this requirement is that the interface cannot support directly elicited responses. That is, there can be no equivalent of the read strobe found in most computer memory and input–output buses because the propagation time through the communication system would probably be much longer than the maximum read cycle time that the computer could tolerate. The equivalent of directly elicited responses can be achieved by sending a control message requesting the desired data. In order to be compatible with computer busses that expect to read data and status directly with minimal delay, the computer interface must contain data and status registers that are updated from the peripheral via the standard interface.

(ix) The cost of the interface implementation should be low, commensurate with the relatively inexpensive peripherals that it is interfacing. This constrains many of the above requirements. Without cost constraints a complicated interface with an extremely high price could be designed which, for the last very good reason alone, would not be used. Consequently, the interface has to do its job as quickly and efficiently as possible while remaining cheap and easy to implement.

3   THE BASIC DIM INTERFACE

The basic DIM interface consists of 22 signal lines comprising 16 bidirectional data–control shift lines, master handshake, and slave handshake control lines, and a dominance line. One side of the interface is termed the *master* and the other side the *slave*.

3.1   DATA TRANSFERS

Sixteen parallel *data lines* DDLØ′–DDL15′ are used to perform the required 16-bit transfers. Of course, devices do not have to use all these lines. For example, terminals usually transmit characters along the 8 least significant lines while analogue-to-digital converters usually use the 8, 10, or 12 most significant lines. These lines are bidirectional to avoid the unnecessary wires, connector pins, and electrical buffering components required if two unidirectional sets of data lines are used.

Transfer of data between a computer and a peripheral is usually divided into command or status information, on the one hand, and actual end-user data on the other. Different registers are associated with each class of transfer.

The *data–control shift line* DDC′ qualifies the data lines to distinguish between these classes during a transfer. If it is true, the data lines contain normal end-user information: *data class*. If it is false, the data lines contain control information: *control class*. Like the data lines it is bidirectional. This enables a single DIM interface to handle the four basic computer–peripheral operations, read data, write data, read status, and write command.

3.2   TRANSFER CONTROL

Two pairs of transfer control lines are provided in the interface. The master handshake lines, *master interrupt* DMI′ and *master acknowledge* DMA′, are used to transfer data from the master to the slave. The slave handshake lines, *slave interrupt* DSI′ and *slave acknowledge* DSA′, are used for transfers from slave to master. When a device wishes to transfer data (of either class), it sets its interrupt line true (the master interrupt if it is the master or the slave interrupt if it is the slave) to indicate that it has data available. After the other device has detected this interrupt and is ready to receive data, it replies by setting its acknowledge line true (the master acknowledge if it is the slave or the slave acknowledge if it is the master). Once the original device has detected the acknowledgment, it then, and only then, enables its data and data class transmitters, placing the data to be transferred onto the interface lines. At the same time it sets its interrupt line false. When the receiver has detected the removal of the interrupt, it first waits for any reflections and cross-talk on the interface cable to die down before quietly accepting the data (usually by strobing it into a buffer register) and then sets the acknowledge line false. When the sender detects the removal of the acknowledgment, it disables the data and data class transmitter and the transfer cycle is complete.

If a noise impulse appears on an interrupt line, there is a danger that it could cause a false transfer of nonsensical or duplicate information or the corruption of data passing in the opposite direction. Similarly, a noise impulse on an acknowledge line could result in the loss of a transfer and/or corruption of data passing in the opposite direction. The designers of the British Standard interface (BS 4421) [7] recognized this danger [8] and protected their interface by the use of low pass filters and large hysteresis receivers on the handshake control lines. DIM uses an alternative method, *validation time-outs*, that is more suitable for large-scale integration. Logic state changes of a control line are not accepted until it has stayed in its new state for a fixed validation interval. If at any time during this interval it returns to its original state, then it is ignored. If, subsequently, it returns to its new state, then the validation timer restarts over. Any noise impulses of width less than the validation interval are therefore rejected no matter how close together they occur.

The noise rejection characteristics of the validation time-out approach may be compared with that of a first-order low pass filter by calculating the impulse height required for acceptance as a function of the pulse width. In order to make the comparison equitable, the time constant τ_0 of the filter is set so that the delay to a normal signal transition is the same as the validation interval t_0, that is,

$$\tau_0 = \frac{t_0}{\log\{\Delta V_s/(\Delta V_s - \Delta V_{th})\}}$$

where ΔV_s is the normal signal swing and ΔV_{th} is the swing required to reach the receiver threshold level. Using a rectangular pulse of height V_p and width τ as the model of a noise impulse, the noise immunity of the filter is given by

$$V_p = (1 - e^{-\tau/\tau_0})^{-1} \Delta V_{th}$$

while that for the time-out circuit is

$$V_p = \infty \qquad \tau < t_0$$
$$= \Delta V_{th} \qquad \tau \geq t_0$$

Perhaps a more realistic noise impulse model is a decaying exponential function of peak height V_p and time constant τ. For this model the noise immunity of the filter circuit can be found, by equating the peak value of the correlation of the noise impulse and the filter impulse response to the threshold voltage, to be

$$
\begin{aligned}
V_{\mathrm{p}} &= \frac{(\tau_0/\tau - 1)\,\Delta V_{\mathrm{th}}}{(\tau/\tau_0)^{\tau/(\tau_0 - \tau)} - (\tau/\tau_0)^{\tau_0/(\tau_0 - \tau)}} \qquad \tau_0 \neq \tau \\
&= e\,\Delta V_{\mathrm{th}} \qquad\qquad\qquad\qquad\quad \tau_0 = \tau
\end{aligned}
$$

The performance of these circuits for both noise models is shown in Figure 1, where $\Delta V_{\mathrm{s}} = 1.5\,\Delta V_{\mathrm{th}}$. It can be seen that the validation time-out approach has superior immunity against short noise pulses of width less than about t_0, while the filter circuit is slightly better for longer pulses. In addition, a series of closely spaced noise spikes cannot "pump up" the time-out circuit as it can the filter circuit, so reducing the latter's immunity.

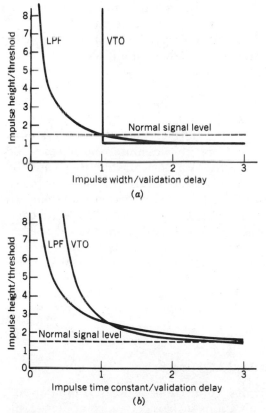

FIGURE 1 Noise immunity of the low pass filter (LPF) and validation time-out (VTO) methods: (a) rectangular noise impulse model; (b) exponential noise impulse model.

Leading and trailing edges of the interrupt and acknowledgment must be validated. Thus there are four separate validations in each transfer cycle. In a noisy environment the validation interval can be increased provided that it does not slow the maximum transfer rate required by the application. It should also be noted that the data lines are not sampled until a settling time interval has elapsed after the last movement of any line on the interface. Thus the danger of cross-talk or reflections in the interface cable is, for all practical purposes, completely removed. In the present realization of the interface the minimum validation interval is approximately 100–200 ns while the minimum settling time before the data are accepted is approximately 300 ns. This gives a maximum of about 10^6 transfers per second.

3.3   MASTER–SLAVE RELATIONSHIP

The master and slave designation of the interfaces principally determines which control lines are used by which party. The master interface has one additional function, however. Since the data lines are bidirectional, it is not possible to perform transfers in both directions at the same time. The master interface arbitrates, on a first-come first-served basis, which transfer goes first. It performs this by ignoring the master acknowledgment if it is already asserting the slave acknowledgment and by delaying the slave acknowledgment if it is already receiving the master acknowledgment. This means that the master designation determines the location of the arbitration logic rather than any priority of data transfer.

By convention, peripherals use slave interfaces and the network uses master interfaces. It is necessary, therefore, for computer interfaces to be masters when they are connected directly to a device, while they must be slaves when connected via the network. To facilitate this without resorting to hardware modification, network interfaces permanently assert true the *dominance line* DDM' in the interface and so are called *dominant masters*. The computer interface (called a *submissive master*) monitors this line and, if it is asserted, acts as a slave interface and, if it is not, acts as a master. Since slaves do not assert the dominance line, a submissive master can be successively connected to either a slave or a dominant master without modification. To prevent the confusion of a transfer by noise on the dominance line, it is heavily low pass filtered by the submissive master.

4   EXTENSIONS OF THE BASIC INTERFACE

The basic interface is sufficient for most connections between computer and peripherals. However, for some applications such as certain computer–computer and computer–network connections there is a need for further qualification of the data transferred. Also some high security applications (e.g., process control) may need some error protection during the data transfer.

These requirements are achieved with DIM by means of address and parity lines.

4.1   THE DIM ADDRESS LINES

Six bidirectional *address lines* DAL∅'–DAL5' are provided that are enabled and can transfer information at the same time as the data lines. These lines can be used to provide subaddressing for links between intelligent devices. For example, two computers could be connected via a single DIM interface, the DIM addresses being mapped into the computers' peripheral address space to form a number of independent virtual connections between processes in the two machines. Another important application arises when a computer is connected to a number of devices via a communications link or network such as MININET. It is plainly absurd to connect the computer to the network via separate interfaces for each remotely connected peripheral, as the transfers would first be demultiplexed to each interface and then re-multiplexed into the network station. Instead a single interface can be used where the address lines discriminate between the different devices addressed. The data would only be demultiplexed when they reached the remote DIM interfaces to the destination devices. Of course, to utilize such an economy a multiple DIM computer interface is required. The advantages and constraints of such an approach are discussed in Section 6.

4.2   PARITY CONTROL

Two bidirectional parity lines are included in the interface: one, the *data parity line* DDP', provides an odd parity check of the data and the data–control shift lines; the other, the *address parity line* DAP', provides an odd parity check on the address lines. In order to enable interfaces without the parity option to be connected to interfaces that are checking parity without the detection of spurious errors, two additional lines are provided. The *master parity available* DMP' and the *slave parity available* DSP' indicate whether the master interface or the slave interface, respectively, is generating parity signals. This is a similar facility to that provided in the BS 4421 interface [7]. Of course, a single bit error on this line could result in the detection of a false error or a failure to detect a parity error because the parity check logic was disabled. This danger is minimized by heavily low pass filtering the parity available lines. These lines, like the dominance line, are quasi-static and would be expected to change state only when the interface connector is physically moved.

5   ELECTRICAL AND MECHANICAL CHARACTERISTICS

Single-ended negative-logic active low signals are used throughout for all interface lines. The electrical characteristics are based on the Unified Bus

Standard [9] for which integrated circuit transmitters and receivers are readily available. The data, address data–control shift and parity lines are terminated at each end by a resistance of 120 Ω in series with a voltage of 3.5 V usually implemented by resistances of 180 Ω connected to a +5 V supply and 390 Ω connected to the common ground (Figure 2a). The drivers may be open collector or tristate with a maximum OFF state leakage current of 100 μA and an ON state voltage of less than 0.7 V. The receivers have high impedance inputs with a maximum input current of 50 μA. They are usually equipped

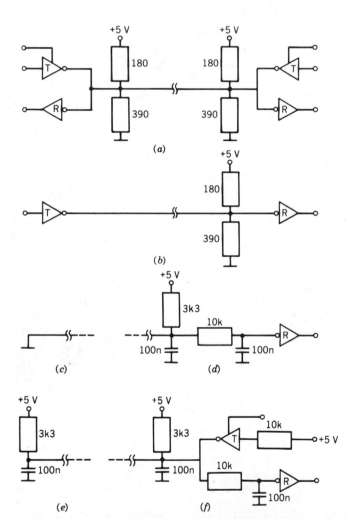

FIGURE 2 DIM interface line termination: (a) line termination for DDL0'–DDL15', DAL0'–DAL5', DDC', DDP', and DAP'; (b) line termination for DMI', DSI', DMA', and DSA'; (c)–(f) line termination for DDM', DMP', and DSP' for (c) transmit true, (d) receive only, (e) transmit false, and (f) transmit and receive.

with hysteresis to improve the interface's noise margin further. The high state threshold voltage is between 1.8 and 2.5 V, and the low state threshold voltage is between 1.05 and 1.55 V. The use of negative logic with this electrical specification has the convenient result that interface lines not used in a particular application are received as logic zero (false). The handshake control lines do not need termination at the transmission end provided that the transmitters have a low impedance active output in the high as well as the low state (Figure 2b). The specifications of the transmitters and receivers are the same as those for the data lines.

The quasi-static lines, that is, the dominance and the two parity available signals, are made active by shorting them to the common return. At their receiving end or if they are not being used, they must be terminated with a 100-nF low inductance capacitor connected to the common ground and a 3.3-kΩ resistor connected to a +5-V supply. We have found that a simple first-order low pass filter consisting of a 10-kΩ resistor and a 100-nF capacitor feeding a standard high input inpedance receiver as described above is quite sufficient for filtering the quasi-static lines (Figures 2c–2f).

A 34-way flat cable is used to implement the interface connection physically (Figure 3). The connector used is compatible with the flat cable and is fastened to the cable by a single crimping operation. Two of the 34 wires are used to form the common return. The connector assignment is shown in Table 1.

6 THE COMPUTER–PERIPHERAL CONVENTION

In order to realize fully the advantages of an intermediate standard, the design of a computer interface must be independent of the device to which it is connected, whether the device be a teletypewriter, an analogue-to-digital converter, or even another computer. Conversely, the device interface design should be independent of the computer to which it is connected. To this end, only the device interface can contain the device-specific circuitry and only the computer interface the circuitry specific to the computer. It is possible to go further as the procedures involved in initializing and maintaining the transfer of information between a device and a computer are remarkably uniform,

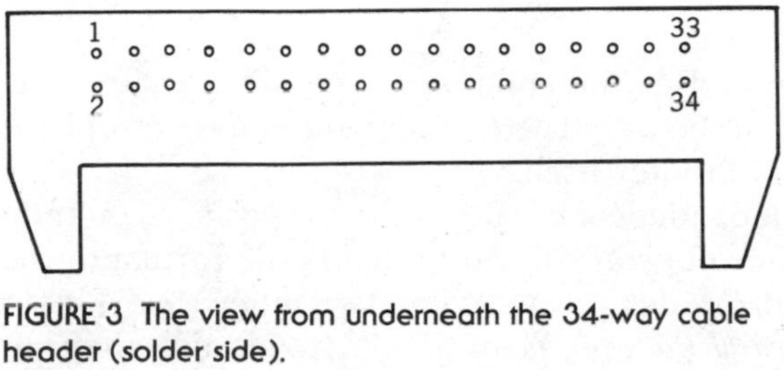

FIGURE 3 The view from underneath the 34-way cable header (solder side).

TABLE 1   DIM Interface Pin Assignment

Pin	Signal	
	Abbreviation	Description
1	GND	Earth
2	DDL0	Data line 0
3	DAL0	Address line 0
4	DDL1	Data line 1
5	DAL1	Address line 1
6	DDL2	Data line 2
7	DAL2	Address line 2
8	DDL3	Data line 3
9	DAL3	Address line 3
10	DDL4	Data line 4
11	DAL4	Address line 4
12	DDL5	Data line 5
13	DAL5	Address line 5
14	DDL6	Data line 6
15	DAP	Address parity (odd)
16	DDL7	Data line 7
17	DDL13	Data line 13
18	DDL8	Data line 8
19	DDL14	Data line 14
20	DDL9	Data line 9
21	DDL15	Data line 15
22	DDL10	Data line 10
23	DDC	Data–control shift
24	DDL11	Data line 11
25	DDP	Data parity (odd)
26	DDL12	Data line 12
27	DMP	Master parity available
28	DSP	Slave parity available
29	DMA	Master acknowledge
30	DSI	Slave interrupt
31	DDM	Dominance
32	DSA	Slave acknowledge
33	DMI	Master interrupt
34	GND	Earth

irrespective both of the device and of the computer type. This enables the bulk of the computer interface to be constructed independently, not only of the peripheral but also of the computer itself.

In order to achieve this independence, it is necessary to specify a common initialization, error and flow control protocol and to define the format of the control class transfers between the device and computer interfaces. This is done with the *DIM computer–peripheral convention* (CPC). In addition the convention allows an intervening communication system such as MININET to report any error condition to the computer interfaces.

This approach facilitates the construction of multiple DIM–computer interfaces such as the *interface processor* shown in Figure 4. Only the part that is concerned with the connection to the computer's peripheral bus needs to be specific to the particular computer type. This *computer personality interface* must conform to the electrical and mechanical standards of the computer bus and may contain logic to map status and command information between the computer's conventional assignments and that of the DIM CPC.

The central part of the interface contains a data-in, a data-out, and a command and a status register for each DIM interface handled. The data-in and data-out registers act as buffers for data class transfers between the computer and the DIM interface. (It should be remembered that directly elicited responses such as a computer bus read operation cannot be extended out to the remote device. Instead the computer reads the data in the buffer register.) The command registers are loaded from the computer. Part of each command register is reserved for control of the internal functions of the interface processor such as the enabling of computer interrupts, and so forth. The remainder of the register is relayed to the device in a control class transfer. Each status register contains information about the health and operational mode of the device and the ready (not-busy) semaphores controlling the flow of data to and from the device. It is updated by events occurring within the interface processor, such as the computer reading or writing to the data registers, and by the receipt of control class transfers from the device or, possibly, from an intervening network.

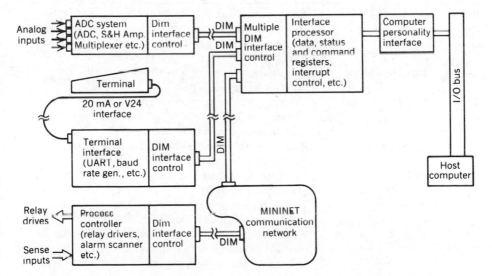

FIGURE 4 Typical uses of a DIM interface processor: ADC, analog-to-digital converter; S & H Amp., sample and hold amplifier; VAR/T, universal asynchronous receiver–transmitter; and I/O, input–output.

0	1	2	3	4	5	6	7	8	9	10	11	12	13	14	15
READ CONT		Device dependent node					Dev dep trigger		INIT		DATA REQ	Res	ENB T T	ENB A A	ENB INT

(a)

0	1	2	3	4	5	6	7	8	9	10	11	12	13	14	15
BRST TRAN		Device dependent node					Dev dep trigger		INIT		DATA REQ	0	0	0	0

(b)

FIGURE 5 DIM CPC command format: (a) command register format; (b) command transfer format. The asterisks indicate that the uses of bits 1 and 8 are restricted to commands to nonintelligent devices only and that otherwise they must be zero.

The DIM control sections in the computer and device interfaces handle the DIM transfer procedure. In the computer interface the DIM section usually maintains a deadman's timer to monitor each transfer, thus preventing a lock-out situation caused because, for example, the device is switched off. The device interface contains circuitry specific to the interfaced device. This may well include a control register that is updated by commands from the computer received via DIM control class transfers, a status register whose contents are relayed to the computer interface whenever a change occurs, and data buffer registers for the data class transfers.

TABLE 2 Explanation of DIM Computer–Peripheral Convention Command Format Abbreviations

Bit	Abbreviation	Description
0	READ CONT BRST TRAN	Read continuous (burst transfer); commands the device to output data in burst mode without waiting for a data request acknowledgment from the computer interface
1–6	Dev Dep Mode	Sets mode of operation of the device; allocation specific to device type
7, 8	Dev Dep Trigger	Triggers the device to perform a device-specific action
9, 10	INIT	Initialization commands; see Table 6
11	DATA REQ	Data request; requests the device to send one word of data to the computer
12	Res	Reserved for future use within the computer interface; must be zero
13	ENB T T	Enable transaction timer; allows the transaction deadman's timer to run
14	ENB A A	Enable auto-acknowledge; causes the computer interface to send a data request message to the device whenever the data-in register is read by the computer
15	ENB INT	Enable interrupt; causes the computer interface to request a computer interrupt whenever an event requiring computer servicing occurs

(a)

0	1	2	3	4	5	6	7	8	9	10	11	12	13	14	15
WRT CONT	REM ERR	Device dependent status						TR ERR	CMD REQ	DEV RDY	WRT RDY	T E Q	D U	READ OVL	READ RDY
*			*	*	*	*	*	*	*			*	*	*	

(b)

0	1	2	3	4	5	6	7	8	9	10	11	12	13	14	15
BRST TRAN	0	Device dependent status						TR ERR	INIT		DATA REQ	0	0	0	0
*			*	*	*	*	*	*	*			*	*	*	

(c)

0	1	2	3	4	5	6	7	8	9	10	11	12	13	14	15
0	1	0	0	0	0	0	0	TR ERR	0	0	0	T E Q	D U	0	0
*			*	*	*	*	*	*	*			*	*	*	

FIGURE 6 DIM CPC status format: (a) status register format; (b) status transfer format (generated by devices; (c) status transfer format (generated by a network). The asterisks indicate exception condition bits.

The overall formats of the command register and transfers are shown in Figure 5 and Table 2 and those of the status register and transfers in Figure 6 and Table 3. The precise mechanisms by which the bits within the command and status registers are set and cleared are detailed in Table 4 and Table 5, respectively.

TABLE 3 Explanation of DIM Computer–Peripheral Convention Status Format Abbreviations

Bit	Abbreviation	Description
0	WRT CONT BRST TRAN	Write continuous (burst transfer); informs the computer interface that the device can accept data in burst mode
1–6	REM ERR	Remote error; see Table 8
2–7	Dev Dep	Allocation specific to device type
8	TR ERR	Transmission error; see Table 8
9	CMD REQ	Command request; indicates that the device has sent an initialize message; see Table 7
10	DEV RDY	Device ready; indicates that the device is ready for operation; see Tables 6 and 7
9, 10	INIT	Initialization messages; see Table 7
11	WRT RDY	Write ready; indicates that the computer can safely write into the data-out register
11	DATA REQ	Data request; requests the computer to send one word of data
12	T E Q	Transmission error qualifier; see Table 8
13	D U	Device unavailable; see Table 8
14	READ OVL	Read overlay; indicates that data have arrived from the device before the previous data have been read by the computer
15	READ RDY	Read ready; indicates that there are new data in the data-in register waiting to be read

TABLE 4 Command Register Control

Bit	Abbreviation	Set	Cleared
0	WRT CONT	By receipt of a command word with the corresponding bit true	By receipt of a command word with the corresponding bit false or after power-up
1–6	Dev Mode		
12	Res		
13	ENB T T		
14	ENB A A		
15	ENB INT		
7, 8	Dev Trig	By receipt of a command word with the corresponding bit true	After the dispatch of a command word towards the device or by receipt of a command word with bit 10 true and the corresponding bit false or after power-up
9, 10	INIT	By receipt of a command word with the corresponding bit true or after power-up	After the dispatch of a command word toward the device
11	DATA REQ	By receipt of a command word with bit 11 true or after the data-in register has been read by the computer if bit 14 of the command register (ENB A A) is set	After the dispatch of a command word toward the device or by receipt of a command word with bit 10 true and bit 11 false or after power-up

6.1   FLOW CONTROL

The transfer of information from the computer to the device is controlled by the WRT RDY and WRT CONT bits in the status register. After initialization, when it is good and ready, the device sends a data request message by means of a control class transfer with bit 11 (DATA REQ) true. Bit 0 (BRST TRAN) of the same transfer determines whether the end-to-end *handshake transfer mode* (bit 0 false) or the *burst transfer mode* (bit 0 true) is to be used for the write operations. The data request message sets the WRT RDY semaphore in the status register and causes a computer interrupt if enabled. By reading the status register the computer can sense when WRT RDY becomes true. It can then safely write data into the data-out register. This event immediately clears WRT RDY in the status register. The data are subsequently dispatched as a data class transfer through the DIM interface towards the device.

TABLE 5 Status Register Control

Bit	Abbreviation	Set	Cleared
0 2, 3	WRT CONT Dev Dep	By receipt of a status word with the corresponding bit true	After power-up or by receipt of a status word with the corresponding bit false or by receipt of a command word with bit 9 true
1 4–7 9	REM ERR Dev Dep CMD REQ	By receipt of a status word with the corresponding bit true	After power-up or by receipt of a status word with the corresponding bit false or by receipt of a command word with bit 10 true
8	TR ERR	By receipt of a status word with bit 8 true or by detection of a parity error in a word received from the device or network	After power-up or by receipt of a status word with bit 8 false or by receipt of a command word with bit 10 true
10	DEV RDY	By receipt of a status word with bit 10 true	After power-up or by receipt of a command word with bit 9 true
11	WRT RDY	If bit 10 of the status register (DEV RDY) is set, by receipt of a status word with bit 11 true or receipt of a status word with bit 10 true and bit 11 true or if bit 0 of the status register (WRT CONT) is set, after dispatch of a data word toward the device	After power-up or by the computer writing to the data-out register or by receipt of a command word with bit 9 true or by receipt of a status word with bit 10 true and bit 11 false

TABLE 5 (*Continued*)

Bit	Abbreviation	Set	Cleared
12	T E Q	By receipt of a status word with bit 12 true or by expiration of the transaction time-out interval or by detection of a parity error in a word received from the device or network	After power-up or by receipt of a status word with bit 12 false or by receipt of a command word with bit 10 true
13	D U	By receipt of a status word with bit 13 true or by expiration of the DIM time-out interval	After power-up or by receipt of a status word with bit 13 false or by receipt of a command word with bit 10 true
14	READ OVL	If bit 15 of the status register (READ RDY) is already set, by the arrival of a data word from the device	After power-up or by receipt of a command word with bit 10 true
15	READ RDY	If bit 10 of the status register (DEV RDY) is already set, by the arrival of a data word from the device	After power-up or by the computer reading the data-in register or by receipt of a command word with bit 10 true

In the end-to-end handshake transfer mode (Figure 7*a*), WRT RDY stays reset until the device is ready to accept another data word, whereupon it dispatches a data request message to the computer interface that sets WRT RDY and so the process continues with alternate data class transfers from computer to device and control class transfers from device to computer. In the burst transfer mode (Figure 7*b*), WRT RDY is set immediately that data are dispatched through the DIM interface and the device sends only the one data request message to start the burst. Thus data are transferred as fast as the computer can write into the data-out register and the interface can dispatch the data. The device must be capable of accepting data at this rate or a loss of data will occur.

It should be noted that, in both modes of transfer, the computer procedures are identical. Thus it is not necessary for the computer to know which mode of transfer is being used. For this reason, when the computer reads the status register the *group error flag* GE, which is the inclusive OR of the exception bits in the status register (bits 4–9 and 12–14), is substituted for WRT CONT in bit 0.

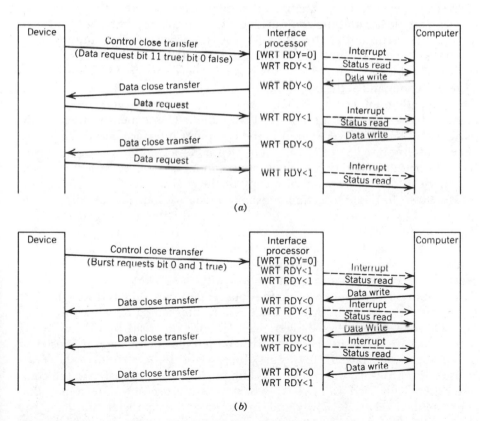

(*a*)

(*b*)

FIGURE 7 Write procedures: (*a*) handshake mode; (*b*) burst mode.

The read process is controlled by READ RDY in the status register and DATA REQ, ENB A A and READ CONT in the command register. If READ CONT (bit 0) is not set by the computer, the data transfer from the device to the computer interface operates in handshake mode. If the computer sets READ CONT the data transfer operates in the burst mode. ENB A A (bit 14) enables the interface to acknowledge automatically the receipt of data by sending a data request message *after* the data have been read by the computer.

To start a read operation, the computer writes a command word with bit 11 (DATA REQ) set. This is dispatched to the device, whereupon the bit is reset in the command register. After receipt of this data request message, the device sends data to the computer. Receipt of the data class transfer by the computer interface causes bit 15 (READ RDY) of the status register to be set and, if enabled, a computer interrupt requested. The computer can detect when the data have arrived by reading the status register and checking bit 15. If it is set, the computer can then read the data-in register, which causes READ RDY to be reset.

Had both READ CONT and ENB A A been false in the command to the device, it is necessary for the computer to output a command with bit 11 true after each read operation (Figure 8a). In this mode there is an explicit handshake between the device and the computer itself. If, in the original command, ENB A A had been true and READ CONT false, the computer interface automatically sets DATA REQ in the command register and dispatches a control class transfer toward the device (Figure 8b). Thus it is only necessary for the computer to issue one command at the beginning of the transfer of a block of information. There is still an implicit handshake between the computer and the device by means of the computer-driven read operation. Consequently, there is no danger that data are overwritten in the computer interface before they have been read. Because DATA REQ cannot be cleared by a computer command with bit 11 false (other than by an initialization command) and because it is immediately cleared after dispatch of a copy of the register (Table 4), commands can be given in the automatic handshake mode without the loss of data request messages before they are transmitted and without the generation of duplicate requests.

The burst mode of transfer is obtained by setting READ CONT as well as DATA REQ in the original command (ENB A A should be reset). The device receives this command as a burst request with BRST TRAN true and so sends data continuously without waiting for data request messages as fast as the data are generated and the interface or a possibly intervening network can transmit. There is therefore a danger that data are not read by the computer before they are overwritten by the arrival of new data in the data-in register of the computer interface. If this occurs, the error bit 14 (READ OVL) of the status register is set to warn the computer. The burst mode of transfer to the computer is basically used with two types of device. It is useful with terminals that generate data relatively slowly and, in any case, the character rate from the terminal to the terminal–DIM interface cannot normally be dynamically con-

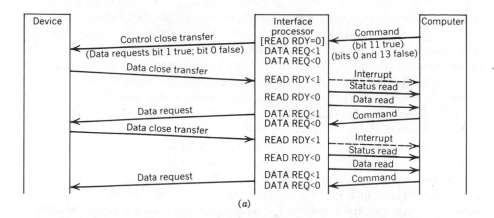

(a)

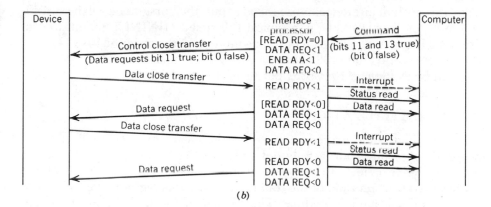

(b)

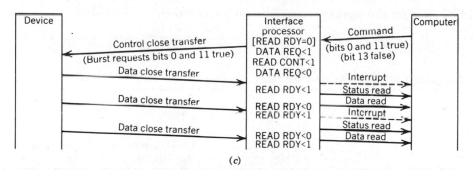

(c)

FIGURE 8 Read procedures: (a) nonautomatic handshake mode (READ CONT = 0 and ENB A A = 0; (b) automatic handshake mode (READ CONT = 0 and ENB A A = 1; (c) burst mode (READ CONT = 1 and ENB A A = 0).

351

trolled. The other common application is where a high speed device such as a multiplexed analogue-to-digital converter is connected via a network and the trans-network delay is too long for the handshake mode. In the latter case the priority of the computer interface and the interrupt latency of the computer can be critical to avoid overlap of data in the computer interface. As far as the computer is concerned, the automatic handshake mode and the burst mode read processes are identical, with the exception of the format of the initial command word.

If the computer and device are separated by a communication link or network such as MININET, then the maximum data transfer rate in the handshake mode is limited by twice the end-to-end transport delay of the network, whereas in the burst mode it is limited by the maximum effective user throughput of the network. Consequently, even with a network such as MININET, which has been specifically designed to minimize the transport delay rather than just to maximize throughput, the burst mode will be considerably faster. For example, in a typical full speed MININET the maximum user data transfer rate in the handshake mode is of the order of $10^4 - 2 \times 10^4$ words per second (160 − 320 kbits/s), whereas in the burst mode the transfer rate can be as high as 10^5 words per second (1.6 Mbits/s).

6.2   INITIALIZATION

The initialization procedure provides the means to reset the device and computer interfaces into a fully defined initial condition with the pathway between the two clear of any pre-existing messages. Bits 9 and 10 of the control class transfers and of the command and status registers are concerned with the initialization procedure. The format of the initialization messages and their effect on the status register when transmitted are shown in Table 6. The effect on the status register when received is shown in Table 7.

TABLE 6 Effect of Initialize Command on Status Register

	Message, Initialize; Code (bit 9, bit 10), 11	Message, Initialize Acknowledge; Code (bit 9, bit 10), 01	Message, None; Code (bit 9, bit 10), 00
DEV RDY (10)	Cleared	Unchanged	Unchanged
WRT RDY (11)	Cleared	Unchanged	Unchanged
READ RDY (15)	Cleared	Cleared	Unchanged
Exception conditions (1, 4−9, 12−14)	Cleared	Cleared	Unchanged
Nonexception conditions (0, 2, 3)	Cleared	Unchanged	Unchanged

TABLE 7 Effect of Received Initialize Message on Status Register

	Message, Initialize; Code (bit 9, bit 10), 11	Message, Initialize Acknowledge; Code (bit 9, bit 10), 01	Message, None; Code (bit 9, bit 10), 00
CMD REQ (9)	Set	Cleared	Cleared
DEV RDY (10)	Set	Set	Unchanged
WRT RDY (11)	a	a	b
READ OVL (14)	Unchanged	Unchanged	Unchanged
READ RDY (15)	Unchanged	Unchanged	Unchanged
Others (0–8, 12, 13)	a	a	a

[a] Jam loaded from corresponding bit in message.
[b] Set if corresponding bit in message is set; otherwise unchanged.

To initialize the interfaces, the computer outputs a command word with bits 9 and 10 true. This clears the status register entirely including the DEV RDY semaphore. If the computer is ready to receive data from the device, bit 11 (DATA REQ) of the command would also be true. If it is not ready, bit 11 of the command word is zero, which clears the corresponding bit in the command word (Table 4). Thus any undispatched data request message is overwritten by the initialize command. After the contents of the command register have been dispatched toward the device, bits 9 and 10 are cleared to avoid duplication.

On receipt of the initialize message the device resets itself, clears any error flags unless the fault condition still exists and halts any ongoing data acquisition or output. After this process is complete, the device sends a control class message back to the computer, which contains the new status of the device and has bit 9 false and bit 10 true to acknowledge the initialize command. The arrival of this acknowledgment at the computer interface causes the DEV RDY semaphore (bit 10) in the status register to be set, thus informing the computer that the device is ready for operation. Any subsequent status transfers from the device informing of exception conditions or carrying a data request must have bits 9 and 10 false. The DEV RDY semaphore remains set, only being cleared by a further initialize command from the computer (Table 5).

In the interim period between the initialize command and its acknowledgment, while DEV RDY is reset, the WRT RDY and READ RDY semaphores remain firmly reset despite the arrival of any data or data requests from the device. Thus any pre-existing data traffic is ignored. However, other status bits, including error flags, can be set and computer interrupts are not suppressed. Consequently, the computer is informed of any error message, which may well originate in an intervening network, concerning the fate of the initialize message.

The device itself can issue an initialize message which indicates that it is initialized but requires a command from the computer as an acknowledgment. This message sets bit 9 (CMD REQ) and bit 10 (DEV RDY) in the status register of the computer interface. The WRT RDY semaphore is jam loaded with the value of bit 11 (DATA REQ) of the status message. The computer must respond with bit 10 of its command word true, normally by sending an initialize acknowledge (bit 9 false, bit 10 true) message. When the computer loads the command register with this message the exception condition bits (i.e. those that set the GE flag) and READ RDY are reset; however, DEV RDY, WRT RDY and the remaining nonexception condition bits (0, 2 and 3) remain unchanged (Table 5) so that if a data or burst request was included with the received initialize message they are not lost.

It is quite acceptable for the computer to respond with an initialize command instead of an initialize acknowledge command. However, this second initialize message would require an acknowledgment from the device before data transfer could begin. Of course, if both sides responded with initialize messages, deadlock would result. Nonintelligent devices usually send an initialize command only after power has been applied. The setting of CMD REQ in the status register therefore informs the computer that the device is now on line or, if in the middle of the transfer of a block of data, that a temporary power interruption has occurred.

Once set by the computer, bits 9 and 10 of the command register are only reset after dispatch to the device (Table 4). Consequently, the initialize commands cannot be inadvertently lost or duplicated by any subsequent commands provided that bits 9 and 10 of these commands are always reset. An initialize command has priority over an initialize acknowledge command in the sense that an initialize command will overwrite an as-yet-undispatched acknowledge message in the command register.

Data or burst requests and other status or command messages may be included with an initialize message or initialize acknowledge message in either direction. These overwrite any flow control information transferred prior to the initialization of the two interfaces. For example, at the beginning of the transfer of a block of data from the device in automatic handshake mode (Figure 8b) the computer can issue a single command with bits 9, 10, 11, and 14 true. Both interfaces are then initialized and the device is informed that the computer is ready for data. The device then acknowledges the initialize command and sends the first data word. After the computer reads this, the next data request is automatically returned to the device with bits 9 and 10 false. In order to avoid the transference of data before the initialize acknowledge has been dispatched, it is important that control class transfers have priority over data class messages when they are being dispatched by either interface. Of course, any communications network being used must handle all transfers on a first-come first-served basis quite independently of the data class. Indeed the initialization procedure depends on the maintenance of a strictly sequential flow of information between the two interfaces by the communication system.

6.3   EXCEPTION CONDITIONS

The status register contains two types of status bits in addition to the three semaphores concerned with initialization and flow control (bits 10, 11, and 15). If any of bits 1, 4–9 or 12–14 are set, an *exception condition* exists and the group error flag GE, which is read by the computer in place of bit 0 of the status register, becomes true. The exception condition may be an error occurring within the device, such as a paper low condition in a printer, or it may be an error arising out of the DIM interface functions themselves (such as a parity error) or it may be an alarm condition such as an overheating bearing in the plant being controlled by the device. It may not be an error condition at all, e.g., an initialize message that causes CMD REQ (bit 9) to be set in the status register. The common characteristic of all these conditions is that the flow of data cannot continue without special action being taken. The form of this recovery procedure must, of course, remain device dependent. However, it typically involves the re-initialization of the interfaces followed by a second attempt at the data transfer, although the condition may well require manual intervention to clear the fault.

The GE flag enables the computer to ascertain whether the device is healthy by testing a single bit. Device service routines normally check that GE is false prior to testing the READ RDY or WRT RDY semaphores.

The nonexception condition bits (0, 2, and 3) of the status register do not set the GE flag and so can be set without affecting the data flow. Bit 0 is reserved for use by the flow control procedures while bits 2 and 3 can be used, for example, to indicate the device's mode of operation.

With the exception of bit 14 (READ OVL), which is wholly controlled from within the computer interface, the exception and nonexception bits are updated, whenever a status message arrives, with the contents of the corresponding bit in the message. It is necessary for the device to send a status message whenever its internal status changes. This status message may be transmitted with a flow control or initialization message at the same time.

An exception condition can be detected and reported by the device, by an intervening network such as MININET or by the computer interface itself. The exception conditions READ OVL (bit 14) and CMD REQ (bit 9) have already been described. Bits 4–7 are used for the exception conditions that arise from within the device. Obviously, their detailed assignment is highly device dependent. Bits 1, 8, 12, and 13 are used to signal exception conditions that are concerned with the general DIM interface functions and are almost completely independent of the device type. The format of these *interface error* group of messages is shown in Table 8.

Some of the error conditions are detected by the computer interface itself. If its DIM interface times out when attempting to output to the device, bit 13 of the status register is set, thus signifying a *local DIM time-out* error. This time-out could arise because the device is physically disconnected or switched off or, if connected through MININET, because the local station is switched off or else there may not be a virtual connection established with the device. If

TABLE 8 Interface Error Codes

Message	Origin	Code for the Following Status Bits			
		1 (REM ERR)	8 (TR ERR)	12 (T E Q)	13 (D U)
Local DIM time-out	Computer interface	0	0	0	1
Transaction time-out	Computer interface	0	0	1	0
Received transmission error	Computer interface	0	1	1	0
Transmitted transmission error	Device	0	1	0	0
Remote DIM time-out	Network	1	0	0	1
Outgoing transmission error	Network	1	1	0	0
Incoming transmission error	Network	1	1	1	0
Link down	Network	1	0	1	1
No error	Network	0	0	0	0

the DIM time-out takes place when the network attempts to deliver a word to the device because the device is not physically connected or is switched off, the network sends a *remote DIM time-out* message back to the computer interface.

If a parity error is detected in an incoming transfer, the damaged word is discarded and bits 8 and 12 of the status register are set to indicate that a *received transmission error* has been detected. If a transmission error in data traveling in the opposite direction is detected by the device, the device can inform the computer by sending a *transmitted transmission error* message. If a parity error is detected by the network or a word is dropped for any other reason, the network can send an *outgoing transmission error* or an *incoming transmission error* message, depending on the direction of the damaged transfer. It should be noted that MININET does not use this mechanism for error recovery in its communication links as this is handled within MININET quite transparently to the user. This type of error message occurs if a parity error is detected by the network in data transferred through the DIM user interface with the network or (hopefully very rarely) because of corruption of the data within a network node. In all cases in which a parity error is detected, the damaged word is never delivered. If, because of node or channel failures, the network cannot deliver or receive any data from the device, the computer is informed by means of a *link-down* message.

All the exception conditions reported by the network are characterized by the REM ERR flag (bit 1). The network must know whether a particular device is capable of accepting these network error messages, i.e., whether the device is intelligent and conforms to DIM CPC as far as the exception condition han-

dling is concerned. Separate flags associated with each DIM interface into the
network indicate whether errors can be reported to the device connected to
the local network interface or, alternatively, whether they can be sent to the
device connected to the remote interface at the other end of the connection.
For transmission errors where both devices are intelligent, the error message
is sent to the device that should have received the damaged word. It is up to
the network to ensure that this error message is inserted in place of the
damaged word with no loss of sequence.

In order to avoid a lock-out situation because a data or data request word
is lost between the device and the computer interface or because of a fault
condition within the device, the computer can request its interface to maintain
a deadman's timer by outputting a command with bit 13 (ENB T T) true. While
this bit is set in the command register, a timer is reset and started whenever a
word of either class is dispatched toward the device. The timer is halted
whenever a data or control class message is received from the device or network, data are dispatched in burst mode or a DIM time-out, or parity error is
detected by the interface itself. If nothing is received, the timer will time out
and bit 12 is set in the status register to indicate that a *transaction time-out* has
occurred.

It can be seen from Table 8 that bit 13 (D U) of the status register is set if
there is a local or remote DIM time-out or if the communication link is down.
This flag therefore serves to inform the computer that the device is unavail-
able. Also bit 8 (TR ERR) indicates that, somewhere, a transmission error has
occurred.

6.4   COMMAND STRUCTURE

The control of the internal operation of the nonintelligent devices is
effected by means of the device-dependent portion (bits 1–8) of the command
word. There are two types of command bits. Most affect the mode of operation of the device and they remain in force until the computer explicitly
changes the mode. These *mode commands* are duplicated every time that a
command word is sent to the device from the computer interface. These
duplications do not have any deleterious effect as they merely update the
control register within the device with the same information that it already
contains. Typical of this type of command function are echo-plex control of a
terminal interface and selection of an external or internal clock in a digital-to-
analogue converter. Device-dependent bits 1–6 are used for mode commands
and bit 0 (READ RDY) is also of this type.

The other type of command triggers a single event or sequence of events
in the device. For example, a single command may trigger an analogue data
acquisition system to perform a scan of its input channels. Device-dependent
bits 7 and 8 are used for these *trigger commands*. The data request and initialization commands are also of this type. Clearly, duplication of a trigger command must be avoided. For this reason the trigger bits (bits 7–11) in the
command register are cleared after dispatch to the device.

Bits 12–15 of the command register are reserved for internal use in the computer interface and are always transmitted as zero toward the device (Figure 4). Bit 15 (ENB INT) is used to enable computer interrupts. If it is set, a computer interrupt request is generated whenever data or status information is received by the interface or when a DIM time-out, parity error, or transaction time-out is detected by the interface, or when data are dispatched toward the device if WRT CONT is set in the status register. The use of bits 13 (ENB T T) and 14 (ENB A A) has already been described. Bit 12 is at present undefined and it should remain zero. It is reserved for future use within the computer interface.

7 OPERATIONAL EXPERIENCE

DIM has been in use at the Polytechnic of Central London and the University of Bologna for over 6 years with interfaces to PDP-11 and Interdata 70 computers. In that time a large amount of equipment has been constructed using the DIM interface including high speed converters for digital audio processing and recording, computer-controlled adaptive filters, and an arbitrary waveform generator as well as the more usual computer peripherals such as terminal interfaces and paper-tape readers. The interface has also been used for resource sharing between microcomputer development systems and a minicomputer, allowing the microcomputer access to the hard discs and fast printers of the minicomputer system. An especially interesting use of DIM has been with "facade" instruments where the device merely acts as a facade for the computer, relaying commands from its front panel to the computer and displaying the resultant waveforms, etc., fed back from the computer.

An interface processor has been produced that is capable of supporting up to 16 DIM interfaces. (Currently it is supporting nine assorted devices.) At its heart is a high speed microprogrammed processor that implements the DIM CPC for each of the 16 independent interfaces. The speed of the processor is such that the computer can (and does) treat the sometimes quite remote devices as if they were separate peripherals directly connected to its input–output bus. Construction of most DIM devices has been greatly eased by the use of a standard DIM interface circuit board that handles the transfer cycle and flow and initialization protocols. It is therefore only necessary to design the device-specific circuitry for each device type. Some 16 device designs have been successfully based on this approach and several more are currently being developed.

ACKNOWLEDGMENTS

The support of the U.K. Science Research Council, the Italian Consiglio Nazionale delle Ricerche, and the North Atlantic Treaty Organization is gratefully acknowledged.

REFERENCES

1 Morling, R. C. S., and G. D. Cain, MININET: a packet-switching minicomputer network for real-time instrumentation, *Proc. AIM Int. Meet. on Minicomputers and Data Communication, Liège, January 1975,* Association des Ingénieurs Electriciens sortis de l'Institut Electrotechnique Montefiore, Liège.

2 Cain, G. D. C., and R. C. S. Morling, MININET: a local area network for real-time instrumentation and control, *Proc. 3rd Conf. on Local Computer Networking, Minneapolis, MN, October 1978,* University of Minnesota, Minneapolis, MN.

3 CAMAC—a modular instrumentation system for data handling, revised description and specification, *CEC Doc. EUR 4100e,* August 1972 (Commission of the European Communities, Luxembourg).

4 An interface system for programmable measuring instruments, *IEC Stand. 625,* 1979 (International Electrotechnical Commission, Geneva, Switzerland).

5 Thornton, J.E., Overview of HYPER channel, *Digest of Papers, COMPCON 79 Spring, San Francisco, CA, February 1979,* IEEE, New York, pp. 262–265.

6 Counterman, W. A., B. B. Bliss, and E. A. Mackey, A ring network IDANET, *Proc. Conf. on New Approaches to Local Computer Networking, St. Paul, MN, September 1976,* University of Minnesota, Minneapolis, MN.

7 A digital input/output interface for data collection systems, *BS Stand. 4421,* 1969 (British Standards Institution, London).

8 Davies, D. W. and D. L. A. Barber, *Communication Networks for Computers,* Wiley, New York, 1973, Appendix.

9 Digital interface bus standards, *Rep.,* 1977 (Motorola Semiconductor Products Inc.).

VME BUS INTERFACING: A CASE STUDY

MARC LOBELLE

Université Catholique de Louvain, Unité d'Informatique, Chemin du Cyclotron 2, B-1348 Louvain-la-Neuve (Belgium)
(Received April 18, 1983).

SUMMARY

VME is a high performance standard bus for multimicroprocessor systems. Its characteristics originate in the 68000 microprocessor's interface signals. Processors with other interface characteristics can, however, also be used in VME systems. The case study of the interfacing of a 6809-based subsystem to the VME bus is presented. A mixture of hardware and software has been used

Reprinted with permission of Marc Lobelle, "VME Bus Interfacing: A Case Study," *Interfaces in Computing,* Vol. 1, No. 3, August 1983, and North-Holland Publishing Company, Amsterdam.

to implement at low cost the matching functions. The 6309-based subsystem can perform all VME-related functions: bus requester in multiprocessor environments, bus master, interrupt handler, bus slave, and interrupter. The last two functions are performed by a dual-port memory included in the subsystem, which is designed to be used as the main processor in small VME systems or as the intelligent peripheral controller in larger VME systems.

1   INTRODUCTION

VME is an emerging new high end standard bus with asynchronous control for microcomputer systems. It features several valuable characteristics [1]. One of these is the fact that, unlike most other minicomputer and microcomputer busses, the VME bus was fully specified before any card was released.

The VME specification was based on the experience gained from previous bus standards and includes several clear and upward compatible options allowing, for example, data transfers 8, 16, or 32 bits wide and the use of 16-, 24-, or 32-bit byte addresses. Strobes specify the width of the data words and a 6-bit address modifier code is included in the bus control signals to define which address width is currently used. This 6-bit code also allows the implementation of sophisticated memory protection mechanisms.

As a consequence, processors with different data path width or addressing ranges can be mixed in the same VME system. The bus control signals will allow them to cooperate efficiently through selected common memory areas. This ability of the VME bus to support nonhomogeneous multiprocessor systems allows the selection of the best-suited processor for each task. All bus arbitration mechanisms have been settled for multimaster architectures and the seven interrupt request lines can be directed to different processors acting as interrupt handlers. This provides the means of system synchronization through interprocessor interrupts.

Other qualities of the VME bus are that it allows high speed data transfers (more than 10 Mwords of 32 bits/s) and that the mechanical and electrical specifications of the cards, connectors and interface circuits to use with the VME bus make VME systems suitable for industrial environments. These mechanical specifications comply with the "double Eurocard" standard (*DIN Stand. 41612* [2] and *DIN Stand. 41494* [3]).

The VME bus is supported by many manufacturers, including large companies (Motorola, Philips–Signetics, Mostek, Thomson–EFCIS) and smaller companies. Many compatible cards have been released and many more are planned for the future. General presentations of the VME bus can be found in [4 and 5]. A brief description of the VME bus signals is given in Table 1. Their locations on the two Eurocard connectors can be found in Table 2.

TABLE 1 VME Bus Signals

Signal	Function	Line[a]
BR0*–BR3* (bus request 0–3)	They indicate that one master in the daisy chain requires access to the bus	OC
BG0IN*–BG3IN* (bus grant in 0–3) BG0OUT*–BG3OUT* (bus grant out 0–3)	They form a daisy-chained bus grant. The bus grant in signal indicates to this board that it may become the next bus master. The bus grant out indicates it to the next board	TP
BBSY* (bus busy)	This is generated by the current master to indicate that it is using the bus	OC
BCLR* (bus clear)	This is generated by the bus arbiter to request the current master to release the bus, in the event that a higher level is requesting it	TP
A01–A23 (address bus bits 1–23)	Optional; only present in expanded systems	TS
A24–A31 (address bus bits 24–31)		TS
AM0–AM5 (address modifier bits 0–5)	They provide additional information about the address bus, such as size, cycle type and/or protection information	TS
AS* (address strobe)	This indicates that a valid address is on the address bus	TS
D00–D15 (data bus bits 0–15)	These are bidirectional data lines providing a data path between a master and a slave	TS
D16–D31 (data bus bits 16–31)	Optional; only present in expanded systems	TS
DS0*, DS1* (data strobes 0, 1)	They indicate during byte and word transfers that a data transfer will occur on data bus lines D00–D07 and D08–D15, respectively	TS
LWORD* (long word)	This indicates that the current transfer is a 32-bit transfer	TS
WRITE* (write)	This indicates that the current transfer is a write operation	TS

361

TABLE 1 (*Continued*)

Signal	Function	Line[a]
DTACK* (data transfer acknowledge)	This is generated by a slave; the falling edge of this signal indicates that valid data are available on the data bus during a read cycle or that data have been accepted from the data bus during a write cycle	OC
BERR* (bus error)	This indicates that an unrecoverable error has occurred and that the bus cycle must be aborted	OC
IRQ1*–IRQ7* (interrupt request 1–7)	These interrupt requests, which are subject to priority, are generated by interrupters; level 7 is the highest priority	OC
IACK* (interrupt acknowledge)	This is a signal from any master processing an interrupt request, as interrupt handler. It is routed via the back plane to slot 1; it is looped back to become slot 1 IACKIN to start the interrupt acknowledge daisy chain. In response to this signal, the interrupt source identification byte is sent on data lines by the interrupter	OC or TS
IACKIN*, IACKOUT*	These are the daisy-chained interrupt acknowledge signals	TP
SYSCLK (system clock)	Constant 16-MHz clock; this can be used for general system timing	TP
SYSFAIL* (system failure)	This indicates a failure in the system; it can be generated by any module on the VME bus	OC
SYSRESET* (system reset)	When low, this causes the system to be reset	OC
ACFAIL* (ac failure)	This indicates that the mains input to the power supply is no longer provided	OC

Signals with an asterisk (*) are in negative logic.
[a] TS, three-state driven line; OC, open collector driven line; TP, totem pole driven line.

TABLE 2 VME Bus Back-Plane Connectors

Pin Number	Row A Signal Mnemonic	Row B Signal Mnemonic	Row C Signal Mnemonic
J1–P1 pin assignments			
1	D00	BBSY*	D08
2	D01	BCLR*	D09
3	D02	ACFAIL*	D10
4	D03	BG0IN*	D11
5	D04	BG0OUT*	D12
6	D05	BG1IN*	D13
7	D06	BG1OUT*	D14
8	D07	BG2IN*	D15
9	GND	BG2OUT*	GND
10	SYSCLK	BG3IN*	SYSFAIL*
11	GND	BG3OUT*	BERR*
12	DS1*	BR0*	SYSRESET*
13	DS0*	BR1*	LWORD*
14	WRITE*	BR2*	AM5
15	GND	BR3*	A23
16	DTACK*	AM0	A22
17	GND	AM1	A21
18	AS*	AM2	A20
19	GND	AM3	A19
20	IACK*	GND	A18
21	IACKIN*	SERCLK[a]	A17
22	IACKOUT*	SERDAT[a]	A16
23	AM4	GND	A15
24	A07	IRQ7*	A14
25	A06	IRQ6[+]	A13
26	A05	IRQ5*	A12
27	A04	IRQ4*	A11
28	A03	IRQ3*	A10
29	A02	IRQ2*	A09
30	A01	IRQ1*	A08
31	−12V	+5V STDBY	+12V
32	+5V	+5V	+5V
J2–P2 pin assignments			
1	User I/O	+5V	User I/O
2	User I/O	GND	User I/O
3	User I/O	RESERVED	User I/O
4	User I/O	A24	User I/O
5	User I/O	A25	User I/O
6	User I/O	A26	User I/O
7	User I/O	A27	User I/O
8	User I/O	A28	User I/O
9	User I/O	A29	User I/O
10	User I/O	A30	User I/O

TABLE 2 (*Continued*)

Pin Number	Row A Signal Mnemonic	Row B Signal Mnemonic	Row C Signal Mnemonic
11	User I/O	A31	User I/O
12	User I/O	GND	User I/O
13	User I/O	+5V	User I/O
14	User I/O	D16	User I/O
15	User I/O	D17	User I/O
16	User I/O	D18	User I/O
17	User I/O	D19	User I/O
18	User I/O	D20	User I/O
19	User I/O	D21	User I/O
20	User I/O	D22	User I/O
21	User I/O	D23	User I/O
22	User I/O	GND	User I/O
23	User I/O	D24	User I/O
24	User I/O	D25	User I/O
25	User I/O	D26	User I/O
26	User I/O	D27	User I/O
27	User I/O	D28	User I/O
28	User I/O	D29	User I/O
29	User I/O	D30	User I/O
30	User I/O	D31	User I/O
31	User I/O	GND	User I/O
32	User I/O	+5V	User I/O

Note: Two *DIN Stand. 41612* 96-pin connectors are used. The standard VME signals run through P1–J1. P2–J2 is optional and used for the signals of expanded VME and for user input–output (I/O).
[a] SERCLK and SERDAT represent provision for a special serial communication bus protocol still being finalized.

In this paper the design of the VME interface of an intelligent local area network controller (LANC) built around a 68B09 microprocessor is presented [6–9]. This LANC is implemented on two VME boards interconnected by a local bus and can act as a VME bus master and interrupt handler. It includes a dual-port memory with interprocessor interrupt facilities. It can be used as a central processor in small VME systems or as an intelligent peripheral controller in larger VME systems.

2   GENERAL DESIGN OPTIONS OF THE LOCAL AREA NETWORK CONTROLLER VME INTERFACE

The interface signals of the 68B09 microprocessor, selected for the VME subsystem presented here, are quite different from those of the VME bus whose design philosophy originates in the 68000 microprocessor's interface signals. All bus control protocols of the 68B09, which are related as well to data transfers as to interrupt handling, have thus to be matched to those of the VME bus. Other functions must be added to the 68B09: those related to bus sharing in multiprocessor environments.

Two solutions would have been possible to reach a full VME bus compatibility. The first solution would have been to add to the 68B09 all the necessary hardware to translate its protocols to those of the VME bus and to perform the missing functions. However, the amount of extra hardware necessary to implement this matching circuit would require a large printed circuit board (PCB) area, which would be lost for the specific functions of the LANC. The second approach is to divide the matching functions into two classes: data transfers; bus sharing and interrupt source identification.

The second class includes the problems related to the transfer of bus mastership between different processors, VME exceptions (e.g., bus errors) handling and interrupt source identification ("vector fetch"). Although it is mandatory to adapt the data transfer mechanism of the VME bus and the 68B09 microprocessor using hardware, this second class of problems can be tackled using a mixture of hardware and software while still respecting the VME bus timing specifications. This approach reduces significantly the amount of hardware related to the VME interface with the advantages of lower cost and of leaving more PCB area available for the local area network control related circuits.

The timing parameters resulting from this mixed approach are acceptable in many applications including the present LANC: (1) there is an increase of a few microseconds in the delay between the arrival of an interrupt request from another VME board and the actual start of the execution of the interrupt routine; (2) there is a similar increase in the time necessary for the LANC to get or release bus mastership. Neither of these inconveniences is thought to have an important effect on global system performance.

3   CHARACTERISTICS OF THE VME INTERFACE OF THE LOCAL AREA NETWORK CONTROLLER SUBSYSTEM

The LANC is a nonexpanded VME subsystem; that is, only address and data lines A01–A23 and D00–D15 are used. As a VME master, the LANC is able to request, receive, and release bus mastership under control of an external bus arbiter circuit. These protocols are summarized in Figure 1. The bus arbitration priority of the LANC is fixed by strapping.

When it has acquired bus mastership, the LANC can transfer data to and from other boards (slaves) on the VME bus. The data transfer protocols are summarized in Figure 2. These transfers occur 8 bits at a time for the LANC. Successive bytes are transferred alternately on VME lines D00–D07 and D08–D15. No internal time-out is generated; an external bus-time-out module must be used to detect the failure of data transfer cycles (the absence of the

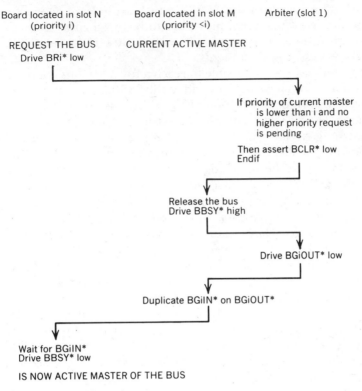

Board located in slot N   Board located in slot M   Arbiter (slot 1)
    (priority i)   (priority <i)

REQUEST THE BUS   CURRENT ACTIVE MASTER
   Drive BRi* low

If priority of current master
is lower than i and no
higher priority request
is pending

Then assert BCLR* low
Endif

Release the bus
Drive BBSY* high

Drive BGiOUT* low

Duplicate BGiIN* on BGiOUT*

Wait for BGiIN*
Drive BBSY* low

IS NOW ACTIVE MASTER OF THE BUS

FIGURE 1 Typical bus arbitration sequence.

slave acknowledgment signal DTACK* or of a BERR* issued by the slave) and to assert the BERR* line. The values of A15–A23 and the address modifier code AM0–AM5 put on the bus by the LANC are defined by straps.

The choice of using straps for address lines A15–A23 and address modifier code AM0–AM5 means that the 68B09 can access one fixed 32k byte window situated anywhere in the VME bus memory space. This window is the part of the 64k-byte 68B09 addressing range that is not used for the memories and peripherals internal to the LANC. It can be used to access other controllers or memories on the VME bus. Another possible choice would have been to use registers instead of straps in order not to restrict the access to one fixed window but to implement one or several movable windows as in the memory management of the PDP/11 [10]. The strap solution is, however, simpler and sufficient for the present application.

The LANC can handle any one of the seven interrupt lines of the VME system. This line is also selected by strapping. The interrupt service protocol is summarized in Figure 3.

As well as its bus master and interrupt handler functions, the LANC also appears to the bus as an 8k-byte dual-port memory and, as such, performs the

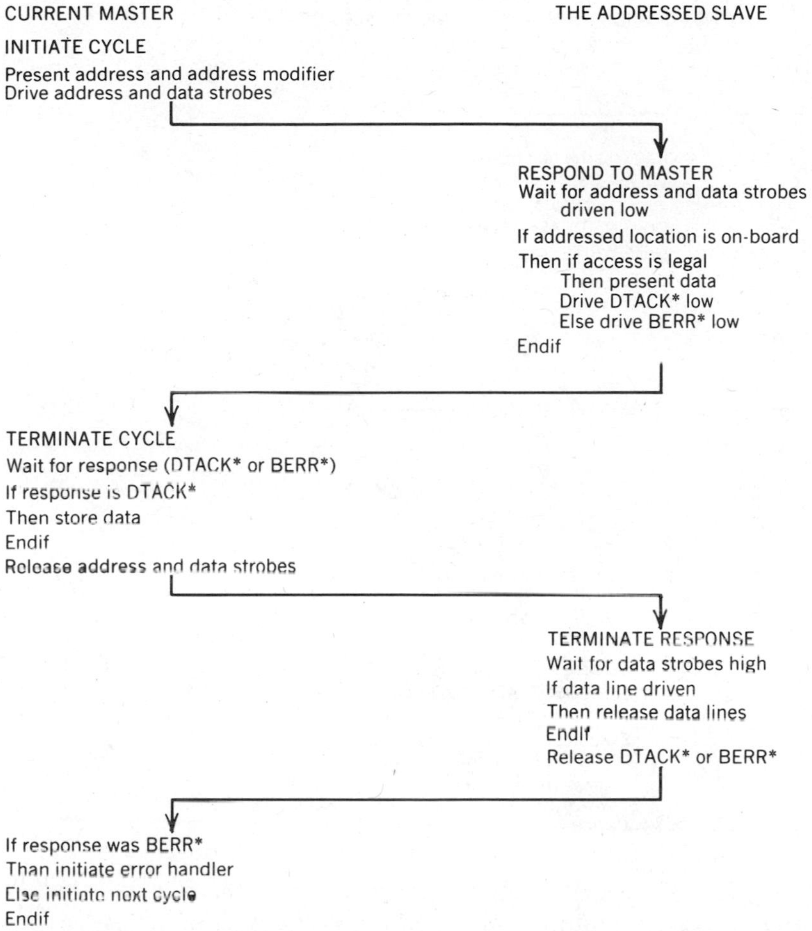

CURRENT MASTER

THE ADDRESSED SLAVE

INITIATE CYCLE

Present address and address modifier
Drive address and data strobes

RESPOND TO MASTER
Wait for address and data strobes
    driven low

If addressed location is on-board

Then if access is legal
        Then present data
        Drive DTACK* low
        Else drive BERR* low
Endif

TERMINATE CYCLE

Wait for response (DTACK* or BERR*)

If response is DTACK*

Then store data

Endif

Release address and data strobes

TERMINATE RESPONSE
Wait for data strobes high
If data line driven
Then release data lines
Endif
Release DTACK* or BERR*

If response was BERR*
Than initiate error handler
Else initiate next cycle
Endif

FIGURE 2 Typical read sequence.

functions of bus slave during data transfers and, as will be seen, of inter-rupter. The access arbitration of this dual-port memory between the VME port and the internal port to the 68B09 local bus is transparent to both sides and is performed in hardware on a cycle-by-cycle first-arrived first-served basis. Access is, however, always granted to the VME port in less than 500 ns, which is the cycle time of the 68B09 synchronous local bus. The transfers between this dual-port memory and the VME bus can occur 8 bits or 16 bits at a time. The VME address lines A12–A24, the address modifier code and LWORD*, the signal specifying whether a 16- or a 32-bit data transfer is per-formed, are compared with strapped values and access to the memory is enabled in the case of match.

This dual-port memory includes two special locations; access to the high-est byte of this memory from the VME port sends an interrupt request signal

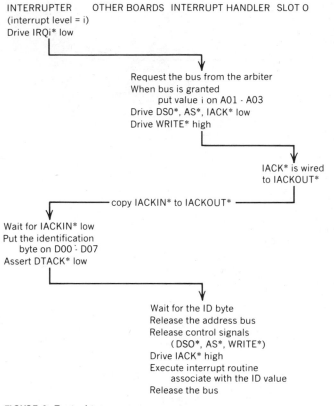

INTERRUPTER   OTHER BOARDS   INTERRUPT HANDLER   SLOT 0
(interrupt level = i)
Drive IRQi* low

Request the bus from the arbiter
When bus is granted
        put value i on A01 - A03
Drive DS0*, AS*, IACK* low
Drive WRITE* high

IACK* is wired
to IACKOUT*

copy IACKIN* to IACKOUT*

Wait for IACKIN* low
Put the identification
    byte on D00 - D07
Assert DTACK* low

Wait for the ID byte
Release the address bus
Release control signals
        (DS0*, AS*, WRITE*)
Drive IACK* high
Execute interrupt routine
        associate with the ID value
Release the bus

FIGURE 3 Typical interrupt sequence.

to the 68B09 processor. This interrupt request can be reset by an access from the internal port to this same location. In contrast, an access to the second-highest byte location from the internal port sends an interrupt request signal to the VME bus. The priority of this request (i.e., on which one of the seven VME interrupt lines it is sent) and the value of the interrupt source identification byte are selected by straps. This interrupt request line is reset by an access from the bus to this second-highest location in the dual-port memory.

4   IMPLEMENTATION OF THE BUS INTERFACE FUNCTIONS

4.1   BUS ARBITRATION CIRCUITS

The circuits related to the bus arbitration mechanism are represented in Figure 4. They are a good illustration of the technique of mixing hardware and software. The signals BNEED*, BNNEED* and RBBSYLOC* are produced by the address decoding circuits of the local bus of the 68B09. They are activated

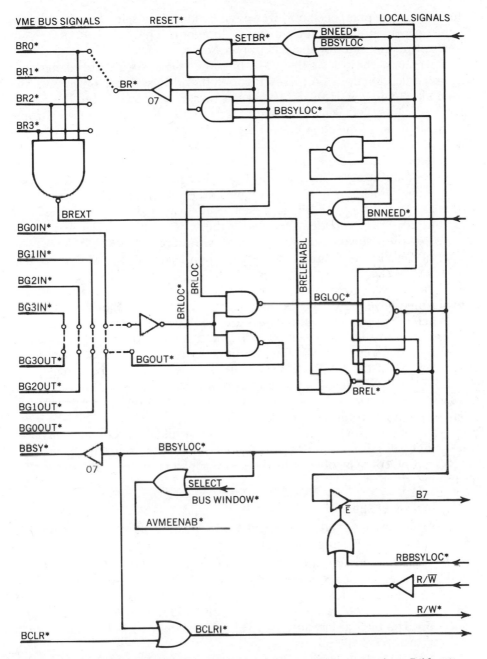

FIGURE 4 Bus requester circuits. All circuits (with the exception of memories) are 74LS series. Only the suffix of 74LS has been noted on the figures. Strap-defined values are denoted by SV.

when the 68B09 puts specific addresses on its address bus, in effect by accesses to the corresponding dummy memory locations. The signal BCLRI* is an interrupt request line for the microprocessor.

To request the use of the VME bus, the processor accesses a memory location called BNEED. This forces, through the address decoding circuits, the signal BNEED* to low. If the VME bus has not yet been granted to the LANC (if it were, BBSYLOC would be high), SETBR* will go low, which will force BRLOC* and one of the BRi* lines of the VME bus (the one strapped to BR*) to low.

When the bus arbiter of the VME bus grants the bus to the LANC, the corresponding BGiIN* line will go low. This signal, gated by BRLOC, makes BGLOC* low, which sets BBSYLOC* to low and resets the BR* signal. BBSYLOC sends BBSY* on the VME bus and enables the use of the VME address and data buffers when an address located in the VME window is selected by the processor (the address decoding circuit then drives AVMEENAB* low). The status of the BBSYLOC signal can be read by the processor as data bit 7 at address RBBSYLOC.

If BGiIN* had gone low while the LANC was not requesting the bus, this signal would have been passed to the next card on the VME bus as BGiOUT*.

When the processor does not need the VME bus any more, it accesses the dummy memory location BNNEED; then BRELENABL will go high and any other requester on the VME bus will be able to force BBSYLOC to false by asserting one of the VME's BRi* signals. If, later, the processor needs the VME bus again, it will again access BNEED, which will reset BRELENABL. If no one else has requested the bus in between, the processor will get it immediately, without arbitration sequence.

If a higher priority module requests the bus while the processor is using it, the arbiter will send a BCLR* signal. This signal will interrupt the processor.

The protocol to be obeyed by the processor when it needs the bus is thus as follows.

1. Access BNEED.
2. Wait until BBSYLOC becomes true by reading BBSYLOC (this implies some "busy waiting," of course) and then use the bus freely.
3. If a BCLR* interruption is received or when the bus is not needed any more, allow the requester circuit to release it by accessing BNNEED. The simplest BCLR interrupt routine would be as follows: Release the bus (access BNNEED); ask for the bus (access BNEED); wait for the bus (monitor data bit 7 at location RBBSYLOC); when the bus is granted back, return from the interrupt.

4.2 BUS MASTER CIRCUITS

Once the LANC has acquired bus mastership, it can transfer data on the VME bus through the circuits of Figure 5.

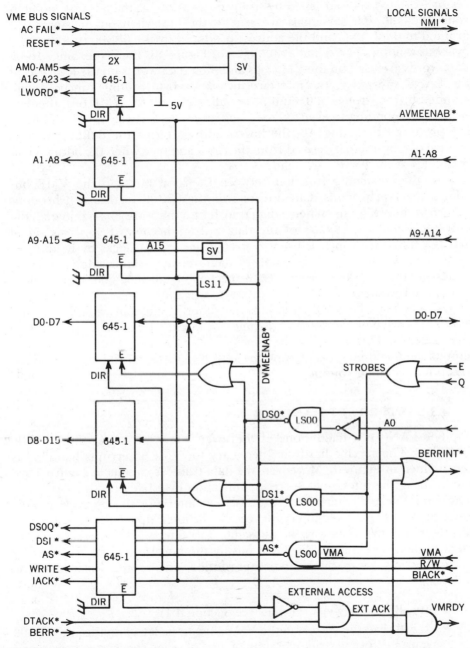

FIGURE 5 Bus master circuits.

The address signals sent to the VME bus are produced by the microprocessor (A01–A14) or specified by straps (A15–A23), as the address modifier code AM0–AM5. These signals are sent to the bus when bus mastership is granted to the LANC and the address produced by the 68B09 is in the 32k-byte bus window. The signal AVMEENAB* of Figure 4 is then true. Since 16-bit data words are used on the VME bus and 8-bit data words on the local bus of the LANC subsystem, the interface includes a byte routing function commanded by the strobes DS0* and DS1* validating, on the VME bus, the data bytes corresponding to even and odd addresses, respectively. DS0* and DS1* are produced by gating A0, the lowest address bit of the 68B09, and the inverse of it by a signal derived from the clock and true when the address and data lines of the 68B09 are valid.

Another matching function between the local bus and the VME bus adapts the synchronous data transfers of the 68B09 to the asynchronous nature of the VME bus. When a data transfer on the VME bus is initiated, the enabling signal DVMEENAB* of the data buffers becomes true. This signal forces VMRDY low, which is used to stretch the 68B09 clocks by holding its MRDY pin low until the reception of the data transfer acknowledgment signal DTACK* of the VME bus. When DTACK* becomes true, MRDY goes high and the clocks are resumed.

If, however, the VME cycle fails (no other module answers by sending DTACK*), a slave or the system bus-time-out module (external to the LANC) asserts BERR*. This signal has the same effect on MRDY as DTACK* has, but it is simultaneously used to interrupt the 68B09 and to initiate the bus error exception handling routine.

4.3   INTERRUPT HANDLER

The LANC can handle one of the seven interrupt request lines of the VME bus. The specific hardware necessary for VME interrupts handling is represented in Figure 6. Moreover, the data transfer circuits of Figure 5 are used during the interrupt source identification byte fetch on the VME bus. The strap-selected interrupt line is used to interrupt the microprocessor. Since the 68B09 has fixed interrupt vectors, the interrupt routine activated by this interrupt line will always be the same. The fetch of the interrupt source identification byte put on the VME bus by the interrupting device must be performed in software by this interrupt routine. This routine will achieve the following.

1. It will request the VME bus (see Section 4.1).
2. It will read the vector on the VME bus. In order to do this, the 68B09 must perform a read operation on one of seven special successive even addresses. When any one of these addresses is accessed, the address decoding circuit asserts INTID*, which, gated by the signal BBSYLOC* (indicating that the LANC has acquired bus mastership) becomes the

VME BUS SIGNALS

LOCAL SIGNALS

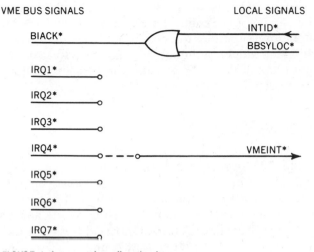

FIGURE 6 Interrupt handler circuits.

interrupt acknowledge signal BIACK* of the VME bus. The VME bus specification requires that simultaneously with IACK* the interrupt handler puts on the address lines A01–A03 the priority of the interrupt being acknowledged. The interrupt source identification byte will then be put on D00–D07 and DTACK* will be asserted by the interrupting device. The function of putting the priority on A01–A03 is performed by using BIACK* to enable the A01–A08 and D00–D07 bus buffers. The A01–A03 bits of the address of the dummy memory location read by the 68B09 to get the interrupt source identification byte will then appear on the VME bus. Since the seven addresses have equivalence for INTID* assertion, the processor must access the specific address for which the A01–A03 bits correspond to the priority of the interrupt being handled by the LANC. In response to this, the interrupting module on the VME bus will put its ID byte on D00–D07. This byte will be read by the microprocessor as if it were the contents of the addressed memory location. The interrupter can thus be identified.

3. It will execute the appropriate service routine.
4. It will release the bus (see Section 4.1).
5. It will return from the interrupt.

4.4   THE 68B09 INTERRUPT HANDLING

Interrupt handling facilities are certainly not the strongest feature of the 68B09; this microprocessor has indeed only two maskable interrupt input lines, and one fixed interrupt vector is associated with each of them. To overcome this limitation, an interrupt vector is associated with each of them. To overcome this limitation, an interrupt controller (6828) has been added to the 68B09 to multiplex eight interrupt lines to one of the interrupt inputs of the 68B09 and to provide independent (although still fixed) vectors to each input. The interrupt request line of the VME bus and the BERR* input are directed to inputs of this interrupt controller.

4.5 THE DUAL-PORT MEMORY

The dual-port memory is built around four 2048-byte static random access memories (Mitsubishi M58725) organized as a 4k word by 16-bit memory. The internal control signals of this memory follow the VME bus asynchronous philosophy. The circuits of the dual-port memory are shown in Figure 7.

The VME interface of this memory has a data bus 16 bits wide and an address bus 24 bits wide. The address lines A13–A23, the address modifier code AM0–AM5 and the control signals LWORD*, IACK*, and AS* are compared with strapped values to produce the ARVME signal that is sent to the shared memory access arbitration circuit to request access to the memory from the VME bus.

The interface between the 68B09 local bus and the dual-port memory routes the 8-bit local data bus of the 68B09 to bits D00–D07 or D08–D15 of the dual-port memory 16-bit data bus. The local address bus interface is 13 bits wide. The dual-port memory has a fixed address in the 68B09 memory map. The ARLOC signal, which is used to request access to the dual-port memory from the local bus, is produced by the address decoding circuits. A signal FMRDY is sent to the MRDY input of the 68B09 as long as access to the memory is requested (ARLOC is true) but not granted (AGLOC* is false). This situation will not be longer than the duration of one access to the shared memory through the VME interface.

The circuit controlling the access to the dual-port memory is built around two nand gates. Each port can send to this circuit an access request signal, the ARVME and ARLOC signals previously mentioned. The first port requesting the access gets it and blocks a request from the other port until the end of the current access. The interface having gained the access receives an access grant signal that enables the address and data bus interfaces (AGVME* and AGLOC*, respectively). The A12,, DS0*, and DS1* signals are then allowed inside the dual-port memory and decoded to yield the chip select signals of the memory chips. The access arbitration circuit's truth table is displayed in Table 3.

It must be noted that the access arbitration circuit grants the access to the dual-port memory for the duration of only one memory operation, that is, as long as the address and other signals decoded to yield the access request signal do not change. In read–modify–write instructions the 68B09 processors release the address bus between the two accesses. The signal ARLOC thus becomes false and the memory can be granted to the other port. The read–modify–write instructions in the shared memory thus cannot be considered as

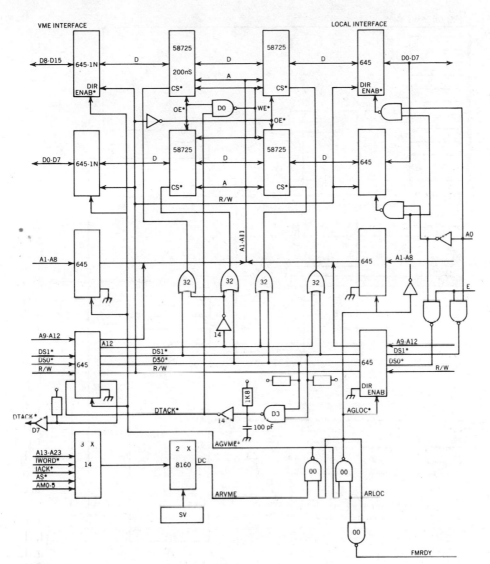

FIGURE 7 Dual-port random access memory.

TABLE 3 Dual-Port Memory
Access Arbitrator Truth Table

AR1	AR2	AG1*	AG2*
0	0	1	1
↑	0	↓	1
1	0	0	1
1	↑	0	1
1	↓	0	1
↓	1	↑	↓

Note: The access arbitrator is an RS flip-flop used
in a somewhat uncommon way. Both access request inputs AR1, AR2 and access grant outputs
AG1, AG2 are equivalent.

375

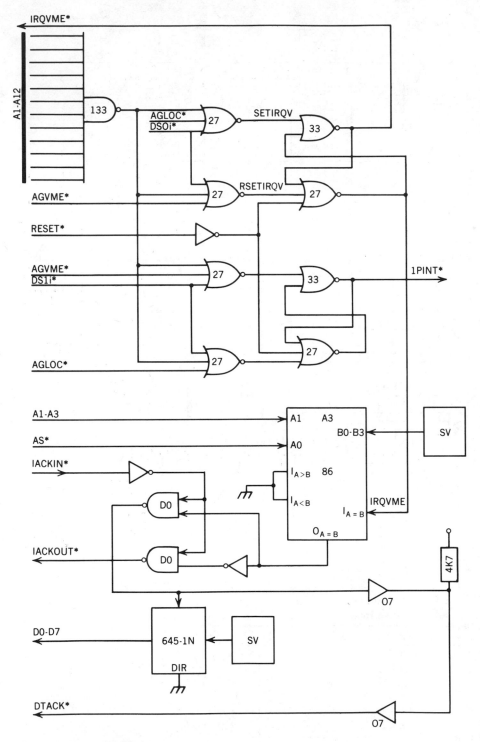

FIGURE 8 Interrupter circuit.

uninterruptible and cannot be used to implement interprocessor locking of nonsharable resources. Then two state variables, each being accessed by only one processor for writing and the other for reading, must be used.

4.6   INTERPROCESSOR INTERRUPTS

The two highest locations of the dual-port memory are used to generate interrupt request signals. As shown in Figure 8, the decoding of the address of each of these locations enables a flip-flop implemented with two nor gates. This flip-flop uses the access grant signals to one of the two ports to set the corresponding interrupt request line and the other to reset it. In response to the interrupt request to the VME bus, the interrupt handler connected to the interrupt request line on which the request was sent replies with an IACK* signal and by sending the priority of the interrupt request on address lines A01-A03. This IACK* enters the LANC through the IACKIN* input. A 74LS86 comparator is used to decide whether this signal must be passed to the next card on the VME bus as IACKOUT* or must let the interrupt source identification byte be put on D00-D07 of this bus. The identification byte is issued if the priority read on A01-A03 is the same as that of the interrupt request line pulled by the LANC.

The connection of the interrupt request signal to one of the VME lines and the specification of the priority of this line to the comparator are both done by strapping.

5   CONCLUSION

Although the design philosophy of the VME bus is similar to that of the 68000 microprocessor's interface signals, other processors with rather different bus organizations can be used as processors in the VME system. An efficient way to perform the necessary matching of these processor's busses to the VME bus is to use a mixture of hardware and software; some functions related to bus arbitration and interrupt handling can economically be performed in software.

ACKNOWLEDGMENTS

The material presented in this paper is part of a joint research project on distributed computer systems and local area networks by the Departement Computer Wetenschappen, Katolieke Universiteit Leuven, and the Unité d'Informatique, Université Catholique de Louvain. All participants in this project are acknowledged.

REFERENCES

1 *VME Bus Specification Manual*, Revision A, Mostek Corporation, Motorola Inc., Signetics–Philips, October 1981.

2 *DIN Stand. 41612*, 1976, 1979 (Deutsches Institut für Normung).

3 *DIN Stand. 41494*, 1972, 1974 (Deutsches Institut für Normung).

4 C. Kaplinsky, Decentralizing microprocessor bus grows easily from 16 to 32 bits, *Electron. Des., 29 (23)*(1981)173–179.

5 M. Rudyk, VME bus, Modulares Koncept für Microprocessor Karten mit Europaformat, *Elektronik, 31*(10)(1982).

6 M. Lobelle, Un réseau local pour l'expérimentation des systèmes distribués, *Actes du Premier Congrès sur la Conception des Systèmes Télématiques, Nice,* June 3–5, *1981,* CITEL, Vallauris, 1981, pp. 249–259.

7 M. Lobelle, The TROUT VME module, *Res. Rep. RR 82-3/2,* May 1982 (Unité d'Informatique, Université Catholique de Louvain).

8 M. Lobelle, The OSI levels, 1, 2, 3 of the TROUT network, *Res. Rep. RR 82-4,* May 1982 (Unité d'Informatique, Université Catholique de Louvain).

9 M. Lobelle, A local area network designed as testbed for distributed operating systems, *Proc. NFWO–FNRS Contact Day on Distributed Systems, Leuven, February 25, 1983,* NFWO–FNRS, Brussels, to be published.

10 R. E. Eckhouse, Jr., and L. R. Morris, *Minicomputer Systems*, Prentice-Hall, Englewood Cliffs, N.J., 2nd ed., 1979.

INDEX

Absolute address, 248
Access control field, 228
Ada, 233, 252, 262
Address, 104
  absolute, 248
  universal, 248
A/D conversion, 192
A/D interfacing, 213
ASCII, 87. *See also Appendix A*
Asynchronous communication, 86

Backbone network, 235
Bandwidth, 220
Baseband, 221
BASIC, 252, 257
Binary exponential back-off, 243, 244
Binary search, 201
Bisync, 103
Bit cell, 243
Block check, 111
Broad band, 221
Broadcast, 219, 248
Bus, 3, 219
  address, 3
  asynchronous, 41, 43
  asynchronous READ, 43
  asynchronous WRITE, 45
  control, 3
  data, 3
  definition, 3
  interfacing, 41
  interfacing example, 74
  semisynchronous, 41, 46
  standards, 48
  synchronous, 41, 45
  synchronous READ, 45, 46
  synchronous WRITE, 45, 46
  WAIT state, 41, 46
Byte erase, 80
Byte write, 80

C, 252, 297
  interfacing example, 303
  program structure, 301
Carrier sense, 225, 241
CAS, 49
CASE, 238
CATV, 221
CCITTV.24, 93
Characteristic impedance, 16
Chip erase, 82
Chip select, 59, 69
CMOS, 13, 72
Coaxial cable, 249
Collision concensus enforcement, 242
Collision detect, 225, 242
Collision interval, 241
Collision window, 241
Common mode, 95
Comparator, analog, 192
Concord Data Systems, 238
Contention, 227
Controller (Ethernet), 246, 248
Counter converter, 200
CSMA/CD, 225, 228, 241
Current loop, 88
Cyclic redundancy check (CRC), 113, 244, 248

Data acquisition systems, 213
Data link, 224
D/A converter, 209
D/A converter interface, 213, 215, 216
DCE, 93
DEC, 227
Delay, 41
  capacitive, 41
  logic, 41
  propagation, 41
  transit time, 41, 43
Demodulator, 220
Differential linearity error, 195

DIM, *see Appendix B*
DTE, 93, 104
Dynamic RAM design, 49, 57

EEPROM, 5, 79
E field, 32
80386 microprocessor, 162
  bus interface, 164
  bus timing, 165
  cache, 179
  description, 163
  example design, 176, 183, 188
  general I/O interface, 187
  interrupt acknowledge, 169
  I/O interfacing, 186
  memory interfacing, 175
  memory refresh, 177
  nonpipelined cycles, 165
  pipelined cycles, 168
  structure, 163
  timing cycles, 172
EMP, 31
EPROM, 5
Errors, A/D, 193, 209
Ether, 242
Ethernet, 227, 241
  controller, 246, 248
  station, 246
Exponential back-off:
  binary, 243
  truncated, 244

Far field, 32
FastBus, 48
FDM, 219, 220
FutureBus, 48
Firmware, 6
Flash converter, 202
FORTRAN, 252, 259
Frame, 103
Frame check sequence, 111, 245
Frequency division multiplexing (FDM), 219,
  220
FTAM, 238
Full duplex, 86

Gain error, 197
General Motors, 233
Generating polynomial, 112
Ground, 33, 36–39

Half-duplex, 86
HDLC, 102
H field, 32

HSCMOS, 72
HSCMOS to NMOS, 73

IBM, 227
IBM PC, 139
  communications adapter, 153
  description, 139
  DMA, 155
  general RAM interface, 158
  general ROM interface, 160
  I/O channel, 143
  printer adapter, 156
  system board, 139, 140
IBM Personal System/2, 160
  description, 160–162
  8086 microprocessor, 162
  80386 microprocessor, 162. *See also* 80386
    microprocessor
  interface, 162
  micro channel, 162
  Models, 30, 50, 60, 80, 162
  ports, 162
INI, 233, 238
Input/Output (I/O):
  basic, 134
  isolated, 133
  linear selection, 133
  memory mapped, 133
  strobed, 135
Integral linearity error, 196
Integrating A/D, 205
  dual slope, 206
  single slope, 205
Interface, 7
Interfacing example, 213
ISO, 102, 223, 225, 238, 240

Jam, 242

LAN, 217
LISP, 252, 264
  atom, 265
  list, 266
  recursion, 267
LLC, 238
Loop, 219
LSTTL, 72

MAC, 238
Manchester encoding, 242, 243, 246
MAP, 233–240
  backbone, 235
  bridge, 235
  CASE, 237, 238

gateway, 235
layers, 235–237
router, 235
TOP, 240
Microcomputer:
  costs, 5
  definition, 3
Microprocessor, definition, 2
MMFS, 238
Modem, 220
Modular programming, 252
Modulator, 220
Monotonicity, 197
MOSFET:
  definition, 7
  depletion mode, 9
  enhancement mode, 9
MOS Memory:
  chip select, 69
  cycles, 54
  interfacing, 49
  READ, 60, 64
  Read-Modify-Write, 54
  refreshing, 54, 60, 69, 177
  timing, 56
  as transmission load, 28
  WRITE, 66
Motorola, 239
MULTIBUS II, 48, 49, 238
Multicast, 248
Multiplexing:
  FDM, 219, 220
  SDM, 219, 220
  TDM, 219, 220, 227

NAND circuit, 11
Near field, 32
Network architecture, 222
NBS, 224, 235
N-channel, 10
NMOS, 10, 13
NMOS to HSCMOS, 72
Nodes, 217
Noise, 31
Noise margin, 74
Noise sources, 31, 34, 35
Nonvolatile memory, 4
NOR circuit, 11
NRZI, 104
NuBus, 48

Offset error, 197
Oscillations, 25
OSI, 223, 225, 234

Packets, 224, 245
Parity, 87
Pascal, 252, 261
Parallel communication, 132
Parallel termination, 25
Pass transistor, 11
P-channel, 10
Phase decoder, 248
Phase-locked loop, 248
Physical level, 224
PL/M-86, 252, 268
  examples, 291
  procedures, 275, 288
  recursion, 281
  reentrant, 281
PMOS, 10, 13
Polling, 227
Preamble, 245, 248
Programming
  modular, 252
  structured, 254
  style, 255
  top-down, 255
Protocol, 222
Pseudobidirectional I/O port, 137

Quantization, 193
Quantization error, 195

RAM, 3, 49, 57
RAS, 49
Ratio logic, 12
Real time operating systems, see Appendix B
Reflection coefficient, 21
Refreshing, 69
RFI, 31
RF transmission line, 14
Ring, 218
RMX-NET, 238
ROM, 3
Router, 235
RS-232-C, 88–95
RS-422A, 95, 96
RS-423A, 96, 97
RS-449, 97–99
RS-485, 99–102
R-2R network, 211

Sample and hold, 201
Serial access, 4
Serial communication, 86
Series termination, 24, 29
SDLC, 102
  address field, 104

SDLC (*Continued*)
CRC, 113
generating polynomial, 112
SDM, 219, 220
Shielding, 32
Simplex, 86
Skew, 41
Slot time, 241
SNA, 103
Software, 252
Star, 218
Station, 217, 246
Steering logic, 12
Structured programming, 254
Stubs, 25
Successive approximation, A/D, 201
Synchronous transmission, 102

TDM, 219, 220, 227
Temperature effects, A/D, 198
Threshold voltage, 7
Token access controller, 230
Token bus, 229
Token passing, 225, 227

Token ring, 228
TOP, 240
Top down design, 255
Transceiver, 246, 249
Transmission, 14
asychronous, 86
serial, 86
synchronous, 102
Transmission line, 25

Unix®, 252

VersaBus, 48
VLSI, 7
VME Bus, 48, 238, *see Appendix B*
Volatile memory, 4

WAN, 225
Wave impedance, 32
Western Digital Corporation, 230
Wirth, Nicholas, 254

Xerox Corporation, 227